Microwave Electron-Tube Devices

Samuel Y. Liao
Professor of Electrical Engineering
California State University, Fresno

Prentice Hall, Englewood Cliffs, N.J. 07632

Library of Congress Cataloging-in-Publication Date

Liao, Samuel Y.
    Microwave electron-tube devices.

    Bibliography: p.
    Includes index.
    1. Microwave tubes.   I. Title.
TK7871.7.L52   1988   621.381'33   88-2530
ISBN 0-13-582073-1

            Editorial/production supervision and
                interior design: Eileen M. O'Sullivan
            Cover design: Wanda Lubelska Design
            Manufacturing buyer: Mary Ann Gloriande

> *The author dedicates this*
> *book to the memory of his*
> *grandparents, Sche ching-Kung*
> *and Wong-Shih for their inspiration.*

 © 1988 by Prentice-Hall, Inc.
A Division of Simon & Schuster
Englewood Cliffs, New Jersey 07632

ISBN   0-13-582073-1

PRENTICE-HALL INTERNATIONAL (UK) LIMITED, *London*
PRENTICE-HALL OF AUSTRALIA PTY. LIMITED, *Sydney*
PRENTICE-HALL CANADA INC., *Toronto*
PRENTICE-HALL HISPANOAMERICANA, S.A., *Mexico*
PRENTICE-HALL OF INDIA PRIVATE LIMITED, *New Delhi*
PRENTICE-HALL OF JAPAN, INC., *Tokyo*
PRENTICE-HALL OF SOUTHEAST ASIA PTE. LTD., *Singapore*
EDITORA PRENTICE-HALL DO BRASIL, LTDA., *Rio de Janeiro*

Contents

Contents

CHAPTER 10   LIGHT IMAGE TUBES   337

CHAPTER 11   INFRARED IMAGE TUBES   391

PREFACE

This book is intended to serve as a text in a course on microwave electron tubes at the senior or beginning graduate level in electrical engineering. Its primary purpose is to provide readers with the quantitative and qualitative analysis of microwave electron-tube devices. It is assumed that readers have had previous courses in electromagnetic theory and electronic circuits. Because the book is to a large extent self-contained, it can also be used as a reference book by electronics engineers working in the area of microwave high-power devices.

It has been observed for years that microwave electron tubes are still the predominant choice for high power sources. Ever since early 1960s prediction has continued to foresee the microwave tubes be displaced by the counterpart solid-state devices. However, this prediction was not materialized except of microwave low-power devices. On the contrary, since spaceborne defense program started in recent years microwave electron tubes have been rearised as the only predominant choice for high power sources.

The text is arranged into twelve chapters:

1. Chapter 1 is introductory.
2. Chapter 2 studies interactions between electrons and electric, magnetic or electro-magnetic field.
3. Chapter 3 describes electron beams and focusing techniques including electron emissions, electron gun design, electron beam modulation, Brillouin flow, confined flow and PM or PPM focusing techniques.
4. Chapter 4 treats electron-tube circuit components including S-parameter theory, cavity resonators, waveguide tees, directional couplers, reentrant cavities, slow-wave structures, circulators and isolators.

5. Chapter 5 deals microwave linear-beam amplifiers including conventional gridded tubes (triodes, tetrodes, and pentodes), klystron amplifier, velocity modulation, bunching process output power and klystrode.

6. Chapter 6 discusses microwave linear-beam oscillators including klystron oscillator, reflex klystron, and electronic admittance.

7. Chapter 7 analyzes microwave traveling-wave tubes including helix traveling-wave tubes, coupled-cavity traveling-wave tubes, gridded-control traveling-wave tubes, voltage stabilization techniques, twystron hybrid amplifier, and backward-wave amplifier (BWA) and backward-wave osicillator (BWO).

8. Chapter 8 investigates microwave crossed-field electron tubes including magnetrons, forward-wave crossed-field amplifier (FWCFA), backward-wave crossed-field amplifier (Amplitron) and oscillator (Carcinotron).

9. Chapter 9 examines microwave fast-wave electron tubes including gyrotrons, phase bunching process, resonance conditions, gyromonotron oscillator, gyro-TWT amplifier, gyroklystron amplifier, Ubitron and Peniotron.

10. Chapter 10 describes the light image tubes including Vidicon, Thermicon, Orthicon, Isoncon, CCI tube, SEC and SIT tube.

11. Chapter 11 covers the infrared image tubes and FLIR image system including FLIR tube, range-finder tube, and shutter tubes.

12. Chapter 12 describes microwave noise measurements and evaluations including noise sources, amplitude modulation noise, phase modulation noise, noise measurement circuits, and noise measurement techniques.

The arrangement of topics is flexible, and the instructor has a choice in the selection or order of the topics to suit either a one-semester or possibly a one-quarter course. Problems for each chapter are included to aid readers in further understanding the subjects discussed in the text. A solutions manual may be obtained from the publisher by instructors who have adopted the book for their courses.

The author is grateful to several anonymous reviewers for their many valuable comments and constructive suggestions, which helped to improve this book. The author would also like to acknowledge his appreciation to Dr. Elden Shaw, Dean of Engineering School, for using his early electron-tube research papers in Stanford University. It is also appropriate to extend thanks to Mr. Bernard M. Goodwin, Executive Editor of Prentice-Hall, for his constant encouragement, and to Ms. Eileen O'Sullivan, Book Production Editor, for her skillful editorial work. Finally, the author is indebted to his wife, Lucia Hsiao-Chuang Lee, and children: Grace in bio-engineering, Kathy in electrical engineering, Gary in electronics engineering, and Jeannie in teacher education, for their valuable collective contributions to this book in many different ways. Therefore, this book is dedicated to them.

CHAPTER 1

Introduction

The purpose of this book is to present the principles, operations, and applications of microwave electron-tube devices. Since the early 1960s, predictions have continued that microwave tubes would have to be displaced by microwave solid-state devices. This displacement has occurred only at the low-power and receiving circuits level of electronic systems. Microwave power tubes continue, however, to perform as the only choice for high-power transmitters and are expected to maintain this dominant role throughout the next generation and beyond. Microwave techniques have been increasingly adopted in many electronic systems, such as airborne radar systems, spaceborne military defense, missile guidance systems, and space communications links. As a result of the accelerating rate of growth of microwave technology in research, design, and development in institutes and industries, students who are preparing for, and electronics engineers working in, the microwave field need to understand the basic structures, operations, and performance of microwave electron tubes for the production of microwave electronic components, modules, and systems.

1-0 MICROWAVE FREQUENCIES

The term *microwave frequencies* traditionally refers to those frequencies from 1 to 300 GHz or to wavelengths measured from 30 cm to 1 mm. However, *microwave* really indicates wavelengths in the micron ranges—that is, microwave frequencies up to infrared and visible-light regions. In this book, microwave frequencies mean those from 1 to 10^6 GHz.

The microwave band designation that resulted from World War II radar security considerations was not officially recognized by any industrial, professional, or government organization. In August 1969, the U.S. Department of Defense, Office of Joint Chiefs of Staff, by message to all services, directed the use of microwave frequency bands, as listed in Table 1-0-1. On May 24, 1970, the Department of Defense adopted another band designation for microwave frequencies as shown in Table 1-0-2. In electronics industries and academic institutes, however, Institute of Electrical and Electronics Engineers (IEEE) microwave frequency bands are commonly used as shown in Table 1-0-3. These three band designations are given in Table 1-0-4 for comparison.

TABLE 1-0-1. U.S. ECM
MICROWAVE FREQUENCY BANDS

Designation	Frequency range, GHz
P band	0.225– 0.390
L band	0.390– 1.550
S band	1.550– 3.900
C band	3.900– 6.200
X band	6.200– 10.900
K band	10.900– 36.000
Q band	36.000– 46.000
V band	46.000– 56.000
W band	56.000–100.000

TABLE 1-0-2. U.S. NEW ECM
MICROWAVE FREQUENCY BANDS

Designation	Frequency range, GHz
A band	0.100– 0.250
B band	0.250– 0.500
C band	0.500– 1.000
D band	1.000– 2.000
E band	2.000– 3.000
F band	3.000– 4.000
G band	4.000– 6.000
H band	6.000– 8.000
I band	8.000– 10.000
J band	10.000– 20.000
K band	20.000– 40.000
L band	40.000– 60.000
M band	60.000–100.000

TABLE 1-0-3.   IEEE FREQUENCY BANDS

Band number	Designation	Frequency	Wavelength
2	ELF(extreme low frequency)	30–300 Hz	10–1 Mm
3	VF(voice frequency)	300–3000 Hz	1–0.1 Mm
4	VLF(very low frequency)	3–30 kHz	100–10 km
5	LF(low frequency)	30–300 kHz	10–1 km
6	MF(medium frequency)	300–3000 kHz	1–0.1 km
7	HF(high frequency)	3–30 MHz	100–10 m
8	VHF(very high frequency)	30–300 MHz	10–1 m
9	UHF(ultrahigh frequency)	300–3000 MHz	100–10 cm
10	SHF(superhigh frequency)	3–30 GHz	10–1 cm
11	EHF(extreme high frequency)	30–300 GHz	1–0.1 cm
12	Decimillimeter	300–3000 GHz	1–0.1 mm
	P band	0.23–1 GHz	130–30 cm
	L band	1–2 GHz	30–15 cm
	S band	2–4 GHz	15–7.5 cm
	C band	4–8 GHz	7.5–3.75 cm
	X band	8–12.5 GHz	3.75–2.4 cm
	Ku band	12.5–18 GHz	2.4–1.67 cm
	K band	18–26.5 GHz	1.67–1.13 cm
	Ka band	26.5–40 GHz	1.13–0.75 cm
	Millimeter	40–300 GHz	7.5–1 mm
	Submillimeter	300–3000 GHz	1–0.1 mm

1-1 MICROWAVE ELECTRON-TUBE DEVICES

In this book, the commonly used conventional vacuum tubes and microwave tubes will be analyzed quantitatively or qualitatively. Conventional vacuum tubes, such as triodes, tetrodes, and pentodes, are still used as signal sources of low output power at the low-microwave frequencies. The most important microwave tubes at the present state of the art are the linear-beam tubes (O-type), as tabulated in Table 1-1-1. The first of the O-type tubes is the two-cavity klystron, and it is followed by the reflex klystron. The helix traveling-wave tube (TWT), the coupled-cavity TWT, the forward-wave amplifier (FWA), and the backward-wave amplifier and oscillator (BWA and BWO) are also O-type tubes, but they have nonresonant periodic structure for electron interactions. The Twystron is a hybrid amplifier using combinations of klystron and TWT components.

TABLE 1-0-4 COMPARISON OF IEEE BANDS, OLD BANDS AND NEW BANDS

Frequency in GHz	0.1	0.15	0.2	0.3	0.4	0.5	0.6	0.75	1	1.5	2	3	4	5	6	7.5	10	15	20	30	40	50	60	75	100
Wavelength in cm	300	200	150	100	75	60	50	40	30	20	15	10	7.5	6	5	4	3	2	1.5	1	0.75	0.6	0.5	0.4	0.3

IEEE bands: VHF, UHF, L, S, C, X, Ku, K, Ka, MILLIMETER, V, W

Old bands: P, L, S, C, X, K, Q, V, W

New bands: A, B, C, D, E, F, G, H, I, J, K, L, M

| Wavelength in cm | 300 | 200 | 150 | 100 | 75 | 60 | 50 | 40 | 30 | 20 | 15 | 10 | 7.5 | 6 | 5 | 4 | 3 | 2 | 1.5 | 1 | 0.75 | 0.6 | 0.5 | 0.4 | 0.3 |
| Frequency in GHz | 0.1 | 0.15 | 0.2 | 0.3 | 0.4 | 0.5 | 0.6 | 0.75 | 1 | 1.5 | 2 | 3 | 4 | 5 | 6 | 7.5 | 10 | 15 | 20 | 30 | 40 | 50 | 60 | 75 | 100 |

4

TABLE 1-1-1. LINEAR BEAM TUBES (O-TYPE)

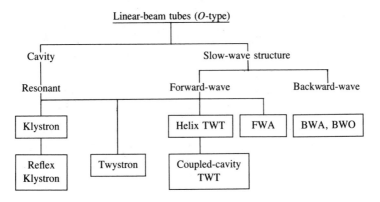

Other equally important microwave tubes are the crossed-field tubes (*M*-type), as tabulated in Table 1-1-2. The magnetron is the oldest and still number one in the family of crossed-field tubes. Others are the forward-wave crossed-field amplifier (FWCFA), the dematron, and the carcinotron (*M*-type backward-wave oscillator).

TABLE 1-1-2. CROSSED-FIELD TUBES (M-TYPE)

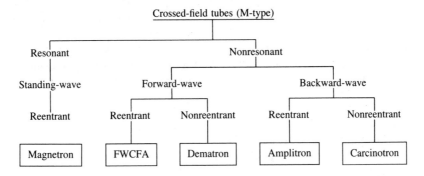

In addition, the following electron tubes are also discussed: high-power gyrotrons including gyromonotron oscillator, gyro-TWT amplifier, and gyroklystron amplifier. These are potential fast-wave devices and are analyzed in Chapter 9. The light-image tubes, including Vidicon, Thermicon, Orthicon, Isocon, CCI, SEC, and SIT, are very useful in the television industry and in military applications, and they are investigated in Chapter 10. The infrared image tubes, such as FLIR tube and range-finder tube, are used for night-vision display, and they are examined in Chapter 11. All these tubes are tabulated in Table 1-1-3.

TABLE 1-1-3. SPECIAL ELECTRON TUBES

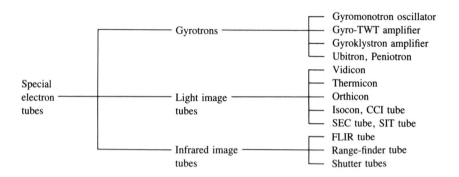

1-2 MICROWAVE ELECTRON-TUBE APPLICATIONS

Microwave electron tubes are vacuum electronic devices used for source generation or signal amplification in the microwave region. Before the invention of solid-state devices, microwave electron tubes were the only active devices available for use in the entire microwave frequency range. Immediately after the invention of the transistor in 1948, it was predicted that microwave electron tubes would have to be displaced by solid-state counterparts. However, up until recent years, microwave electron tubes are still the predominant choices for high-power, high-frequency, and wide-bandwidth applications. The following electronic systems are predominant examples:

1. Military radars
2. Electronic communications
3. Electronic countermeasure (ECM)

In addition, the future applications will be extended into:

1. Microwave cooking wares
2. Microwave medical equipment
3. Nuclear fusion

Table 1-2-1 lists the applications of microwave electron-tube devices.

TABLE 1-2-1. APPLICATIONS OF MICROWAVE ELECTRON-TUBE DEVICES

Device	Operational mechanism	Applications

Gridded tubes
   Triode
   tetrode
   Pentode
	Linear beam	Low microwave frequencies
		low power, low cost

Klystron	Linear beam	Up to 20 GHz medium gain, medium power, TV and broadcasting
Reflex klystron	Linear beam	Medium power, local oscillator
Traveling-wave tube	Linear beam	High gain, high power, wide bandwidth
Magnetron	Crossed field	High power, narrow bandwidth, source generator
Voltage tunable magnetron	Crossed field	High power, wide bandwidth
Amplitron	Crossed field	Medium gain, high power, ECM
Carcinotron	Crossed field	Medium power, wide bandwidth, jammers, swept generator

Gyrotron
   Gyromonotron oscillator,
   Gyro-TWT amplifier,
   gyroklystron amplifier,
   Ubitron, Peniotron
	Linear beam
	
	Free electron
		High gain, high power, radar,
		   nuclear fusion
		
		Radar, nuclear fusion

Light image tubes
   Vidicon, Thermicon, Orthicon,
   Isocon, CCI tube, SEC tube,
   SIT tube
	Photothermic effect	Military and commercial applications

Infrared image tubes
   FLIR image tube
   Range-finder tube
	Thermal-effect	Military and commercial applications

1-3 MICROWAVE ELECTRON-TUBE DEVICES FOR THE FUTURE

As described previously microwave electron tubes are and will be the predominant devices for high-power, high-frequency, and wide-bandwidth applications. For example: the Ubitron uses a free electron laser mechanism to generate high power at 30 GHz and above; and the gyrotron employs plasma to reduce the electron beam velocity spread due to Coulomb repulsion force in order to increase its output power up to 20-MW (megawatt) pulses at 11 GHz. Figure 1-3-1 shows the predicted peak power for Ubitrons and gyrotrons in comparison to laser devices in the years ahead. For spaceborne military applications (Strategic Defense Initiative-SDI, or Star Wars), the subject of microwave electron tubes must remain as the first priority in research, development, and production in order to achieve the projected goal in the future.

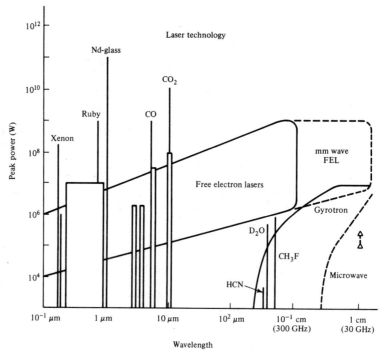

Figure 1-3-1   Comparison of microwave electron tubes [1] (Reprinted by permission of Prentice-Hall, Inc.).

REFERENCE

[1] COLEMAN, JAMES T., *Microwave Devices*. Prentice-Hall, Inc., Englewood Cliffs, N.J., 1982.

CHAPTER 2

Interactions Between Electrons and Fields

2-0 INTRODUCTION

In this chapter we are concerned with electron-field interactions. The motion of the electron-beam is assumed to be in a uniform electric field, or a uniform magnetic field, or a uniform electromagnetic field because the inhomogeneous differential equations governing the motion of an electron-beam in a field involve three dimensions and their solution in a nonuniform field are, in most cases, extremely difficult to obtain and usually cannot be determined exactly. On the other hand, fortunately, all present microwave devices employ uniform field for the electron-field interactions.

Our primary purpose here is to provide the reader with a background for understanding the electron-field interactions in microwave electron-tube devices that will be discussed in later chapters.

2-1 ELECTRON MOTION IN AN ELECTRIC FIELD

In describing fields and electron-field interactions, certain experimental laws of electricity and magnetism are discussed first. The fundamental force law of charges is Coulomb's law, which states that between two charges there exists either an attractive or a repulsive force, depending on whether the charges are of opposite or like sign. That is,

$$\mathbf{F} = \frac{Q_1 Q_2}{4\pi\epsilon_0 R^2}\, \mathbf{u}_{R_{12}} \qquad \text{newtons} \qquad (2\text{-}1\text{-}1)$$

where Q = charge in coulombs

ϵ_0 = $8.854 \times 10^{-12} \simeq \dfrac{1}{36\pi} \times 10^{-9}$ F/m is the permittivity of free space

R = separation between the charges in meters

u = unit vector

It should be noted that since the MKS system is used throughout this text, a factor of 4π appears in the preceding equation.

The electric field intensity produced by the charges is defined as the force per unit charge—that is,

$$\mathbf{E} \equiv \frac{\mathbf{F}}{Q} = \frac{Q}{4\pi\epsilon_0 R^2} \, \mathbf{u}_R \qquad \text{volts/meters} \qquad (2\text{-}1\text{-}2)$$

If there are n charges, the electric field becomes

$$\mathbf{E} = \sum_{N=1}^{n} \frac{Q_N}{4\pi\epsilon_0 R_N^2} \, \mathbf{u}_{R_N} \qquad (2\text{-}1\text{-}3)$$

In order to determine the path of an electron in an electric field, the force must be related to the mass and acceleration of the electron by Newton's second law of motion. Hence,

$$\mathbf{F} = -e\mathbf{E} = m\mathbf{a} = m\frac{d\mathbf{v}}{dt} \qquad (2\text{-}1\text{-}4)$$

where m = 9.109×10^{-31} kg, mass of electron

a = acceleration in m/s^2

v = vector velocity of electron in m/s

e = 1.602×10^{-19} C, charge of electron that is negative

It can be seen that the force is in the opposite direction of the field because the electron has a negative charge. Thus, when an electron moves in an electric field **E**, it experiences a force $-e\mathbf{E}$ newtons. The differential equations of motion for an electron in an electric field in rectangular coordinates are given by

$$\frac{d^2x}{dt^2} = -\frac{e}{m} E_x \qquad (2\text{-}1\text{-}5a)$$

$$\frac{d^2y}{dt^2} = -\frac{e}{m} E_y \qquad (2\text{-}1\text{-}5b)$$

$$\frac{d^2z}{dt^2} = -\frac{e}{m} E_z \qquad (2\text{-}1\text{-}5c)$$

where $\dfrac{e}{m}$ = 1.759×10^{11} coul/kg is the ratio of charge to mass of electron and E_x, E_y,

and E_z are the components of **E** in rectangular coordinates.

In many cases, the equations of motion for electrons in an electric field in cylindrical coordinates are useful. The cylindrical coordinates (r, ϕ, z) are defined as in Fig. 2-1-1.

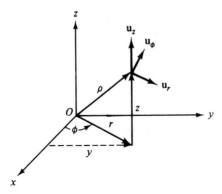

Figure 2-1-1 Cylindrical coordinates.

It can be seen that

$$x = r \cos \phi \tag{2-1-6a}$$

$$y = r \sin \phi \tag{2-1-6b}$$

$$z = z \tag{2-1-6c}$$

and conversely,

$$r = (x^2 + y^2)^{1/2} \tag{2-1-7a}$$

$$\phi = \tan^{-1}\left(\frac{y}{x}\right) = \sin^{-1}\frac{y}{(x^2 + y^2)^{1/2}} = \cos^{-1}\frac{x}{(x^2 + y^2)^{1/2}} \tag{2-1-7b}$$

$$z = z \tag{2-1-7c}$$

A system of unit vectors, $\mathbf{u}_r$, $\mathbf{u}_\phi$, $\mathbf{u}_z$, in the directions of increasing r, ϕ, z, respectively, is also shown in the same diagram. While $\mathbf{u}_z$ is constant, $\mathbf{u}_r$ and $\mathbf{u}_\phi$ are functions of ϕ—that is,

$$\mathbf{u}_r = \cos \phi \, \mathbf{u}_x + \sin \phi \, \mathbf{u}_y \tag{2-1-8a}$$

$$\mathbf{u}_\phi = -\sin \phi \, \mathbf{u}_x + \cos \phi \, \mathbf{u}_y \tag{2-1-8b}$$

Differentiation of Eqs. (2-1-8) with respect to ϕ yields

$$\frac{d\mathbf{u}_r}{d\phi} = \mathbf{u}_\phi \tag{2-1-9a}$$

$$\frac{d\mathbf{u}_\phi}{d\phi} = -\mathbf{u}_r \tag{2-1-9b}$$

The position vector $\boldsymbol{\rho}$ can be expressed in cylindrical coordinates in the form

$$\boldsymbol{\rho} = r\mathbf{u}_r + z\mathbf{u}_z \tag{2-1-9c}$$

Differentiation of Eq. (2-1-9c) with respect to t once for velocity and twice for acceleration yields

$$\mathbf{v} = \frac{d\boldsymbol{\rho}}{dt} = \frac{dr}{dt}\mathbf{u}_r + r\frac{d\mathbf{u}_r}{dt} + \frac{dz}{dt}\mathbf{u}_z = \frac{dr}{dt}\mathbf{u}_r + r\frac{d\phi}{dt}\cdot\frac{d\mathbf{u}_r}{d\phi} + \frac{dz}{dt}\mathbf{u}_z$$

$$= \frac{dr}{dt}\mathbf{u}_r + r\frac{d\phi}{dt}\mathbf{u}_\phi + \frac{dz}{dt}\mathbf{u}_z \tag{2-1-10}$$

$$\mathbf{a} = \frac{d\mathbf{v}}{dt} = \left[\frac{d^2r}{dt^2} - r\left(\frac{d\phi}{dt}\right)^2\right]\mathbf{u}_r + \left[r\frac{d^2\phi}{dt^2} + 2\frac{dr}{dt}\frac{d\phi}{dt}\right]\mathbf{u}_\phi + \frac{d^2z}{dt^2}\mathbf{u}_z$$

$$= \left[\frac{d^2r}{dt^2} - r\left(\frac{d\phi}{dt}\right)^2\right]\mathbf{u}_r + \frac{1}{r}\frac{d}{dt}\left(r^2\frac{d\phi}{dt}\right)\mathbf{u}_\phi + \frac{d^2z}{dt^2}\mathbf{u}_z \tag{2-1-11}$$

Therefore, the equations of motion for electrons in an electric field in cylindrical coordinates are given by

$$\frac{d^2r}{dt^2} - r\left(\frac{d\phi}{dt}\right)^2 = -\frac{e}{m}E_r \tag{2-1-12a}$$

$$\frac{1}{r}\frac{d}{dt}\left(r^2\frac{d\phi}{dt}\right) = -\frac{e}{m}E_\phi \tag{2-1-12b}$$

$$\frac{d^2z}{dt^2} = -\frac{e}{m}E_z \tag{2-1-12c}$$

where E_r, E_ϕ, and E_z are the components of $\mathbf{E}$ in cylindrical coordinates.

From Eq. (2-1-4), the work done by the field in carrying a unit positive charge from point A to point B is

$$-\int_A^B \mathbf{E}\cdot d\ell = \frac{m}{e}\int_{\mathcal{V}_A}^{\mathcal{V}_B} \mathcal{V}\, d\mathcal{V} \tag{2-1-13}$$

where $\mathcal{V}$ is the scalar velocity of the electron.

However, by definition, the potential V of point B with respect to point A is the work done against the field in carrying a unit positive charge from A to B. That is,

$$V \equiv -\int_A^B \mathbf{E}\cdot d\ell \tag{2-1-14}$$

Substitution of Eq. (2-1-14) in Eq. (2-1-13) and integration of the resultant yield

$$eV = \frac{1}{2}m(\mathcal{V}_B^2 - \mathcal{V}_A^2) \tag{2-1-15}$$

The left side of Eq. (2-1-15) is the potential energy, and the right side represents the change in kinetic energy. The unit of work or energy is called the electron volt (eV), which means that if an electron falls through a potential of 1 V, its kinetic energy will increase 1 eV. That is,

$$1 \text{ eV} = (1.60 \times 10^{-19} \text{ C})(1 \text{ V}) = 1.60 \times 10^{-19} \text{ J} \qquad (2\text{-}1\text{-}16)$$

If an electron starts from rest and is accelerated through a potential rise of V volts, its final velocity is

$$\mathcal{V} = \left(\frac{2eV}{m}\right)^{1/2} = 0.593 \times 10^6 \sqrt{V} \qquad \text{meters/seconds} \qquad (2\text{-}1\text{-}17)$$

Since $d\ell$ is the increment of distance in the direction of an electric field E, the change in potential dV over the distance $d\ell$ can be expressed as

$$dV = Ed\ell = \mathbf{E} \cdot d\ell \qquad (2\text{-}1\text{-}18)$$

In vector notation, the electric field can be expressed as the negative gradient of a voltage and it is

$$\mathbf{E} = -\nabla V \qquad (2\text{-}1\text{-}19)$$

where the vector operator ∇ is given by

$$\nabla = \frac{\partial}{\partial x}\mathbf{x} + \frac{\partial}{\partial y}\mathbf{y} + \frac{\partial}{\partial z}\mathbf{z} \qquad \text{(cartesian)}$$

$$\nabla = \frac{\partial}{\partial r}\mathbf{r} + \frac{\partial}{r\partial\phi}\boldsymbol{\phi} + \frac{\partial}{\partial z}\mathbf{z} \qquad \text{(cylindrical)}$$

$$\nabla = \frac{\partial}{\partial r}\mathbf{r} + \frac{1}{r}\frac{\partial}{\partial \theta}\boldsymbol{\theta} + \frac{1}{r\sin\theta}\frac{\partial}{\partial\phi}\boldsymbol{\phi} \qquad \text{(spherical)}$$

The minus sign implies that the field is directed from regions of higher potential to those of lower potential. Equation (2-1-19) is valid in regions in which there is space charge as well as regions that are free of charge.

2-2 ELECTRON MOTION IN A MAGNETIC FIELD

A charged particle in motion in a magnetic field of flux density **B** is experimentally found to experience a force that is directly proportional to the charge Q, its velocity $\mathcal{V}$, the flux density **B**, and the sine of the angle between the vectors **v** and **B**. The direction of the force is perpendicular to the plane of both **v** and **B**. Therefore the force exerted on the charged particle by the magnetic field may be expressed in vector form as

$$\mathbf{F} = Q\mathbf{v} \times \mathbf{B} \qquad (2\text{-}2\text{-}1)$$

Since the electron has negative charge,

$$\mathbf{F} = -e\mathbf{v} \times \mathbf{B} \qquad (2\text{-}2\text{-}2)$$

The motion equations of an electron in magnetic field in rectangular coordinates can be written as

$$\frac{d^2x}{dt^2} = -\frac{e}{m}\left(B_z\frac{dy}{dt} - B_y\frac{dz}{dt}\right) \tag{2-2-3a}$$

$$\frac{d^2y}{dt^2} = -\frac{e}{m}\left(B_x\frac{dz}{dt} - B_z\frac{dx}{dt}\right) \tag{2-2-3b}$$

$$\frac{d^2z}{dt^2} = -\frac{e}{m}\left(B_y\frac{dx}{dt} - B_x\frac{dy}{dt}\right) \tag{2-2-3c}$$

Since

$$\mathbf{v} \times \mathbf{B} = (B_z r \mathcal{V}_\phi - B_\phi \mathcal{V}_z)\mathbf{u}_r + (B_r \mathcal{V}_z - B_z \mathcal{V}_r)\mathbf{u}_\phi + (B_\phi \mathcal{V}_r - B_r r \mathcal{V}_\phi)\mathbf{u}_z \tag{2-2-4}$$

the equations of motion for electrons in magnetic field for cylindrical coordinates may be given by

$$\frac{d^2r}{dt^2} - r\left(\frac{d\phi}{dt}\right)^2 = -\frac{e}{m}\left(B_z r\frac{d\phi}{dt} - B_\phi\frac{dz}{dt}\right) \tag{2-2-5a}$$

$$\frac{1}{r}\frac{d}{dt}\left(r^2\frac{d\phi}{dt}\right) = -\frac{e}{m}\left(B_r\frac{dz}{dt} - B_z\frac{dr}{dt}\right) \tag{2-2-5b}$$

$$\frac{d^2z}{dt^2} = -\frac{e}{m}\left(B_\phi\frac{dr}{dt} - B_r r\frac{d\phi}{dt}\right) \tag{2-2-5c}$$

Consider next an electron moving with a velocity of $\mathcal{V}_x$ to enter a constant uniform magnetic field that is perpendicular to $\mathcal{V}_x$ as shown in Fig. 2-2-1. The velocity of the electron is assumed as

$$\mathbf{v} = \mathcal{V}_x\mathbf{u}_x \tag{2-2-6}$$

where $\mathbf{u}_x$ is a unit vector in the x direction.

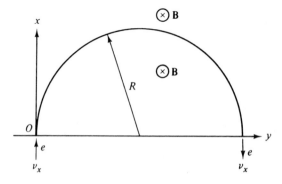

Figure 2-2-1 Circular motion of an electron in a transverse magnetic field.

Since the force exerted on the electron by the magnetic field is normal to the motion at every instant, no work is done on the electron and its velocity remains constant. The magnetic field is assumed to be

$$\mathbf{B} = B_z \mathbf{u}_z \tag{2-2-7}$$

Then the magnetic force at the instant when the electron just enters the magnetic field is given by

$$\mathbf{F} = -e\mathbf{v} \times \mathbf{B} = e\mathcal{V}B\,\mathbf{u}_y \tag{2-2-8}$$

This means that the force remains constant in magnitude but changes the direction of motion because the electron is pulled by the magnetic force in a circular path. This type of magnetic force is analogous to the problem of a mass tied to a rope and twirled around with constant velocity. The force in the rope remains constant in magnitude and is always directed toward the center of the circle, and thus is perpendicular to the motion. At any point on the circle, the outward centrifugal force is equal to the pulling force. That is,

$$\frac{m\mathcal{V}^2}{R} = e\mathcal{V}B \tag{2-2-9}$$

where R is the radius of the circle.

From Eq. (2-2-8), the radius of the path is given by

$$R = \frac{m\mathcal{V}}{eB} \qquad \text{meters} \tag{2-2-10}$$

The cyclotron angular frequency of the circular motion of the electron is

$$\omega_c = \frac{\mathcal{V}}{R} = \frac{eB}{m} \qquad \text{radians/seconds} \tag{2-2-11}$$

The period for one complete revolution is expressed by

$$T = \frac{2\pi}{\omega_c} = \frac{2\pi m}{eB} \qquad \text{seconds} \tag{2-2-12}$$

It should be noted that the radius of the path is directly proportional to the velocity of the electron, but that the angular frequency and the period are independent of velocity or radius. This means that faster-moving electrons or particles will traverse larger circles in the same time that a slower-moving particle moves in a smaller circle. This very important result is the basis of operation of microwave devices such as magnetic-focusing apparatus.

2-3 ELECTRON MOTION IN AN ELECTROMAGNETIC FIELD

If both electric and magnetic fields exist simultaneously, the motion of the electrons will depend upon the orientation of the two fields. If the two fields are in the same or in opposite directions, the magnetic field exerts no force on the electron, and the electron motion depends only upon the electric field which has been described in Section 2-1. Linear-beam tubes (*O*-type devices) use a magnetic field whose axis coincides with that of the electron-beam to hold the beam together as it travels the length of the tube. In these tubes, the electrons receive the full potential energy of the electric field but are not influenced by the magnetic field.

When the electric field **E** and the magnetic field **B** are at right angle to each other, a magnetic force is exerted on the electron-beam. This type of field is called a *crossed field*. In a crossed-field tube (*M*-type device), electrons emitted by the cathode are accelerated by the electric field and gain velocity, but the greater their velocity, the more their path is bent by the magnetic field. The Lorentz force acting on an electron due to the presence of both the electric field **E** and the magnetic flux **B** is given by

$$\mathbf{F} = -e(\mathbf{E} + \mathbf{v} \times \mathbf{B}) = m\frac{d\mathbf{v}}{dt} \tag{2-3-1}$$

The equations of motion for electrons in a crossed field are expressed in rectangular coordinates and cylindrical coordinates, respectively, as

$$\frac{d^2x}{dt^2} = -\frac{e}{m}\left(E_x + B_z\frac{dy}{dt} - B_y\frac{dz}{dt}\right) \tag{2-3-2a}$$

$$\frac{d^2y}{dt^2} = -\frac{e}{m}\left(E_y + B_x\frac{dz}{dt} - B_z\frac{dx}{dt}\right) \tag{2-3-2b}$$

$$\frac{d^2z}{dt^2} = -\frac{e}{m}\left(E_z + B_y\frac{dx}{dt} - B_x\frac{dy}{dt}\right) \tag{2-3-2c}$$

$$\frac{d^2r}{dt^2} - r\left(\frac{d\phi}{dt}\right)^2 = -\frac{e}{m}\left(E_r + B_z r\frac{d\phi}{dt} - B_\phi\frac{dz}{dt}\right) \tag{2-3-3a}$$

$$\frac{1}{r}\frac{d}{dt}\left(r^2\frac{d\phi}{dt}\right) = -\frac{e}{m}\left(E_\phi + B_r\frac{dz}{dt} - B_z\frac{dr}{dt}\right) \tag{2-3-3b}$$

$$\frac{d^2z}{dt^2} = -\frac{e}{m}\left(E_z + B_\phi\frac{dr}{dt} - B_r r\frac{d\phi}{dt}\right) \tag{2-3-3c}$$

where $\dfrac{d\phi}{dt} = \omega_c = \dfrac{e}{m}B$ is the cyclotron frequency.

It is, of course, very difficult to solve these equations for solutions in three dimensions. In microwave devices and circuits, however, only one dimension is involved in most cases. So the equations of motion become simple and can easily be solved. An example may show how to solve some of the preceding equations.

Example 2-3-1: Electron Motion in an Electromagnetic Field

The inner cylinder of radius a is the cathode and the outer shell with radius b is the anode. A dc voltage V_0 is applied between the anode and cathode, and a magnetic flux density **B** is into the page as shown in Fig. 2-3-1. The problem is to adjust the applied voltage V_0 and the magnetic flux density **B** to such levels that the electrons emitted from the cathode will just graze the anode and travel in space between the cathode and the anode only.

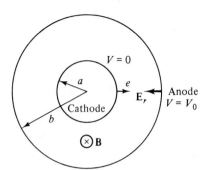

Figure 2-3-1 Electron motion in an electromagnetic field.

Solution:

1. Write the equations of motion for electrons in cylindrical coordinates.

(a) $$\frac{d^2r}{dt^2} - r\left(\frac{d\phi}{dt}\right)^2 = +\frac{e}{m}E_r - \frac{e}{m}r\frac{d\phi}{dt}B_z$$

(b) $$\frac{1}{r}\frac{d}{dt}\left(r^2\frac{d\phi}{dt}\right) = \frac{e}{m}B_z\frac{dr}{dt}$$

2. From (b)

$$\frac{d}{dt}\left(r^2\frac{d\phi}{dt}\right) = \frac{1}{2}\omega_c\frac{d}{dt}(r^2) \qquad \left(\text{where } \omega_c = \frac{e}{m}B_0\right)$$

$$r^2\frac{d\phi}{dt} = \frac{1}{2}\omega_c r^2 + \text{constant}$$

3. Applications of the boundary conditions: At $r = a$,

$$a^2\frac{d\phi}{dt} = \frac{1}{2}\omega_c a^2 + \text{constant}$$

$$\frac{d\phi}{dt} = 0 \qquad \text{constant} = -\frac{1}{2}\omega_c a^2$$

Hence,

$$r^2\frac{d\phi}{dt} = \frac{1}{2}\omega_c(r^2 - a^2)$$

4. The magnetic field does no work on the electrons:

$$\frac{1}{2}m\mathcal{V}^2 = eV$$

$$\mathcal{V}^2 = \frac{2e}{m}V = \mathcal{V}_r^2 + \mathcal{V}_\phi^2 = \left(\frac{dr}{dt}\right)^2 + \left(r\frac{d\phi}{dt}\right)^2$$

5. For grazing the anode,

$$r = b, \qquad V = V_0, \qquad \frac{dr}{dt} = 0$$

$$b^2 \left(\frac{d\phi}{dt} \right)^2 = \frac{2e}{m} V_0 \quad \text{and} \quad b^2 \frac{d\phi}{dt} = \frac{1}{2} \omega_c (b^2 - a^2)$$

$$b^2 \left[\frac{1}{2} \omega_c \left(1 - \frac{a^2}{b^2} \right) \right]^2 = \frac{2e}{m} V_0$$

6. The Hull cutoff voltage is

$$V_{0c} = \frac{e}{8m} B_0^2 b^2 \left(1 - \frac{a^2}{b^2} \right)^2 \qquad (2\text{-}3\text{-}3)$$

This means that if $V_0 < V_{0c}$ for a given B_0, the electrons will not reach the anode. Conversely, the cutoff magnetic field can be expressed in terms of V_0

$$B_{0c} = \frac{(8V_0 m/e)^{1/2}}{b \, (1 - a^2/b^2)} \qquad (2\text{-}3\text{-}4)$$

This implies that if $B_0 > B_{0c}$ for a given V_0, the electrons will not reach the anode.

2-4 ELECTRIC AND MAGNETIC WAVE EQUATIONS

The principles of electromagnetic plane waves are based on the relationships between electricity and magnetism. A changing magnetic field will induce an electric field, and a changing electric field will induce a magnetic field. Also, the induced fields are not confined but ordinarily extend outward into space. The time variation of a sinusoidal wave causes energy to be interchanged between the magnetic and electric fields in the direction of propagation of the wave.

A plane wave has a plane front, a cylindrical wave has a cylindrical front, and a spherical wave has a spherical front. In the far field of free space, electric and magnetic waves are always perpendicular to each other, and both are normal to the direction of propagation of the wave. This type of wave is known as the *transverse electromagnetic* (TEM) wave. If the electric wave is transverse to the direction of wave propagation, the wave is called TE-mode wave. That means there is no component of the electric wave in the direction of propagation. In TM modes only magnetic wave is transverse to the direction of wave propagation.

The electric and magnetic wave equations can be basically derived from Maxwell's equations, which in time domain are expressed as

$$\nabla \times \mathbf{E} = -\frac{\partial \mathbf{B}}{\partial t} \qquad (2\text{-}4\text{-}1)$$

$$\nabla \times \mathbf{H} = \mathbf{J} + \frac{\partial \mathbf{D}}{\partial t} \qquad (2\text{-}4\text{-}2)$$

$$\nabla \cdot \mathbf{D} = \rho_v \tag{2-4-3}$$

$$\nabla \cdot \mathbf{B} = 0 \tag{2-4-4}$$

It should be noted that the boldface Roman letters indicate vector or complex quantities. The units of these field variables are

$\mathbf{E}$ = *electric field intensity* in volts per meter,
$\mathbf{H}$ = *magnetic field intensity* in amperes per meter,
$\mathbf{D}$ = *electric flux density* in coulombs per square meter
$\mathbf{B}$ = magnetic flux density in webers per square meter or in tesla
    (1 tesla = 1 weber/m^2 = 10^4 gauss = 3×10^{-6} ESU)
$\mathbf{J}$ = *electric current density* in amperes per square meter
ρ_v = *electric volume charge density* in coulombs per cubic meter

The electric current density includes two components. That is,

$$\mathbf{J} = \mathbf{J}_c + \mathbf{J}_0 \tag{2-4-5}$$

where $\mathbf{J}_c = \sigma\mathbf{E}$ is called the conduction current density
    $\mathbf{J}_0$ = the *impressed current density,* which is independent of the field

In addition to Maxwell's four equations, the characteristics of the medium in which the fields exist are needed to specify the flux in terms of the fields in a specific medium. These constitutive relationships are

$$\mathbf{D} = \epsilon\mathbf{E} \tag{2-4-6}$$

$$\mathbf{B} = \mu\mathbf{H} \tag{2-4-7}$$

$$\mathbf{J}_c = \sigma\mathbf{E} \tag{2-4-8}$$

$$\epsilon = \epsilon_r\epsilon_0 \tag{2-4-9}$$

$$\mu = \mu_r\mu_0 \tag{2-4-10}$$

where ϵ = permittivity or capacitivity of the medium in farad per meter
    ϵ_r = relative permittivity or dielectric constant (dimensionless)
    ϵ_0 = $8.854 \times 10^{-12} \simeq 1/36\pi \times 10^{-9}$ F/m is permittivity of vacuum or free space
    μ = permeability or inductivity of the medium in H/m
    μ_r = the relative permeability or magnetic constant (dimensionless)
    μ_0 = $4\pi \times 10^{-7}$ H/m is the permeability of vacuum or free space
    σ = conductivity of the medium in Siemens or mhos per meter

If a sinusoidal time function in the form of $e^{j\omega t}$ is assumed, $\partial/\partial t$ can be replaced by $j\omega$. Then Maxwell's equations in frequency domain are given by

$$\nabla \times \mathbf{E} = -j\omega\mu\mathbf{H} \tag{2-4-11}$$

$$\nabla \times \mathbf{H} = (\sigma + j\omega\epsilon)\mathbf{E} \tag{2-4-12}$$

$$\nabla \cdot \mathbf{D} = \rho_\nu \tag{2-4-13}$$

$$\nabla \cdot \mathbf{B} = 0 \tag{2-4-14}$$

Taking the curl of Eq. (2-4-11) on both sides yields

$$\nabla \times \nabla \times \mathbf{E} = j\omega \, \mu \nabla \times \mathbf{H} \tag{2-4-15}$$

Substitution of Eq. (2-4-12) for the right-hand side of Eq. (2-4-15) gives

$$\nabla \times \nabla \times \mathbf{E} = j\omega\mu(\sigma + j\omega\epsilon)\mathbf{E} \tag{2-4-16}$$

The vector identity for the curl of the curl of a vector quantity $\mathbf{A}$ is expressed as

$$\nabla \times \nabla \times \mathbf{A} = -\nabla^2 \mathbf{A} + \nabla(\nabla \cdot \mathbf{A}) \tag{2-4-17}$$

In free space, the space-charge density is zero, and in a perfect conductor time-varying or static fields do not exist. Hence,

$$\nabla \cdot \mathbf{D} = \rho_\nu = 0 \tag{2-4-18}$$

$$\nabla \cdot \mathbf{E} = 0 \tag{2-4-19}$$

Substitution of Eq. (2-4-17) for the left-hand side of Eq. (2-4-16) and replacement of Eq. (2-4-19) yield the electric wave equation as

$$\nabla^2 \mathbf{E} = \gamma^2 \mathbf{E} \tag{2-4-20}$$

where $\gamma = \sqrt{j\omega\mu(\sigma + j\omega\epsilon)} = \alpha + j\beta$ is called the intrinsic propagation constant of a medium,

$\alpha = $ *attenuation constant* in nepers per meter, and

$\beta = $ *phase constant* in radians per meter.

Similarly, the magnetic wave equation is given by

$$\nabla^2 \mathbf{H} = \gamma^2 \mathbf{H} \tag{2-4-21}$$

It should be noted that the "double del" or "del squared" is a scalar operation—that is,

$$\nabla \cdot \nabla = \nabla^2 \tag{2-4-22}$$

which is a second-order operator in three different coordinate systems.

In rectangular (cartesian) coordinates,

$$\nabla^2 \equiv \frac{\partial^2}{\partial x^2} + \frac{\partial^2}{\partial y^2} + \frac{\partial^2}{\partial y^2} \tag{2-4-23}$$

In cylindrical (circular) coordinates,

$$\nabla^2 \equiv \frac{1}{r}\frac{\partial}{\partial r}\left(r\frac{\partial}{\partial r}\right) + \frac{1}{r^2}\frac{\partial^2}{\partial \phi^2} + \frac{\partial^2}{\partial z^2} \tag{2-4-24}$$

In spherical coordinates,

$$\nabla^2 \equiv \frac{1}{r^2} \frac{\partial}{\partial r} \left(r^2 \frac{\partial}{\partial r} \right) + \frac{1}{r^2 \sin \theta} \frac{\partial}{\partial \theta} \left(\sin \theta \frac{\partial}{\partial \theta} \right) + \frac{1}{r^2 \sin^2 \theta} \frac{\partial^2}{\partial \phi^2} \qquad (2\text{-}4\text{-}24a)$$

Also, the solutions of Eqs. (2-4-1) and (2-4-2) solved simultaneously yield the electric and magnetic wave equations in the time domain as

$$\nabla^2 \mathbf{E} = \mu\sigma \frac{\partial \mathbf{E}}{\partial t} + \mu\epsilon \frac{\partial^2 \mathbf{E}}{\partial t^2} \qquad (2\text{-}4\text{-}25)$$

$$\nabla^2 \mathbf{H} = \mu\sigma \frac{\partial \mathbf{H}}{\partial t} + \mu\epsilon \frac{\partial^2 \mathbf{H}}{\partial t^2} \qquad (2\text{-}4\text{-}26)$$

At what rate will electromagnetic energy be transmitted through free space or any medium, be stored in the electric and magnetic fields, and be dissipated as heat? From the standpoint of complex power in terms of the complex field vectors, the time-average power density of any two complex vectors is equal to the real part of the product of one complex vector multiplied by the complex conjugate of the other vector. Hence the time-average power density of an instantaneous Poynting vector in steady state is given by

$$\langle P \rangle = \langle \mathbf{E} \times \mathbf{H} \rangle = \tfrac{1}{2}\text{Re}(\mathbf{E} \times \mathbf{H}^*) \qquad \text{w/m}^2 \qquad (2\text{-}4\text{-}27)$$

where the notation $\langle \ \rangle$ stands for the average and the factor of 1/2 appears in the equation for complex power when peak values are used for the complex quantities $\mathbf{E}$ and $\mathbf{H}$. Re represents the real part of the complex power, and the asterisk indicates the complex conjugate.

From the field theory the electric and magnetic field components can be written in the following forms if the direction of the wave propagation is in the positive z direction

$$\mathbf{E} = E_{xo} e^{-\gamma z} e^{j\omega t} \mathbf{x} = E_{xo} e^{-\alpha z} e^{j(\omega t - \beta z)} \ \mathbf{x} \qquad (2\text{-}4\text{-}28)$$

$$\mathbf{H} = H_{yo} e^{-\gamma z} e^{j\omega t} \mathbf{y} = H_{yo} e^{-\alpha z} e^{j(\omega t - \beta z)} \ \mathbf{y} \qquad (2\text{-}4\text{-}29)$$

Alternatively, they can be expressed as

$$\mathbf{E} = E_{xo} e^{-\alpha z} \cos (\omega t - \beta z) \ \mathbf{x} \qquad (2\text{-}4\text{-}30)$$

$$\mathbf{H} = H_{yo} e^{-\alpha z} \cos (\omega t - \beta z) \ \mathbf{y} \qquad (2\text{-}4\text{-}31)$$

Example 2-4-1: Wave Propagation

An electric-field wave is given by

$$\mathbf{E} = 50 \cos (10^{10} t - 60 z) \ \mathbf{x} \qquad \text{V/m}$$

Determine:

1. The direction of wave propagation
2. The direction of the electric field
3. The direction of the magnetic field

4. The frequency in GHz
5. The wave velocity in m/s
6. The dielectric constant of the material
7. The magnetic field H
8. The average power density

Solution:

1. Positive z direction
2. Positive x direction
3. Positive y direction
4. The frequency is

$$f = \frac{10^{10}}{2\pi} = 1.59 \text{ GHz}$$

5. The phase velocity is

$$\mathcal{V}_{ph} = \frac{10^{10}}{60} = 1.667 \times 10^8 \text{ meters/seconds}$$

6. The dielectric constant is

$$\epsilon_r = \left(\frac{3 \times 10^8}{1.67 \times 10^8} \right)^2 = 3.22$$

7. The magnetic field is

$$\mathbf{H} = \frac{50}{377/\sqrt{3.22}} \cos (10^{10}t - 60z) \, \mathbf{y}$$
$$= 0.239 \cos (10^{10}t - 60z) \, \mathbf{y}$$

8. The average power density is

$$P_{av} = \frac{1}{2} \text{Re}(\mathbf{E} \times \mathbf{H}^*)$$
$$= \frac{1}{2} \times 50 \times \frac{50}{209.4}$$
$$= 5.97 \text{ W/m}^2$$

2-5 RADAR EQUATION AND SIGNAL-TO-NOISE RATIO

The word RADAR is an acronym for RAdio Detection And Ranging and it signifies a means of employing radio waves to detect and locate some material object or target. A target is located by determining the distance and direction from the radar site to the target. The determination of a target location requires, in general, the measurement of three coordinates—usually the range, angle of azimuth, and angle of elevation. The radar detection depends upon the reflection of microwaves from a target. The microwaves reflected from a target are usually called *microwave echoes*.

2-5-1 Radar Cross Section

The radar cross section σ of a target is defined as the area intercepting the amount of electromagnetic power which when reradiated isotropically by the target produces an echo at the source of radiation equal to that observed from the target, as shown in Fig. 2-5-1.

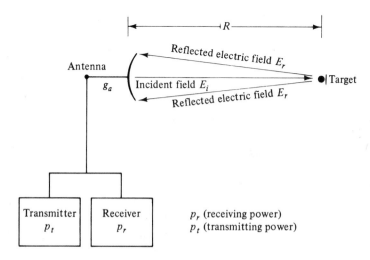

Figure 2-5-1 Power radiation for radar.

    Let the electric field incident upon the target be E_i and the echo electric field at the radar reflected by the target be E_r. Then the radar cross section σ can be expressed as

$$\sigma = 4\pi R^2 \frac{|E_r|^2}{|E_i|^2} \tag{2-5-1}$$

where R is the range between the radar and the target in meters. For a matched antenna, the power received in a matched load is

$$P_i = \frac{|E_i|^2}{\eta_0} A_e \tag{2-5-2}$$

where $\eta_0 = 120\pi$ ohms is the free space impedance

    $A_e = \dfrac{\lambda^2}{4\pi} g_a$ is the effective antenna aperture or area

    g_a = antenna gain

When the target reradiates its received power isotropically, the power density at the radar is

$$P_{di} = \frac{P_i g_a}{4\pi R^2} \tag{2-5-3}$$

Then the electric field at the radar reflected from the target is

$$E_r = \sqrt{120\pi P_{di}} = \frac{1}{R}\sqrt{30 P_i g_a} \tag{2-5-4}$$

Substitution of Eqs. (2-5-2) and (2-5-3) in Eq. (2-5-1) yields the radar cross section σ as

$$\sigma = \frac{\lambda^2}{4\pi} g_a^2 \qquad \text{in } m^2 \tag{2-5-5}$$

For instance, the very short-dipole antenna has a gain of 1.5 and its radar cross section is

$$\sigma = \frac{\lambda^2}{4\pi} (1.5)^2 = 0.18 \, \lambda^2 \tag{2-5-6}$$

2-5-2 Radar Equation

A simple radar consists of a transmitter and a receiver. The transmitter generates electrical power for radiation and the receiver receives electrical power from radiation. Both the transmitter and the receiver use the same antenna as shown in Fig. 2-5-1.

Let P_t be the power of the radar transmitter. Then the power density P_d at a distance R from the transmitting antenna with a gain of g_t is

$$P_d = \frac{P_t g_t}{4\pi R^2} \tag{2-5-7}$$

When the signal impinges the target, the power density reradiated by the target toward the receiving antenna—i.e., the transmitting antenna in this case, is

$$P_{rd} = \frac{P_d \sigma}{4\pi R^2} = \frac{P_t g_t \sigma}{(4\pi R^2)^2} \tag{2-5-8}$$

The power received by the receiving antenna with an effective aperture A_{er} is given by

$$P_r = \frac{P_t g_t \sigma}{(4\pi R^2)^2} A_{er} \tag{2-5-9}$$

If the transmitting and the receiving antennas are the same, then

$$g_t = \frac{4\pi}{\lambda^2} A_{et} = \frac{4\pi}{\lambda^2} A_{er} = g_r = g = \frac{4\pi}{\lambda^2} A_e \tag{2-5-10}$$

and the received power may be expressed as

$$P_r = \frac{P_t g^2 \lambda^2 \sigma}{(4\pi)^3 R^4} = \frac{P_t A_e^2 \sigma}{4\pi \lambda^2 R^4} \qquad (2\text{-}5\text{-}11)$$

The maximum radar range R_{max} is the distance beyond which the target can no longer be detected. It occurs when the received echo signal power P_r is just equal to a minimum detectable signal power P_{min}, and therefore the maximum range is

$$R_{max} = \left[\frac{P_t g^2 \lambda^2 \sigma}{(4\pi)^3 P_{min}} \right]^{1/4} = \left[\frac{P_t A_e^2 \sigma}{4\pi \lambda^2 P_{min}} \right]^{1/4} = \left[\frac{P_t g \sigma A_e}{(4\pi)^2 P_{min}} \right]^{1/4} \qquad (2\text{-}5\text{-}12)$$

where P_t = power generated by the transmitter in watts
 g = antenna gain (numeric value)
 σ = radar cross section in square meters
 λ = wavelength in meters
 A_e = effective antenna aperture in square meters
 P_{min} = minimum detectable signal power in watts

Maximum range.   For a radar set and an isolated object in space, the maximum range of detection depends upon the transmitting power, the minimum detectable echo-pulse signal power, the effective antenna aperture, the square of the wavelength, and the radar cross section of the object. The minimum detectable echo-pulse power depends mainly on the quality of the receiver input circuit, the pulse repetition frequency, and the number of pulses returned from the object as the antenna beam scans past it.

Minimum range.   Objects at a very short range, as well as those at an extremely great range, cannot be detected by radar. One reason for the minimum-range limitation is that most transmitters can not end a pulse with a perfect suddenness. Another effect that limits minimum range is the recovery time of the duplexer. This is because a transmitting-receiving device functions as a switch to connect the antenna to the transmitter during the time of the pulse width and to the receiver during the remainder of the repetition period.

Example 2-5-2:   Radar Range

A certain radar has a transmitter power of 100 MW (peak) at a frequency of 3 GHz and a minimum detectable power of 1.00 μW. The radar antenna has a power gain of 20 dB. (The receiver power gain is not considered).

(a) Calculate the radar cross section in square meters.
(b) Compute the effective antenna aperture in square meters.
(c) Determine the maximum radar range in meters.

Solution:

(a) The radar cross section is

$$\sigma = \frac{\lambda^2}{4\pi} g_a^2 = \frac{[3 \times 10^8/(3 \times 10^9)]^2}{4\pi} (100)^2 = 7.96 \text{ m}^2$$

(b) The effective antenna aperture is

$$A_e = \frac{\lambda^2}{4\pi} g_a = \frac{[3 \times 10^8/(3 \times 10^9)]^2}{4\pi} (100) = 7.96 \times 10^{-2} \text{ m}^2$$

(c) The maximum radar range is

$$R_{max} = \left[\frac{P_t g \sigma A_e}{(4\pi)^2 P_{min}} \right]^{1/4}$$

$$= \left[\frac{100 \times 10^6 \times 100 \times 7.96 \times 7.96 \times 10^{-2}}{(4\pi)^2 \times 1 \times 10^{-6}} \right]^{1/4}$$

$$= 2.52 \text{ km} = 1.56 \text{ miles}$$

2-5-3 Duty Cycle and Signal-to-Noise Ratio

Duty cycle.   Duty cycle (DC) is defined as the ratio of pulse duration over repetition period for a pulse train and it may be expressed as

$$\text{Duty cycle} = \frac{\text{pulse duration}}{\text{pulse repetition period}} = \frac{\tau}{T} = \tau f \qquad (2\text{-}5\text{-}13)$$

$$= \frac{\text{average power}}{\text{peak power}} = \frac{P_{av}}{P_{pk}}$$

Figure 2-5-2 shows a pulse train for determination of the duty cycle.

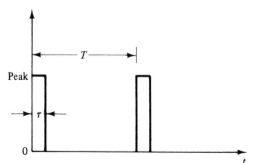

Figure 2-5-2   Pulse train.

Example 2-5-3:   Duty Cycle and Peak Power

The pulse duration of a radar signal is 1 μs and its pulse period is 2 ms.

(a) Calculate the duty cycle.

(b) Compute the peak power if its average power is 50 W.

Solution:

(a) From Eq. (2-5-13), the duty cycle is

$$\text{Duty cycle} = \frac{10^{-6}}{2 \times 10^{-3}} = 0.0005$$

(b) The peak power is

$$P_{\text{pk}} = \frac{50}{0.0005} = 100 \text{ kW}$$

Signal-to-noise ratio.   The signal-to-noise ratio at the output within a narrow frequency band can be expressed as

$$\text{SN} = \frac{S_0}{N_0} \tag{2-5-14}$$

The noise N_0 at the output of the network within the same frequency band arises from amplification of the input noise and the noise N_n generated within the network and it may be written as

$$N_0 = N_n + GkTB \tag{2-5-15}$$

where kTB is the available noise power of the standard source in a band-width B at the room temperature T of 290°K and G the available power gain of the network at the frequency of the band considered. The signal-to-noise at the output is

$$\text{SN} = \frac{S_i G}{N_n + GkTB} \tag{2-5-16}$$

where S_i is the input signal power and $S_i G$ is the power of the amplified input signal.

Since noise does limit the receiver sensitivity, the input circuits of the receiver are the most critical parts of the entire radar system. To make possible the detection of a given object and a maximum range, the input circuits must utilize the smallest echo signal as effectively as possible and at the same time must combine with the signal the least possible amount of noise. That is, the highest obtainable signal-to-noise ratio is desired in order that the signal at the indicator may be detectable. When the signal falls below the level required by the signal-to-noise ratio the signals are obscured. An increase of amplification is then useless, because signal and noise are amplified together. Radar receivers are always designed with enough stages so that large-amplitude noise voltages appear in the output with full gain; therefore the limit of sensitivity is therefore set by the noise.

2-5-4 Doppler Frequency

When the radar transmitter and receiver are colocated, a moving target along the direction of wave propagation causes each frequency component of transmitted wave that strikes the target to be shifted by an amount

$$f_{d1} = \frac{\mathcal{V}_r}{c} f_t \qquad (2\text{-}5\text{-}17)$$

where $\mathcal{V}_r$ = relative or radial velocity of the target with respect to radar
f_t = frequency of transmitted signal
$c = 3 \times 10^8$ m/s is the velocity of light in vacuum

When this signal is reflected or reradiated from the moving target back to the radar, the total Doppler frequency shift of each component is

$$f_d = \frac{2\mathcal{V}_r}{c} f_t \qquad (2\text{-}5\text{-}18)$$

This situation is shown in Fig. 2-5-3. Thus, when the relative velocity of the target is 300 m/s and the transmitted frequency is 10 GHz, the Doppler frequency is 20 kHz.

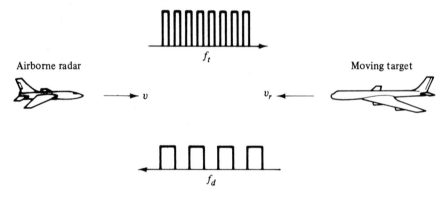

Figure 2-5-3 Doppler effect.

Alternatively, when a moving target is approaching toward a radar with a relative velocity of $\mathcal{V}_r$ with respect to the radar, the transmitted wave from the radar will travel a speed of $c - \mathcal{V}_r$ toward the target and the reflected wave from the target will travel a speed of $c + \mathcal{V}_r$ back to the radar. Then the Doppler frequency is given by

$$f_d = f_t \frac{c + \mathcal{V}_r}{c - \mathcal{V}_r} - f_t \doteq \frac{2\mathcal{V}_r}{c} f_t \qquad (2\text{-}5\text{-}19)$$

On the contrary,

$$f_d = f_t - f_t \frac{c - \mathcal{V}_r}{c + \mathcal{V}_r} \doteq \frac{2\mathcal{V}_r}{c} f_t \qquad (2\text{-}5\text{-}20)$$

In an airborne radar, the radar transmits electromagnetic energy toward the ground and utilizes the Doppler shift of the received energy to determine two or three of the velocity components of the aircraft. This situation is illustrated schematically in Fig. 2-5-4.

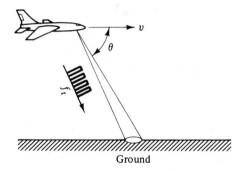

Ground   **Figure 2-5-4**   Doppler beam toward ground.

The basic Doppler equation is given by

$$f_d = \frac{2\mathcal{V} f_t}{c} \cos \theta \tag{2-5-21}$$

where $\mathcal{V}$ = velocity of the aircraft

θ = the angle between the velocity vector and the direction of propagation

Since the antenna beam has a finite width and since the scattering from the ground is randomlike, the information from the ground is not a single frequency, but instead is in the form of a noiselike frequency spectrum. A certain amount of smoothing time of about 1 sec is required to determine the quasi-instantaneous velocity for a given frequency.

Example 2-5-4:   Doppler Frequency

An airborne radar has a velocity of 700 m/s and a transmitted signal of 9 GHz. A moving-target airplane has a relative velocity of 200 m/s with respect to the radar and a stationary target on the ground with an angle of 60° between the radar velocity vector and the direction of radar-beam propagation.

(**a**) Determine the Doppler frequency when the radar sees the moving target in the air.

(**b**) Find the Doppler frequency when the radar sees the stationary target on the ground.

Solution:

(**a**) From Eq. (2-5-18), the Doppler frequency for the moving target in the air is

$$f_d = \frac{2\mathcal{V}_r}{c} f_t = \frac{2 \times 200}{3 \times 10^8} (9 \times 10^9) = 12 \text{ kHz}$$

(**b**) From Eq. (2-5-21), the Doppler frequency for the stationary target on the ground is

$$f_d = \frac{2\mathcal{V} f_t}{c} \cos \theta = \frac{2 \times 700 \times 9 \times 10^9}{3 \times 10^8} \cos 60°$$

$$= 21 \text{ kHz}$$

2-5-5 Radar System

A radar system contains two major generators. One is the transmitting oscillator, which provides the high-frequency powerful pulses needed for echo detection. The other is the local oscillator of the superheterodyne receiver. This oscillator produces a low-power continuous oscillation of a frequency nearly equal to the transmitting frequency. In a microwave radar system, the transmitting oscillator is usually a magnetron, and the receiving oscillator is a reflex klystron. The traveling-wave tube (TWT) is used as an amplifier in the receiver set. The transmitting antenna is also used as the receiving antenna. A radar system is shown in Fig. 2-5-5.

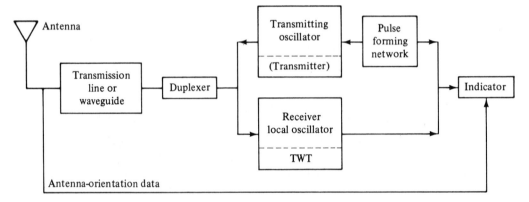

Figure 2-5-5 Radar system.

A radar system should have the following characteristics:

1. *Reliability of Detection:* Reliability of detection includes not only the maximum detection range but also the probability or percentage of the time that the desired target will be detected at any range. Since detection is inherently a statistical problem, this measure of performance must also include the probability of mistaking an unwanted target or noise for a true target.

2. *Accuracy:* Accuracy is measured with respect to the target parameter estimates. These parameters include the target range and the angular coordinates.

3. *Certainty:* The third quality of a radar system is the extent to which the accuracy parameters can be measured without ambiguity or, alternatively, the difficulty encountered in resolving any ambiguities that may be present.

4. *Resolution:* Resolution is the degree to which two or more targets may be separated in one or more spatial coordinates, in radial velocity, or in acceleration. In a simple sense, resolution measures the ability of a radar system to distinguish the radar echoes from similar aircraft in a formation or to distinguish a missile from a possible decoy. In a more sophisticated sense, resolution in a ground mapping radar includes the separation of a multitude of targets with widely divergent radar-echoing areas without cross talk between the various reflectors.

5. *Discrimination:* Discrimination is the ability to detect or to track (or acquire) a target echo in the presence of environmental echoes. It is convenient to include here the discrimination of a missile or an aircraft from man-made dipoles or decoys, and the ability to separate the echoes from a reentry body from its wake.

6. *Countermeasure Immunity:* Countermeasure immunity is the selection of a transmitted signal to give the enemy the least possible information. It includes those processing techniques to make the least use of the identifying characteristics of the desired signals. In some cases, the information received by two or more receivers, or derived by two or more complete systems of different locations or utilizing different principles, parameters, or operation may be compared (correlated) to provide useful discrimination between the desired and undesired signals.

7. *Immunity to Radio Frequency Interference (RFI):* Immunity to radio-frequency (RF) interference means the ability of a radar system to perform its mission in close proximity to other radar systems. It includes both the ability to inhibit detection or display of the transmitted signals from another radar and the ability to detect desired targets in the presence of another radar signal. The RF interference is also called electromagnetic compatibility (EMC).

SUGGESTED READINGS

[1] GEWARTOWSKI, J. W., and H. A. WATSON, *Principles of Electron Tubes.* Van Nostrand Company, Princeton, N.J., 1965.

[2] KRAUS, JOHN D., and K. R. CARVER, *Electromagnetics,* 2nd ed. McGraw-Hill Book Company, New York, 1973.

[3] SKITEK, G. G., and S. V. MARSHALL, *Electromagnetic Concepts and Applications,* 2nd ed. Prentice-Hall, Inc., Englewood Cliffs, N.J., 1987.

PROBLEMS

2-1. At time $t = t_0$ an electron is emitted from a planar diode with zero initial velocity and the anode voltage is $+15$ volts. At time $t = t_1$ the electron is midway between the plates and the anode voltage changes discontinuously to -30 volts.
  (a) Determine which electrode the electron will strike.
  (b) Compute the kinetic energy of the electron in electron volts (eV) when it strikes the electrode.

2-2. A cylindrical cavity is constructed of a center conductor and an outer conductor. The inner center conductor has a radius of 1 cm and is grounded. The outer conductor has a radius of 10 cm and is connected to a power supply of $+10$ kV. The electrons are emitted from the cathode at the center conductor and move toward the anode at the outer conductor.
  (a) Determine the magnetic flux density **B** in webers/m² in the axial direction so that the electrons just graze the anode and return to the cathode again.

(b) If the magnetic flux density B is fixed at 4 milliwebers/m^2, find the required supply voltage so that the electrons just graze the anode and return to the cathode again.

2-3. An electric-field wave is given by

$$\mathbf{E} = 40 \cos (10^{11}t - 50x) \, \mathbf{y} \qquad \text{volts/meters}$$

(a) Determine the direction of wave propagation.

(b) Find the direction of the electric field.

(c) Decide the direction of the magnetic field.

(d) Compute the frequency in GHz.

(e) Calculate the wave velocity in m/s.

(f) Find the dielectric constant of the material.

(g) Determine the magnetic field $\mathbf{H}$ in A/m.

(h) Decide the average power density in W/m^2.

2-4. The distance between a target and a transmitter is 3 km. If the target requires a minimum peak power density of 10 mW/m^2 for detection determine the minimum radiated peak power from the transmitter for proper operation.

2-5. If the duty cycle of a signal is 0.001, what is the average power for a peak power of 2 MW?

2-6. If the transmitted frequency is 9 GHz and the relative velocity of a target is 400 m/s, determine the Doppler frequency.

2-7. A radar antenna has a power gain of 10 dB and is operating at a frequency of 10 GHz. Calculate the radar cross section.

2-8. The minimum power $P_{\min}$ for a radar is 0.001 mW and the transmitting power is 100 MW(peak). The radar antenna has a power gain of 20 dB and is operating at a frequency of 10 GHz. Determine the maximum radar range.

CHAPTER 3

Electron Emissions and Beam Focusing Techniques

3-0 INTRODUCTION

In many microwave electron-tube devices, a fine pencil-like electron beam with a low voltage and high current density must be employed. The problems of forming such a beam by a specific cathode-anode configuration, focusing the beam to a given diameter by means of magnetic-focusing method, and maintaining this beam diameter over a long path by using periodic-permanent-magnet (PPM) focusing shall be discussed in this chapter.

3-1 ELECTRON EMISSIONS

Electron emission can be produced as a result of free electrons within a metal acquiring sufficient energy to overcome the potential-energy barriers existing on the metal surface. According to the Fermi-Dirac distribution function of electrons in metal there is a basic concept of the Fermi energy level in the metal. The Fermi-Dirac distribution function is given by

$$f(E) = \left[\exp \left(\frac{E - E_F}{kT} \right) + 1 \right]^{-1} \tag{3-1-1}$$

where E = energy level in electron volts (eV)
   E_F = Fermi energy level in eV
   k = 1.38×10^{-23} W-s/°K is the Boltzmann's constant
   T = absolute temperature in degrees Kelvin
   kT = $8.63 \times 10^{-5} \, T$ eV

The Fermi energy level in eV is defined as a statistical occupancy of electrons in the metal. The distribution function $f(E)$ is equal to unity for $E < E_F$ at $T = 0°K$ and becomes 0.5 for $E = E_F$ at $T > 0°K$. For instance, at the absolute temperature $(T = 0°K$ or $-273°C)$ all energy levels below the Fermi level are occupied by the electrons, and all energy levels greater than the Fermi level are empty. At $T > 0°K$, the Fermi level is defined as the energy level that has 50 percent occupancy by the electrons, independent of temperature. In other words, 50 percent of the energy level above the Fermi level is occupied by the electrons and the other 50 percent of the energy level below the Fermi level is occupied by the holes. The distribution function is symmetrical with respect to the Fermi level. Figure 3-1-1 shows the Fermi-Dirac distribution function at different temperatures.

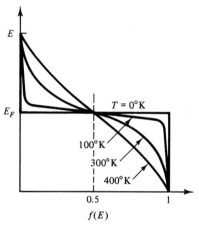

Figure 3-1-1 Fermi-Dirac distribution function at different temperatures.

Each metal has a work function and the work function is defined as the minimum energy in electron volts (eV) required for an electron to escape from the Fermi energy level in a metal to the vacuum surface of the metal at rest.

Electron emissions from a metallic cathode into vacuum can be formed in various ways, such as thermionic emission, photoelectric emission, high-field emission, and secondary emission.

3-1-1 Thermionic Emission

Thermionic emission is the one in which the electrons are emitted from the cathode by heating the cathode metal to high temperature. The current density for thermionic emission is a function of temperature T and the metal work function ϕ_m as

$$J = A_0 T^2 \exp\left(-\frac{\phi_m}{kT}\right) \qquad \text{A/cm}^2 \qquad (3\text{-}1\text{-}2)$$

where $A_0 = 4\pi mek^2h^{-3} = 120$ A/cm^2/°K^2 is the thermionic emission constant
$m = 9.11 \times 10^{-31}$ kg is the electron mass
$e = 1.60 \times 10^{-19}$ coulomb is the electron charge

$k = 1.38 \times 10^{-23}$ J/°K is the Boltzmann's constant

$h = 6.626 \times 10^{-34}$ J-s is the Planck's constant

ϕ_m = metal work function in electron volts (eV)

T = absolute temperature in °K

$kT = 1.38 \times 10^{-23} T/(1.6 \times 10^{-19}) = 8.63 \times 10^{-5} T$ eV

This expression is known as the Richardson-Dushman equation (see Appendix I). The value of the current density is influenced most by the exponential factor and is accordingly a sensitive function of the metal work function and the absolute temperature. Table 3-1-1 lists the thermionic emission data for several commonly used metals.

TABLE 3-1-1.   THERMIONIC EMISSION DATA.

Metal	Emission constant A_0 (A/cm^2/°K^2)	Work function ϕ_m (eV)	Melting temperature (°C)
Tungsten (W)	60	4.50	3370
Tantalum (Ta)	55	4.10	3000
Platinum (Pt)	32	5.30	
Nickel (Ni)	30	4.67	
Chromium (Cr)	48	4.62	
Cesium (Cs)	160	1.80	
	26	4.48	
Molybdenum (Mo)	55	4.20	
Barium on W (Ba on W)	1.5	1.56	
Cesium on W (Cs on W)	3.2	1.36	

It is observed from Table 3-1-1 that the experimental thermionic emission constant for tungsten is only about half of the theoretical value as predicted by Eq. (3-1-2). This discrepancy is due to crystal surface imperfection and the poor emission efficiency.

Tungsten is a good emitter in spite of its high work function (4.5 eV), because it can be heated up to a high temperature (3370°C) without melting and its evaporation rate is also low. On the other hand, copper has a lower work function (4.1 eV), but its melting temperature is too low (1083°C).

Tantalum is also a good emitter because its work function is 4.1 eV lower than tungsten and its emission efficiency is even better than tungsten. Another attractive feature of tantalum is that it is available in sheets and can be formed in special shapes.

Thoriated tungsten and porous tungsten (L-cathode) are also used as thermionic emitters. A thin layer of thorium with a work function of 3.4 eV forms at the tungsten surface and the composite structure can reduce the work function to 2.6 eV. The porous tungsten is impregnated with an alkaline-earth metal compound and reduces the work function for electron emission.

The most extensively used emitting surface is a mixture of barium and strontium oxides. Such an oxide coating has an effective work function of the order of 1.0 to 1.5 eV and an emission constant of the order of 1.20. Because of the very low equivalent work function, an emission current density of 100 mA/cm^2 can be obtained at a temperature of 1000°K with an efficiency of 20 mA/W of heating current. The oxide coatings may be applied either to an indirectly heated cathode surface or directly to a filament.

The most recent cathode for klystron is made of porous tungsten impregnated with barium, calcium, and aluminum oxides; and then coated with a layer of osmium ruthenium alloy. The metal work function is considerably lowered and the temperature is significantly decreased. As a result, the life of the cathode is extended.

Example 3-1-1: Thermionic Emission

Calculate the thermionic electron-emission current density from the surface of a tungsten cathode at a temperature of 2000°K.

Solution: From Eq. (3-1-2), we have

$$J = A_0 T^2 \exp\left(-\frac{\phi_m}{kT}\right) \qquad \text{A/cm}^2$$

where $A_0 = 120$ A/cm^2/°K^2
$T^2 = (2 \times 10^3)^2 = 4 \times 10^6$
$\phi_m = 4.5$ eV for tungsten
$kT = 8.63 \times 10^{-5} \times 2000 = 1.726 \times 10^{-1}$ eV

The current density is

$$J = 120 \times 4 \times 10^6 \exp\left(-\frac{4.5}{0.1726}\right)$$
$$= 4.8 \times 10^8 \times 4.8 \times 10^{-12}$$
$$= 2.3 \text{ mA/cm}^2$$

3-1-2 Photoelectric Emission

The ejection of electrons from metal under the effect of light is known as the photoelectric emission or photoemission. When a light beam is incident on a metal, each photon in the incident light beam is assumed to give up all its photon energy $h\nu$ to an electron in the metal. In order for this electron to get out of the metal, it must expend some energy against the potential barrier of the metal. This energy is called the work function of the metal. The Einstein photoelectric equation is given by

$$h\nu = \phi_m + \frac{1}{2} m \mathcal{V}^2 \qquad (3\text{-}1\text{-}3)$$

where $h\nu =$ photon energy in eV
$h = 6.626 \times 10^{-34}$ J-s is the Planck's constant
$\nu =$ frequency in Hz
$\mathcal{V} =$ electron velocity in m/s

The minimum energy for an electron to emit from metal with zero velocity at the metal surface is the work function ϕ_m of the metal. In other words, the minimum frequency of light, known as the threshold frequency f_{th}, that can be used to cause photoelectric emission can be found from Eq. (3-1-3) by setting the velocity equal to zero. The result is

$$f_{th} = \frac{\phi_m}{h} \qquad \text{Hz} \qquad (3\text{-}1\text{-}4)$$

Then the threshold wavelength, beyond which photoelectric emission cannot take place, is expressed by

$$\lambda_{th} = \frac{c}{f_{th}} = \frac{1.242}{\phi_m} \qquad \mu m \qquad (3\text{-}1\text{-}5)$$

where $c = 3 \times 10^8$ m/s is the velocity of light in vacuum
ϕ_m = metal work function in eV

For response over the entire visible light region, 0.429 to 0.698 MGHz or 0.70 to 0.43 μm, the work function of the photosensitive surface must be less than 1.77 eV.

Phototubes and multiplier phototubes are commonly used for nuclear-radiation detection, television pickup devices, colorimetry, astronomy, and many industrial processes.

Example 3-1-2: Photoelectric Emission

The surface of a barium-strontium oxide is commonly used as a photoelectric emitter. Calculate

(a) The threshold light frequency
(b) The velocity of the emitted electron if the infrared light frequency is 4.29×10^{14} Hz

Solution:

(a) The barium-strontium oxide has a work function of 1.2 eV. From Eq. (3-1-4) the threshold light frequency is

$$f_{th} = \frac{\phi_m}{h} = \frac{1.2 \times 1.6 \times 10^{-19}}{6.626 \times 10^{-34}} = 2.90 \times 10^{14} \text{ Hz near infrared}$$

(b) From Einstein photoelectric Equation (3-1-3), we have

$$\frac{1}{2} m \mathcal{V}^2 = h\nu - \phi_m$$

$$= 6.626 \times 10^{-34} \times 4.29 \times 10^{14} - 1.20 \times 1.6 \times 10^{-19}$$

$$= 0.93 \times 10^{-19} \text{ J}$$

then the velocity is

$$\mathcal{V} = \sqrt{\frac{2 \times 0.93 \times 10^{-19}}{9.1 \times 10^{-31}}}$$

$$= 4.52 \times 10^5 \text{ m/s}$$

or the velocity is found from Eq. (2-1-7) as

$$\mathcal{V} = 0.593 \times 10^5 \times (0.58)^{1/2}$$
$$= 4.52 \times 10^5 \text{ m/s}$$

where $h\nu - \phi_m = \dfrac{6.626 \times 10^{-34} \times 4.29 \times 10^{14}}{1.60 \times 10^{-19}} - 1.20$

$$= 1.78 - 1.20$$
$$= 0.58 \text{ eV}$$

3-1-3 High-Field Emission

Electron emission can be formed by using the Schottky-barrier lowering effect or quantum-mechanical tunneling effect when an adequate electric field is applied to the cathode surface. This type of electron emission is called high-field emission or cold cathode emission, or autoelectronic emission. If the accelerating electric field at the surface of a "cold" cathode is high enough the variation of the emission current density with the strength of the electric field intensity at the metal surface can be expressed as [1]

$$J = CE^2 \exp(-D/E) \qquad \text{A/m}^2 \qquad (3\text{-}1\text{-}6)$$

where $C = \dfrac{6.2 \times 10^{-6}}{E_a} \left(\dfrac{E_F}{\phi_m}\right)^{1/2} \qquad \text{A/V}^2$

$D = 6.8 \times 10^9 \, \phi_m^{3/2} \qquad$ V/m

E = electric field between cathode and the point in consideration in V/m

E_a = applied anode potential energy in eV

E_F = Fermi energy level in eV

ϕ_m = metal work function in eV

The electric-field intensity at an electrode with a sharp point or edge geometry may be very high even if the applied voltage is moderate. Hence, if the high-field emission is to be eliminated, it is very important to shape the electrodes in a tube properly so that a concentration of electrostatic lines of flux does not take place on any metallic surface. The cold-cathode emission has been used to provide several thousand amperes in an X-ray tube used for high-speed radiography.

Example 3-1-3: High-Field Emission

The tungsten emitter has a work function of 4.5 eV and its Fermi energy level is assumed to be 1 eV. The applied anode potential energy E_a is 200 eV and the electric field on the cathode surface is 2700 kV/mm. Find the emission current density.

The constants C and D are

$$C = \frac{6.2 \times 10^{-6}}{200} \left(\frac{1}{4.5}\right)^{1/2} = 1.46 \times 10^{-8} \text{ A/V}^2$$

$$D = 6.8 \times 10^9 \, (4.5)^{3/2} = 6.487 \times 10^{10} \text{ V/m}$$

The emission current density is

$$J = 1.46 \times 10^{-8} \, (27 \times 10^8)^2 \exp\left(-\frac{6.487 \times 10^{10}}{27 \times 10^8}\right)$$
$$= 1.064 \times 10^{11} \times 3.678 \times 10^{-11}$$
$$= 3.913 \text{ A/m}^2$$

3-1-4 Secondary Emission

Secondary emission can be produced from a cathode by bombarding the cathode surface with a primary beam. The most striking aspect of secondary emission is that the number of secondary electrons can be several times greater than the number of primary electrons. It is equally important that it occurs for very low impact energies of primary electrons down to a few electron volts. As a result, secondary electron emission is virtually always present in any electron tubes. Under many conditions the secondary electrons will not go anywhere and will have little effect on the primary beam. However, under some conditions they may have a serious effect on the operation of the electron tube.

3-1-5 Space-Charge-Limited Effect

When the cathode of a parallel-plane electrode structure is heated to a temperature T at a fixed anode voltage the emission current density will follow the Richardson-Dushman's law. This means that the current density is operated under the space-charge-limited mode as shown in Fig. 3-1-2.

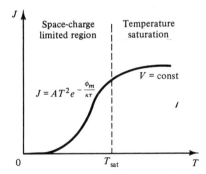

Figure 3-1-2 Emission current density as a function of temperature.

The voltage distribution between cathode and anode is influenced by the presence of electrons in that region. In two parallel plates, the potential varies linearly between cathode and anode voltages in the absence of electrons as shown in curve 1 in Fig. 3-1-3.

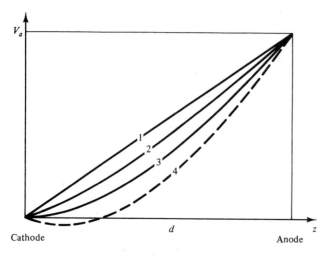

Figure 3-1-3 Potential distribution between parallel plates .

When the cathode is heated, the negative charge of the emitted electrons depresses the potential at the interelectrode region as shown in curve 2. When the operating temperature is increased, more electrons are emitted and they depress the potential to zero slope at the cathode as shown in curve 3. Further increase in temperature depresses further the slope to be negative at the cathode as shown in curve 4.

It is clearly indicated that not all the emitted electrons, but only those that have sufficient velocity to overcome the retarding force, can reach the anode. Some electrons are turned back and reenter the cathode. Further increase in temperature does not increase the emission current. The emission current density is then saturated to the value with the temperature as shown in curve 3. The method of operating the tube above the saturation temperature is called the *space-charge-limited operation*. The current curve follows Richardson's equation up to the saturation temperature for constant voltage, and then it breaks off gradually and becomes almost constant. This effect is also applied to the concentric cylinders and concentric spheres.

3-1-6 Temperature-Limited Effect

Most microwave tubes are operated in the space-charge-limited mode. In this mode of operation, the emission current density is mainly determined by the accelerating voltage and is rather insensitive to small temperature changes. Based on the conservation of energy concept, when an electron is just emitted from the cathode through a potential V the electron has a kinetic energy

$$\frac{1}{2} m \mathcal{V}^2 = eV \tag{3-1-7}$$

The emission current density is

$$J = -\rho \mathcal{V} \qquad \text{A/m}^2 \tag{3-1-8}$$

where ρ = magnitude of the electronic volume charge density in C/m^3 and it is negative for electrons.

From Poisson's equation, the V-ρ relation is given by

$$\nabla^2 V = - \rho/\epsilon_0 \qquad (3\text{-}1\text{-}9)$$

Then

$$\frac{d^2V}{dz^2} = \frac{J}{\epsilon_0 \sqrt{\dfrac{2e}{m} V}} = \frac{J}{\epsilon_0} \sqrt{\frac{m}{2e}} V^{-1/2} \qquad (3\text{-}1\text{-}10)$$

where the $+z$ direction is assumed to be the current flow direction.

Integration of Eq. (3-1-10) with boundary conditions $V = 0$, $\dfrac{dV}{dz} = 0$ at $z = 0$ yields the current density to be

$$J = \frac{4\sqrt{2}}{9z^2} \epsilon_0 \sqrt{\frac{e}{m}} V^{3/2}$$

$$= \frac{2.33 \times 10^{-6}}{z^2} V^{3/2} \qquad \text{A/m}^2 \qquad (3\text{-}1\text{-}11)$$

This is the well-known Langnuir-Child's law. It is also commonly called the *three-half-power equation for parallel-plane geometry*. If the temperature is held constant and the voltage is increased above a certain value, the anode attracts all the emitted electrons from the cathode and a further increase in voltage does not increase the current significantly. This phenomenon is called *voltage saturation,* and operating the tube above the saturation voltage is known as *temperature-limited operation*. Figure 3-1-4 shows the emission current density as a function of voltage under the temperature-limited mode.

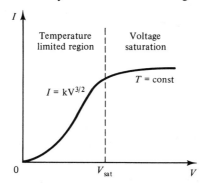

Figure 3-1-4 Emission current density as a function of voltage .

The emission current density follows the three-half-power law up to the saturation voltage for constant temperature, and then it breaks off gradually and becomes almost constant.

In conclusion, Richardson's equation applies to the anode current only when all emitted electrons reach the anode under temperature-limited operation, whereas Child's equation gives only the current when the supply of electrons exceeds the demand under space-charge-limited operation.

3-2 ELECTRON GUN DESIGN

Most microwave tubes are operated under the space-charge-limited mode and their beam current depends only on the voltage between cathode and anode for constant temperature. The operation of an electron gun under such mode may be described by a figure of merit, perveance, which is defined as

$$K = \frac{I}{V^{3/2}} \tag{3-2-1}$$

If an electron gun produces 2 mA at 100 V its perveance is $K = 2 \times 10^{-6}$ A/V$^{3/2}$. The factor 10^{-6} is usually omitted; and it can be said to have a perveance of 2. The perveance of an electron gun is independent of the actual size of the emitting structure and depends only on the electrode shape of the electron gun.

3-2-1 Pierce Gun

Since most microwave tubes require a long and thin rectilinear electron beam, the electron beam needed usually requires a fairly large current with a relatively low voltage. Pierce developed a cathode structure that obtains a rectilinear current flow between two electrodes of finite size. His method is known as Pierce gun and is shown in Fig. 3-2-1 [2].

The Pierce gun employs a concave emitting surface as a cathode. The emitting curvature compensates for the space-charge repulsion force. In order to ensure rectilinear flow between the electrodes, the potential between the electrodes must satisfy Laplace's equation ($\nabla^2 V = 0$) in the charge-free region outside the beam, reduce to the correct value along the beam edge given by Eq. (3-1-11), and yield zero electric field or potential gradient normal to the boundary $[E_\perp = (-\nabla V)_\perp = 0]$.

From the current density equation as shown in Eq. (3-1-11), the potential along the beam edge at $x = 0$ can be expressed as

$$V(z,0) = \left(\frac{9J}{4\epsilon_0 \sqrt{2e/m}} \right)^{2/3} z^{4/3} \tag{3-2-2}$$

Outside the beam, the potential must satisfy Laplace's rather than Poisson's equation, because the charge density there is zero. Pierce has shown that the potential distribution is given by

$$V = \left(\frac{9J}{4\epsilon_0 \sqrt{2e/m}} \right)^{2/3} [\text{Re}(z + jx)^{4/3}] \tag{3-2-3}$$

and that Eq. (3-2-3) satisfies the Laplace equation $\nabla^2 V = 0$ outside the beam, reduces to Eq. (3-2-2) at $x = 0$, and has a zero potential gradient at the beam edge—that is, $V/x = 0$ at $z = 0$.

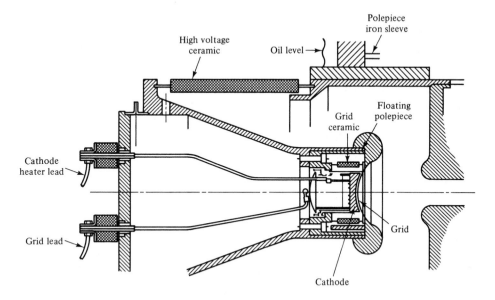

(a) Pierce gun layout

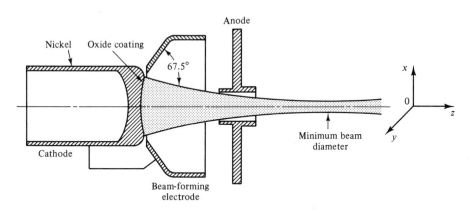

(b) Pierce gun equivalent diagram

Figure 3-2-1 Pierce electron gun: (a) Pierce gun layout, (b) Pierce gun equivalent diagram.

Substitution of various values of constant V in Eq. (3-2-3) yields the shapes of the equipotentials outside the beam region as expressed as

$$(z^2 + x^2)^{2/3} \cos\left(\frac{4\theta}{3}\right) = K \tag{3-2-4}$$

where $\theta = \arctan(x/z)$ is the curvature angle

$$K = V\left(\frac{9J}{4\epsilon_0\sqrt{2e/m}}\right)^{-2/3} \qquad \text{is a constant}$$

The zero equipotential is a plane whose intersection with the x-z plane is given by

$$x = z \tan 67.5° \tag{3-2-5}$$

The angle of 67.5° (or $3/4 \times 90°$) arises out of the exponent 4/3 that occurs in Eq. (3-2-2).

In order to obtain a rectilinear flow between a plane cathode and anode, it is only necessary to confine the region by metal surfaces shaped according to Eq. (3-2-4). For a practical design, the two surfaces chosen are those corresponding to the cathode and anode potentials as shown in Fig. 3-2-1. The external electrode at cathode potential is usually separated from the cathode by a small gap to improve the thermal insulation and to provide the possibility for biasing voltage adjustment.

Example 3-2-1: Electron Gun

A Pierce electron gun has the following parameters:

> Potential $V = 2.5$ kV
> Perveance $K = 2 \times 10^{-6}$ A/V$^{3/2}$

Determine the current I.

Solution: The current is

$$I = KV^{3/2} = 2 \times 10^{-6} \times (2.5 \times 10^3)^{3/2}$$
$$= 0.25 \text{ A}$$

3-3-2 Gridded Control Beam

In electron gun design, a third electrode was introduced by DeForest in 1906 in the front of the cathode surface between the cathode and anode, and this third electrode is called the *control grid*. Figure 3-2-2 shows a sectional view of a gridded gun.

The grid is always biased negatively with respect to the cathode and consequently the current emitted from the cathode comes from a small circular area opposite the grid aperture. Equipotential lines are plotted in Fig. 3-2-3.

If the grid is made sufficiently negative, the beam current is cut off. The cutoff condition clearly will prevail when the off-cathode potential gradient at a point on the cathode surface directly opposite the center of the grid aperture is zero or negative. The

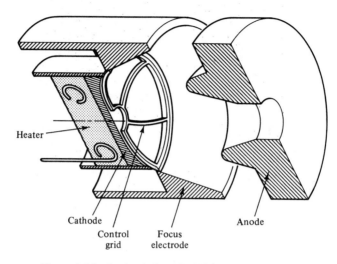

Figure 3-2-2   Sectional view of gridded gun.

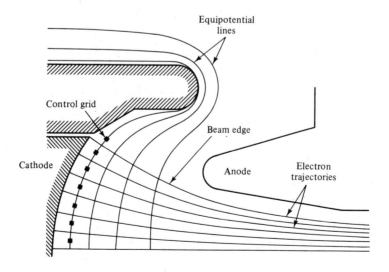

Figure 3-2-3   Equipotential lines in triode gun.

difference between the applied grid voltage and the cutoff voltage is called the grid drive voltage. As the grid is made more positive than cutoff, the current density drawn from the cathode surface is increased. As a result, the triode tube can be used as an amplifier by simply applying a signal to the grid electrode.

3-2-3 Other Cathode Configurations

In addition to the parallel-plane emitter described earlier, two other electron emitters such as concentric cylinders and concentric spheres are also used for electron emission.

Concentric cylinders.   A concentric-cylinder emitter is shown in Fig. 3-2-4.

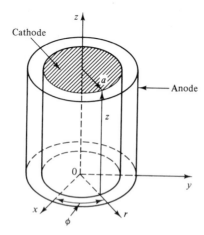

Figure 3-2-4   Concentric-cylinder emitter.

The Poisson's equation for the cylindrical coordinates can be written from the general Poisson's equation of Eq. (3-1-9) to be

$$\frac{d^2V}{dr^2} + \frac{1}{r}\frac{dV}{dr} = \frac{J}{\epsilon_0}\sqrt{\frac{m}{2e}}\, V^{-1/2} \qquad (3\text{-}2\text{-}6)$$

The current per unit length of Eq. (3-2-6) is given by

$$J = \frac{8\pi\epsilon_0}{9\gamma\beta^2}\sqrt{\frac{m}{2e}}\, V^{3/2} \qquad \text{A/m} \qquad (3\text{-}2\text{-}7)$$

where $\gamma = \ell n(r/r_c)$
       r_c = cathode radius in meters

$$\beta = \gamma - \frac{2}{5}\gamma^2 + \frac{11}{120}\gamma^3 + \frac{47}{3300}\gamma^4 + \cdots$$

Values of β^2 have been calculated by Langmuir and Blodgett [3] both for the cases where the cathode is the inner cylinder and where it is the outer cylinder (see Appendix

C). In the latter case, the approximate function is designated as $(-\beta)^2$ for convergent-beam configuration in which the cathode is a cylindrical sector. If θ is the half angle of the sector at the center of the cylinder, the current per unit length is expressed as

$$J = \frac{14.648 \times 10^{-6}}{r\,(-\beta)^2}\left(\frac{\theta}{180°}\right) V^{3/2} \qquad \text{A/m} \qquad (3\text{-}2\text{-}8)$$

Example 3-2-2: Cylinder Emitter

A concentric-cylinder emitter has the following parameters:

Potential at point P outside cathode	$V = 3$ kV
Ratio of radii	$r/r_c = 2$

Compute the current J per unit length.

Solution: From Eq. (3-2-7), we have

$$J = \frac{8\pi\epsilon_0}{9\gamma\beta^2}\sqrt{2e/m}\ V^{3/2}$$

$$= \frac{8 \times 3.1416 \times 8.854 \times 10^{-12}}{9\ \ell n(2)(0.2793)}(2 \times 1.759 \times 10^{11})^{1/2}(3 \times 10^3)^{3/2}$$

$$= 12.448 \ \text{A/m}$$

Concentric spheres.   Figure 3-2-5 shows a concentric sphere emitter.

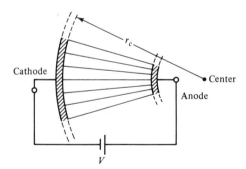

Figure 3-2-5   Concentric-sphere emitter.

Poisson's equation for the spherical coordinates system can be derived from Eq. (3-1-9) as

$$\frac{d^2V}{dr^2} + \frac{2}{r}\frac{dV}{dr} = \frac{J}{\epsilon_0}\sqrt{\frac{m}{2e}}\ V^{-1/2} \qquad (3\text{-}2\text{-}9)$$

The current per unit length is given by

$$J = \frac{16\pi\epsilon_0}{9\alpha^2}\sqrt{2e/m}\ V^{3/2} = \frac{29.297 \times 10^{-6}}{\alpha^2}\ V^{3/2} \qquad \text{A/m} \qquad (3\text{-}2\text{-}10)$$

where $\gamma = \ell n(r/r_c)$
$\quad\quad r_c =$ cathode radius in meters
$\quad\quad \alpha = \gamma - 0.3 \gamma^2 + 0.075 \gamma^3 - 0.0143182 \gamma^4 + \ldots$

Values of α^2 have been computed by Langmuir and Blodgett (see Appendix D) for the potential point outside the cathode and the values $(-\alpha)^2$ indicate the case where the cathode is the outer sphere.

Spherical caps are being used in klystrons and traveling-wave tubes (TWTs) for convergent electron beam. If θ is the half angle of the cone that forms the cap, the current equation of Eq. (3-2-10) must be multiplied by a factor of $1/2(1 - \cos \theta)$ and then it becomes

$$J = \frac{14.648 \times 10^{-6}(1 - \cos \theta)}{(-\alpha)^2} V^{3/2} \quad\quad (3\text{-}2\text{-}11)$$

$$= \frac{29.297 \times 10^{-6}}{(-\alpha)^2} \sin^2 (\theta/2) \, V^{3/2} \quad\quad \text{A/m}$$

Example 3-2-3: Sphere Emitter

A concentric-sphere emitter has the following parameters:

$\quad\quad\quad$ Potential at point P inside cathode $\quad\quad V = 2.5$ kV
$\quad\quad\quad$ Ratio of radii $\quad\quad\quad\quad\quad\quad\quad\quad\quad\quad r_c/r = 2$

Calculate the current J.

Solution: From Eq. (3-2-10), we obtain

$$J = \frac{29.297 \times 10^{-6}}{(-\alpha)^2} V^{3/2}$$

$$= \frac{29.297 \times 10^{-6}}{0.750} (2.5 \times 10^3)^{3/2}$$

$$= 4.88 \text{ A/m}$$

3-3 ELECTRON BEAM MODULATION

Electron beam modulation can be achieved either at the modulating anode or at the input cavity. The former is called the *pulse modulation* and the latter is called the *velocity modulation*.

3-3-1 Pulse Modulation

The anode modulator applies a highly regulated positive grid drive voltage with respect to the cathode to turn the electron beam on for RF amplification. An unregulated negative grid bias voltage with respect to the cathode is used to cut the beam off. Thus, the anode modulator acts as a pulse switch for the electron beam. This technique is frequently used in traveling-wave tubes (TWTs) for power amplification. Figure 3-3-1 shows a diagram of anode modulation.

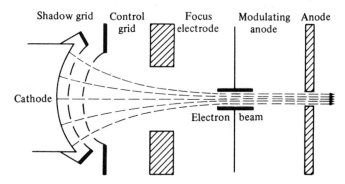

Figure 3-3-1   Pulse modulation at anode.

3-3-2 Velocity Modulation

When the gap of the input cavity of a klystron is energized by an ac input signal, the electrons entering the gap at different times will experience different forces and the electrons leaving from the gap will form a velocity modulated beam. Let's consider three different electrons entering the gap at different times and with different ac signals. For instance: when electron A enters the gap at some time before t_0 with a negative ac signal at the entering grid its velocity is slowed down, when electron B enters the gap at time t_0 with zero ac signal at the grid its velocity remains the same, and when electron C enters the gap at some time after t_0 with a positive signal at the grid its velocity is speeded up. Consequently, these three electrons will be bunched together at some distance at time t_1. This process is called the velocity modulation and it will be discussed in detail in Chapter 5 for klystrons. The modulated velocity is given by

$$\mathcal{V}(t_1) = \mathcal{V}_0 \left[1 + \frac{\beta_i V_1}{2V_0} \sin \left(\omega t_0 + \frac{\theta_g}{2} \right) \right] \tag{3-3-1}$$

where $\mathcal{V}_0 = \sqrt{2e/m\ V_0} = 0.593 \times 10^6 \sqrt{V_0}$ is the dc velocity of electron

V_0 = dc beam voltage in volts

$\beta_i = \dfrac{\sin (\theta_g/2)}{\theta_g/2}$ is the beam coupling coefficient

V_1 = signal voltage in volts

$\theta_g = \dfrac{\omega d}{\mathcal{V}_0}$ is the average gap transit angle

d = cavity gap of the klystron

3-4 BRILLOUIN FLOW

Most microwave electron tubes require a thin electron beam for electron and field interactions. If the drift tube diameter is larger than the beam diameter at the anode aperture, the beam can be transmitted through the tube for a small distance without appreciable loss of electrons to the walls of the drift tube due to the repulsion force such as in the reflex klystron. However, if the beam is going to travel over an extended distance beyond the anode aperture such as in the traveling-wave tube (TWT) and/or if the current density is sufficiently high, some form of beam focusing structure beyond the anode is required to hold the beam diameter constant. When the electron beam travels in free space the electrons have a tendency to diverge their paths due to their coulomb repulsion force and their centrifugal force outward from the center of rotation. In order to confine the beam in a uniform diameter for an extended distance, these two divergent forces must be counterbalanced by a magnetic centripetal force inward toward to the center of rotation—that is,

$$\begin{array}{ccc} \text{Coulomb repulsion} \\ \text{force} \end{array} + \begin{array}{c} \text{Electron centrifugal} \\ \text{force} \end{array} = \begin{array}{c} \text{Magnetic centripetal} \\ \text{force} \end{array}$$

or   (3-4-1)

$$F_{cou} \qquad + \quad F_{cen} \qquad = F_{mc}$$

These three forces are to be determined. The electron beam which travels under the conditions of Eq. (3-4-1) is defined as the Brillouin flow and the magnetic flux density B_0 is known as the Brillouin field. The effect of a magnetic field on moving electrons may be described as follows: the paths remain unchanged by a magnetic field parallel to the path; the paths become circular when they enter a magnetic field perpendicular to the path; and the paths become helical if the magnetic field and the path intersect at an intermediate angle of less than 90°. Figure 3-4-1 shows diagrams of an electron beam flowing through a permanent magnetic field in a cylindrical (r, ϕ, z) coordinates system.

3-4-1 Coulomb Repulsion Force

From Maxwell's electric flux divergence equation

$$\mathbf{\nabla} \cdot \mathbf{D} = \rho_v \tag{3-4-2}$$

where ρ_v = volume charge density

and Gauss's theorem

$$\oint_s \mathbf{D} \cdot d\mathbf{s} = \int_v \mathbf{\nabla} \cdot \mathbf{D} \, dv \tag{3-4-3}$$

the radial electric force is given by

$$(-e)\mathbf{E}_r = \frac{e\rho_\ell}{2\pi\epsilon_0 a} \mathbf{r} \tag{3-4-4}$$

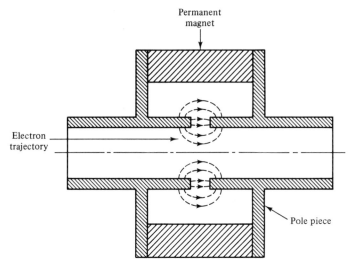

(a) Permanent magnet focusing

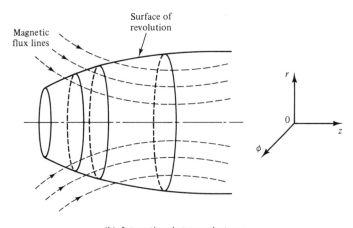

(b) Interactions between electrons
and magnetic field

Figure 3-4-1   Permanent magnet focusing: (a) Permanent magnet focusing, (b) Interactions between electrons and magnetic field.

where ρ_ℓ = linear charge density in coulomb per meter
    ϵ_0 = 8.854×10^{-12} F/m is permittivity of free space
    a = radius of electron beam in meters
    e = 1.602×10^{-19} C is the electron charge

The charge density can be expressed as

$$\rho_\ell = \frac{I_0}{\mathscr{V}_0} = \frac{I_0}{\sqrt{2e/m\ V_0}} \qquad \text{C/m} \qquad (3\text{-}4\text{-}5)$$

where $\mathscr{V}_0$ = dc electron velocity in meters per second
    I_0 = dc beam current in Amperes
    V_0 = dc beam voltage in volts
    e/m = 1.759×10^{11} C/kg is the ratio of charge to mass of an electron

Then the coulomb repulsion force of electron is given by

$$\mathbf{F}_{\text{cou}} = (-e)\mathbf{E}_r = \frac{eI_0}{2\pi\epsilon_0 a\sqrt{2e/m\ V_0}}\ \mathbf{r} \qquad (3\text{-}4\text{-}6)$$

Example 3-4-1: Electron Beam

    An electron beam has the following parameters:

        Beam voltage $V_0 = 2$ kV
        Beam current $I_0 = 2$ A

Calculate the charge density.

Solution: From Eq. (3-4-5), we have the charge density

$$\rho_\ell = \frac{2}{\sqrt{2 \times 1.759 \times 10^{11} \times 2 \times 10^3}}$$
$$= 7.54 \times 10^{-6} \text{ C/m}$$

3-4-2 Electron Centrifugal Force

When an electron rotates in a circular path about the center of rotation the electron will experience a centrifugal force which is outward from the center of rotation. This force is given by

$$\mathbf{F}_{\text{cen}} = ma\dot{\phi}^2\ \mathbf{r} \qquad (3\text{-}4\text{-}7)$$

where $\mathscr{V}_T\dot{\phi} = a\dot{\phi}^2$ is the angular acceleration in radians/s^2
    $\dot{\phi}$ = angular velocity in radians per second
    a = radius of electron beam in meters
    $\mathscr{V}_T = a\dot{\phi}$ is the tangential linear velocity in m/s

3-4-3 Magnetic Centripetal Force

When an electron beam travels through the magnetic field the interaction of the axial electron velocity with the radial component of the magnetic field yields a tangential velocity. That is

$$-e\mathcal{V}_z \, \mathbf{z} \times B_0 \, \mathbf{r} = -e\mathcal{V}_z B_0 \boldsymbol{\phi} \tag{3-4-8}$$

where B_0 = magnetic flux density in Wb/m^2 in $+z$ direction
$\mathcal{V}_z$ = axial electron velocity in m/s

In turn, the same electron interacting with the field in the axial region produces an inward focusing force. Then the magnetic centripetal force is expressed as

$$\mathbf{F}_{mc} = -e\mathcal{V}_\phi \boldsymbol{\phi} \times B_0 \, \mathbf{z} = -e\mathcal{V}B_0 \, \mathbf{r} = -ea\dot{\phi}B_0 \, \mathbf{r} \tag{3-4-9}$$

where $\mathcal{V}_\phi$ = ϕ-component of electron velocity in m/s

3-4-4 Brillouin Flow Equation

Applying Newton's second law to the radial direction the sum of the coulomb repulsion force and the electron centrifugal force must be equal to the magnetic centripetal force as

$$\frac{eI_0}{2\pi\epsilon_0 a \sqrt{2e/m \, V_0}} + ma\dot{\phi}^2 = ea\dot{\phi}B_0 \tag{3-4-10}$$

This equation is known as the Brillouin flow equation.

As the electron crosses the magnetic flux lines, it experiences a force in the $+\phi$ direction. From Eq. (2-2-5b), we can write

$$\frac{d}{dt}(r^2\dot{\phi}) = -\frac{e}{m} r (B_r\dot{z} - B_z\dot{r}) \tag{3-4-11}$$

where $\dot{\phi} = d\phi/dt$, $\dot{z} = dz/dt$, and $\dot{r} = dr/dt$.

Multiplication of Eq. (3-4-11) by dt yields

$$d(r^2\dot{\phi}) = +\frac{e}{m} r(B_z dr - B_r dz) \tag{3-4-12}$$

Consider that the electron advances a distance dz in the z direction and a distance dr in the r direction at an incremental length of trajectory. The magnetic flux that crosses the portion of the surface of revolution corresponding to the axial length dz can be expressed as

$$d\psi = 2\pi r (B_z dr - B_r dz) \tag{3-4-13}$$

where the differential magnetic flux $d\psi$ is considered to be positive if the flux within the surface of revolution increases as z increases. Combining Eq. (3-4-12) with Eq. (3-4-13), we obtain

$$d(r^2\dot{\phi}) = \frac{e}{2\pi m}\,d\psi \tag{3-4-14}$$

where ψ = magnetic flux in webers.

Integration of Eq. (3-4-14) along the axis from a point at $\dot{\phi} = \psi = 0$ to a point within the region of field at ϕ and ψ yields

$$r^2\dot{\phi} = \frac{e}{2\pi m}\,\psi \tag{3-4-15}$$

For small r, ψ is related to the axial magnetic flux density by $B_z = \psi/(\pi r^2)$, so that

$$\dot{\phi} = \frac{e}{2m}\,B_0 = \frac{\omega_c}{2} \tag{3-4-16}$$

where $\omega_c = \dfrac{e}{m}\,B_0$ is the cyclotron angular frequency

$B_0 = B_z$ is the magnetic flux density in $+z$ direction

Thus, the angular velocity of the electron at a given point on its trajectory is proportional to the z component of magnetic field at that point; and when the electron has traveled beyond the region of field, its angular velocity is reduced to zero. Substitution of Eq. (3-4-16) into Eq. (3-4-10) yields

$$B_0 = (e/m)^{-3/4}\,\sqrt{1.414\,I_0/\epsilon_0 \pi a^2\,V_0^{1/2}} = 8.31 \times 10^{-4}/a\,\sqrt{I_0/V_0^{1/2}} \tag{3-4-17}$$

The magnetic flux density B_0 given by Eq. (3-4-17) is known as the Brillouin field and is the minimum value of flux density required to confine an electron beam of dc current I_0 and dc voltage V_0 for a constant diameter a [4]. The dc beam current can be expressed as

$$I_0 = (e/m)^{3/2}\,\epsilon_0 \pi a^2/\sqrt{2}\,B_0^2\,V_0^{1/2} \tag{3-4-18}$$

Example 3-4-1: Electron-Beam Flow

A Brillouin-flow electron beam has the following operating parameters:

Beam voltage	$V_0 = 10$ kV
Beam current	$I_0 = 1$ A
Beam radius	$a = 3$ mm

Determine:

(a) The required magnetic flux density

(b) The cyclotron frequency

(c) The angular velocity

(d) The coulomb repelsion force

(e) The electron centrifugal force

(f) The magnetic centripetal force

(g) The verification of Brillouin-flow equation

Solution:

(a) The required magnetic flux density is

$$B_0 = \frac{8.31 \times 10^{-4}}{3 \times 10^{-3}} \left[\frac{1}{(10 \times 10^3)^{1/2}} \right]^{1/2} = 0.027 \ \text{Wb/m}^2$$

(b) The cyclotron frequency is

$$\omega_c = 1.759 \times 10^{11} \times 0.027 = 4.749 \times 10^9 \ \text{rad/s}$$

(c) The angular velocity is

$$\dot{\phi} = 4.749 \times 10^9/2 = 2.375 \times 10^9 \ \text{rad/s}$$

(d) The coulomb repulsion force is

$$F_{\text{cou}} = \frac{1.6 \times 10^{-19} \times 1}{6.2832 \times 8.854 \times 10^{-12} \times 3 \times 10^{-3} \times 5.93 \times 10^7}$$

$$= 0.16 \times 10^{-13} \ \text{newton}$$

where $\sqrt{2e/m \ V_0} = (2 \times 1.759 \times 10^{11} \times 10^4)^{1/2} = 5.93 \times 10^7 \ \text{m/s}$

(e) The electron centrifugal force is

$$F_{\text{cen}} = 9.1 \times 10^{-31} \times 3 \times 10^{-3} \times (2.375 \times 10^9)^2$$

$$= 0.15 \times 10^{-13} \ \text{newton}$$

(f) The magnetic centripetal force is

$$F_{mc} = 1.6 \times 10^{-19} \times 3 \times 10^{-3} \times 2.375 \times 10^9 \times 0.027$$

$$= 0.31 \times 10^{-13} \ \text{newton}$$

(g) The Brillouin-flow equation is verified:

$$F_{\text{cou}} + F_{\text{cen}} = F_{mc}$$

Example 3-4-2: Electron-Beam Current

An electron beam has a radius a of 5 mm and the magnetic flux density B_0 is 0.02 Web/m². The dc beam voltage V_0 is 10 kV. Calculate the dc beam current.

Solution: The dc beam current is

$$I_0 = (1.759 \times 10^{11})^{3/2} \times \frac{8.854 \times 10^{-12} \times 3.1416}{1.414}$$

$$\times (5 \times 10^{-3})^2 \times (0.02)^2 \times (10^4)^{1/2}$$

$$= 73.77 \times 10^{15} \times 19.67 \times 10^{-12} \times 25 \times 10^{-6}$$

$$\times 4 \times 10^{-4} \times 100$$

$$= 1.45 \text{ A}$$

3-5 CONFINED FLOW

Another method of focusing an electron beam is to have the electrons entirely immersed into an axial magnetic field. To achieve this, the permanent magnet or solenoid must be longer than the cathode-collector distance so that the entire electron-flow path is confined in the axial magnetic field. As the electron beam tends to spread, the radial component of velocity $\mathcal{V}_r$ introduces a force in the ϕ direction. That is

$$\mathbf{F}_\phi = -e\mathcal{V}_r \mathbf{r} \times B_0 \mathbf{z} = e\mathcal{V}_r B_0 \boldsymbol{\phi} \qquad (3\text{-}5\text{-}1)$$

where B_0 is assumed in $+z$ direction.

Equating the torque of rF_ϕ about the z-axis to the time rate of change of angular momentum, we have

$$er\dot{r}B_0 = \frac{d}{dt}(mr^2\dot{\phi}) \qquad (3\text{-}5\text{-}2)$$

where $\dot{r} = \mathcal{V}_r = dr/dt$ is replaced.

Integration of Eq. (3-5-2) with respect to time t yields

$$\dot{\phi} = \frac{\omega_c}{2} + \frac{C}{r^2} \qquad (3\text{-}5\text{-}3)$$

where $\omega_c = \dfrac{e}{m} B_0$ is the cyclotron frequency

$$C = \text{constant}$$

If initial angular velocity $\dot{\phi}$ of electrons at the cathode is zero, then $C = -\omega_c r_c^2/2$, where r_c is the radius of the cathode. Hence, Eq. (3-5-3) becomes

$$\dot{\phi} = \frac{\omega_c}{2}\left(1 - \frac{r_c^2}{r^2}\right) \qquad (3\text{-}5\text{-}4)$$

In turn, the $\dot{\phi}$ component intersecting with the axial magnetic field creates a restoring force in the r direction, which is

$$\mathbf{F}_r = -er\dot{\phi}\boldsymbol{\phi} \times B_0 \mathbf{z} = -er\dot{\phi}B_0 \mathbf{r} \qquad (3\text{-}5\text{-}5)$$

Substitution of Eq. (3-5-4) into Eq. (3-5-5) gives

$$\mathbf{F}_r = -\frac{\mathrm{r}e^2}{2m}B_0^2\left(1-\frac{r_c^2}{r^2}\right)\mathbf{r} \tag{3-5-6}$$

It can be seen from Eq. (3-5-6) that if the beam is well focused so that r is kept close to r_c, the magnetic flux density B_0 must be very large to provide the necessary restoring force. This is the disadvantage of using confined rather than Brillouin flow. However, the former usually yields a better defined constant-diameter beam than the latter does.

In practice, Brillouin flow focusing is not often used because of the high sensitivity of the electron beam to perturbations. Instead, confined flow focusing is commonly employed to confine the electron beam along the axis. Figure 3-5-1 shows the comparison of Brillouin flow, confined flow and space-charge balanced flow [5].

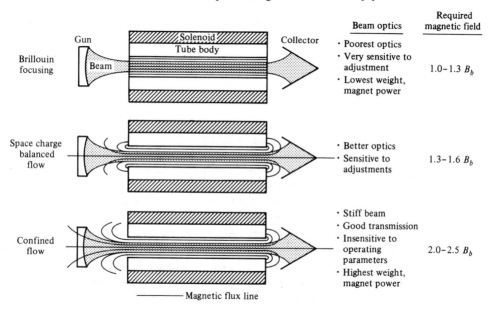

Figure 3-5-1   Comparison of electron flows (Reprinted by permission of the IEEE, Inc.).

3-6 PM OR PPM FOCUSING TECHNIQUES

In microwave tube design, confined flow requires a very large magnetic field for good focusing while Brillouin flow needs relatively large but still finite fields for perfect focusing. This field must be supplied by permanent magnet (PM) or solenoids as shown in Fig. 3-6-1.

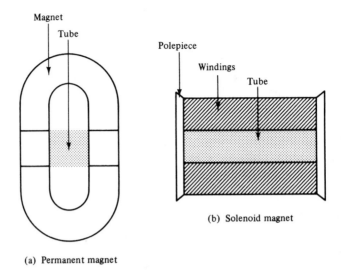

(a) Permanent magnet

(b) Solenoid magnet

Figure 3-6-1 PM or Solenoid focusing structures: (a) Permanent magnet, (b) Solenoid magnet.

In either focusing method, the magnet structure is inconveniently heavy, typically weighing about 40 lb for Brillouin flow. If a solenoid is used, an additional amount of electrical power must be invested to obtain the magnetic force.

The better and inexpensive focusing method is to use a periodic-permanent magnet (PPM), as shown in Fig. 3-6-2.

The PPM focusing structure reduces the magnet weight to $1/N^2$ where N is the number of magnets used in the periodic structure [6]. If the values of magnetic flux density B_0, beam voltage V_0, and magnetic cell length L of the PPM are adjusted in such a way that nearly perfect or 99-percent focusing can be reached. For mathematic simplicity, it is assumed that the magnetic field is given by

$$B_z = B_0 \cos\left(\frac{2\pi z}{L}\right) \tag{3-6-1}$$

and that the electric field is expressed as

$$\mathbf{E}_\phi = -e\mathcal{V}_r \, \mathbf{r} \times B_z \, \mathbf{z} = e\mathcal{V}_r B_0 \boldsymbol{\phi} \tag{3-6-2}$$

The Langrangian equation [7] for an electron in an electric and magnetic field is expressed as

$$\mathcal{L} = \frac{m}{2}(\dot{r}^2 + r^2\dot{\phi} + \dot{z}^2) + e\,(\mathbf{v} \cdot \mathbf{A}) - eV \tag{3-6-3}$$

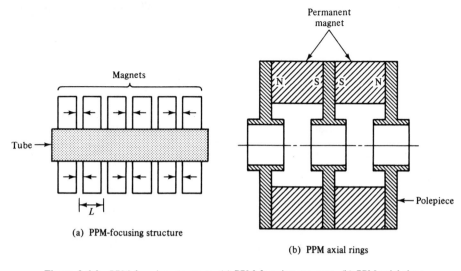

Figure 3-6-2   PPM focusing structures: (a) PPM focusing structure, (b) PPM axial rings.

where $\dot{z} = \mathcal{V}_0$ is dc beam velocity in m/s

  A = vector magnetic potential in webers per meter

  V = scalar electric potential in volts

  $\mathbf{v}$ = vector velocity of electron in m/s

Applying the relationships of $\mathbf{H} = \nabla \times \mathbf{A}$ and $\mathbf{H} = H_z\mathbf{z}$, we obtain the component equations

$$\frac{1}{r}\frac{\partial A_z}{\partial \phi} - \frac{\partial A_\phi}{\partial z} = 0 \tag{3-6-4a}$$

$$\frac{\partial A_r}{\partial z} - \frac{\partial A_z}{\partial r} = 0 \tag{3-6-4b}$$

$$\frac{1}{r}\frac{\partial (rA_\phi)}{\partial r} - \frac{1}{r}\frac{\partial A_r}{\partial \phi} = 0 \tag{3-6-4c}$$

These equations can be satisfied if

$$A_r = A_z = 0$$

$$A_\phi = rH_z/z \tag{3-6-5}$$

From Eq. (3-4-16), the angular velocity for Brillouin flow in a cathode is given by

$$\dot{\phi} = \frac{e}{2m}B_z = \frac{\omega_c}{2} \tag{3-6-6}$$

From Maxwell's diavergence equation $\nabla \cdot \mathbf{D} = \rho$, the potential at the edge of the beam is

$$V = \frac{r_0^2 \rho}{2r\epsilon_0}$$

(3-6-7)

where r_0 = beam radius in meters at the entrance of the focusing structure
ρ = volume charge density in coulombs per cubic meter

Substitution of Eqs. (3-6-5), (3-6-6) and (3-6-7) into Eq. (3-6-3) yields

$$\mathcal{L} = \frac{m}{2}\left(\dot{r}^2 - \frac{e^2 B_z}{4m^2}r^2 + \dot{z}^2\right) + \frac{er_0^2\rho}{2\epsilon_0 r}$$

(3-6-8)

The equation of electron motion in terms of the Langrangian equation is

$$\frac{\partial \mathcal{L}}{\partial q_k} - \frac{d}{dt}\left(\frac{\partial \mathcal{L}}{\partial q_k}\right) = 0$$

(3-6-9)

where $q_k = r, \phi, z$.

Substitution of Eq. (3-6-8) into Eq. (3-6-9) gives

$$\ddot{r} + \left(\frac{eB_z}{2m}\right)^2 r - \frac{e\rho}{2m\epsilon_0}\frac{r_0^2}{r} = 0$$

(3-6-10)

By change of variables, Eq. (3-6-10) becomes

$$\ddot{\sigma} + \alpha[1 + \cos(2T)]\sigma - \frac{\beta}{\sigma} = 0$$

(3-6-11)

where $\ddot{\sigma} = \dfrac{d^2\sigma}{dT^2}$ $\alpha = \dfrac{1}{2}\left(\dfrac{\omega_L}{\omega}\right)^2$ $\omega = \dfrac{2\pi \mathcal{V}_0}{L}$ V_0 = dc voltage

$\sigma = \dfrac{r}{r_0}$ $\beta = \dfrac{1}{2}\left(\dfrac{\omega_\rho}{\omega}\right)^2$ $\omega_\rho^2 = \dfrac{\rho e r_0}{\epsilon_0 m}$

$T = \omega t$ $\omega_L = \dfrac{1}{2}\dfrac{e}{m}B_0$ $\mathcal{V}_0 = \sqrt{\dfrac{2e}{m}V_0}$

Equation (3-6-11) is a nonlinear equation and has been solved by Mendel [4] with the aid of a computer. Figure 3-6-3 shows the magnetic fields for three different cases.

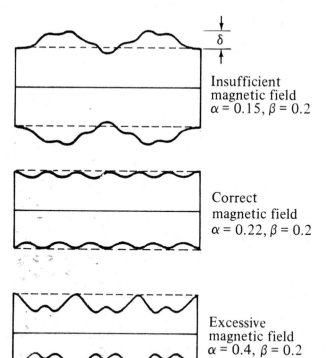

Insufficient
magnetic field
$\alpha = 0.15, \beta = 0.2$

Correct
magnetic field
$\alpha = 0.22, \beta = 0.2$

Excessive
magnetic field
$\alpha = 0.4, \beta = 0.2$

Figure 3-6-3 Magnetic fields of PPM focusing (Reprinted by permission of the IEEE, Inc.).

It can be seen from Fig. 3-6-3 that the optimum focusing occurs when $\alpha = \beta$. In addition, Eq. (3-6-11) can be written as

$$I_0 = \left(\frac{e}{m}\right)^{3/2} \frac{\epsilon_0 \pi r_0^2}{\sqrt{2}} \frac{B_0^2}{\sqrt{2}} V_0^{1/2} \qquad (3\text{-}6\text{-}12)$$

In comparison of Eq. (3-6-12) with Eq. (3-4-18), the two equations are similar for Brillouin flow except of the rms B_0 and average diameter r_0 for PPM method. These two equations also are applicable only if electrons enter the focusing structure with no tranverse velocities.

Example 3-6-1: Electron-Beam Current

An electron beam has the following parameters:

Beam voltage	$V_0 = 10$ kV
Magnetic flux density	$B_0 = 0.02$ Web/m^2
Radius of electron beam	$r_0 = 5$ mm

Calculate the beam current.

Solution: From Eq. (3-6-12), we have the beam current

$$I_0 = (1.759 \times 10^{11})^{3/2} \times \frac{8.854 \times 10^{-12} \times 3.1416 \times (5 \times 10^{-3})^2}{2^{1/2}}$$

$$\times \frac{(0.02)^2}{\sqrt{2}} \times (10^4)^{1/2}$$

$$= 73.77 \times 10^{15} \times 491.79 \times 10^{-18} \times 2.83 \times 10^{-4} \times 100$$

$$= 1.03 \text{ A}$$

REFERENCES

[1] FOWLER, R. H., and L. NORDHEIM, Electron emission in intense electric fields. *Proc. Roy. Soc.* (London), **119,** pp. 173–81, May 1928.

[2] PIERCE, J. R., Rectilinear flow in beams. *J. Appl. Phys.* **11,** pp. 548–54, (1940).

[3] LANGMUIR, I., and K. B. BLODGETT, Currents limited by space-charge between coaxial cylinders. *Phys. Rev.,* **22,** pp. 347–56, (1922).

[4] CHODOROW, MARVIN and CHARLES SUSSKIND, Fundamentals of Microwave Electronics. McGraw-Hill Book Company, New York, 1964. p. 48.

[5] STAPRANS, ARMAND and others, High-power linear-beam tubes. *Proc. IEEE,* vol. 61, no. 1, pp. 299–330, March 1973.

[6] MENDEL, J. T., and et al., *Proc. Inst. Radio Engrs.* **42,** p. 800, (1954).

[7] MACLACHLAN, N. W., *Theory and Applications of Mathieu Functions.* Oxford University Press, New York, 1947.

PROBLEMS

3-1. A tungsten cathode is operating at a temperature of 2500 °K. Calculate the thermionic electron-emission current density from the tungsten surface.

3-2. A tungsten cathode is operating at a temperature of 1800 °K. Compute the thermionic electron-emission current density from the emitting surface.

3-3. The barium on tungsten oxide has a work function of 1.56 eV and operates as a photoelectric emitter.
(a) Calculate its threshold light frequency
(b) Compute the velocity of the emitted electron if the infrared light frequency is 4.10×10^{14} Hz.

3-4. The cesium on tungsten oxide has a work function of 1.36 eV and is used as a photoelectric emitter.
(a) Compute its threshold light frequency.
(b) Calculate the velocity of the emitted electron if the infrared light frequency is 3.50×10^{14} Hz.

3-5. The tungsten emitter has a work function of 4.50 eV and its Fermi energy level is 1 eV. The applied anode potential energy E_a is 150 eV and the electric field on the cathode surface is 2600 kV/mm.

(a) Compute the constant C.

(b) Calculate the constant D.

(c) Find the emission current density.

3-6. An electron gun has a perveance of 3×10^{-6} A/V$^{3/2}$ and the applied cathode voltage is 2000 V. Determine the electron beam current.

3-7. A concentric-cylinder emitter has a potential of 2 kV at point P outside cathode and a radius ratio of 3. Calculate the current J per unit length.

3-8. A concentric-sphere emitter has a potential of 3 kV at point P inside cathode and a radius ratio of 2.50. Compute the current J in amperes per meter.

3-9. An electron beam has the following parameters:

$$\text{Beam voltage} \quad V_0 = 3.0 \text{ kV}$$
$$\text{Beam current} \quad I_0 = 2.5 \text{ A}$$

Compute the charge density.

3-10. A Brillouin-flow electron beam has the following operating parameters:

$$\text{Beam voltage} \quad V_0 = 12 \text{ kV}$$
$$\text{Beam current} \quad I_0 = 1.2 \text{ A}$$
$$\text{Beam radius} \quad a = 3 \text{ mm}$$

Compute:

(a) The required magnetic flux density in Wb/m^2

(b) The cyclotron angular frequency in rad/s

(c) The angular velocity in rad/s

(d) The coulomb repelling force in newtons

(e) The electron centrifugal force in newtons

(f) The magnetic centrifugal force in newtons

(g) The Brillouin-flow equation

3-11. A Brillouin-flow electron beam has the following operating parameters:

$$\text{Beam voltage} \quad V_0 = 30 \text{ kV}$$
$$\text{Beam current} \quad I_0 = 10 \text{ A}$$
$$\text{Beam radius} \quad a = 4 \text{ mm}$$

Calculate:

(a) The required magnetic flux density in Wb/m^2

(b) The cyclotron angular frequency in rad/s

(c) The angular velocity in rad/s

(d) The coulomb repelsion force in newtons

(e) The electron centrifugal force in newtons

(f) The magnetic centrifugal force in newtons

(g) The Brillouin-flow equation

3-12. An electron beam has a radius of 4 mm and the magnetic flux density B_0 is 0.03 Wb/m^2. The dc beam voltage V_0 is 20 kV. Compute the beam current I_0.

3-13. Derive Eq. (3-5-6) for confined flow.

3-14. An electron beam has the following parameters:

Beam voltage	$V_0 = 40$ kV
Beam radius	$r_0 = 8$ mm
Magnetic flux density	$B_0 = 0.08$ Wb/m^2

Compute the beam current in amperes.

3-15. Derive Eq. (3-4-4) from the linear charge density ρ_ℓ in Coulomb per meter.

CHAPTER 4

Microwave Circuit Components

4-0 INTRODUCTION

Microwave circuit components are often used in the electron-tube devices or modules for energy transmission. For instance, the cavity resonators are used in klystrons, the slow-wave structures in traveling-wave tubes (TWTs), and the reentrant cavities in magnetrons. In this chapter, we will study these microwave passive components plus waveguide tees, directional couplers, circulators, and isolators, which are used in the later chapters.

4-1 S-PARAMETER THEORY

4-1-1 S Parameters

From network theory, a two-port device as shown in Fig. 4-1-1 can be described by a number of parameter sets, such as the H, Y, and Z parameters.

$$H \text{ parameters} \qquad V_1 = h_{11}I_1 + h_{12}V_2 \qquad (4\text{-}1\text{-}1)$$

$$I_2 = h_{21}I_1 + h_{22}V_2 \qquad (4\text{-}1\text{-}2)$$

$$Y \text{ parameters} \qquad I_1 = y_{11}V_1 + Y_{12}V_2 \qquad (4\text{-}1\text{-}3)$$

$$I_2 = y_{21}V_1 + y_{22}V_2 \qquad (4\text{-}1\text{-}4)$$

and

$$\text{Z parameters} \qquad V_1 = z_{11}I_1 + z_{12}I_2 \qquad\qquad \text{(4-1-5)}$$

$$V_2 = z_{21}I_1 + z_{22}I_2 \qquad\qquad \text{(4-1-6)}$$

All of these network parameters relate total voltages and total currents at each of the two ports. For instance,

$$h_{11} = \frac{V_1}{I_1}\bigg|_{V_2 = 0} \qquad \text{short circuit} \qquad\qquad \text{(4-1-7)}$$

$$h_{12} = \frac{V_1}{V_2}\bigg|_{I_1 = 0} \qquad \text{open circuit} \qquad\qquad \text{(4-1-8)}$$

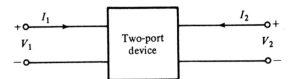

Figure 4-1-1 Two-port network.

However, if the frequencies are in the microwave range, the H, Y, and Z parameters can not be measured. This is because

1. Equipment is not readily available to measure total voltage and total current at the ports of the network.
2. Short and open circuits are difficult to achieve over a broad band of frequencies.
3. Active devices, such as power transistors and tunnel diodes, very often will not be short or open circuit stable.

Consequently, some new method of characterization is to be found to overcome these problems. The logical variables to use at the microwave frequencies are traveling waves rather than total voltages and total currents. These are the S parameters that are expressed as

$$b_1 = S_{11}a_1 + S_{12}a_2$$

$$b_2 = S_{21}a_1 + S_{22}a_2 \qquad\qquad \text{(4-1-9)}$$

Figure 4-1-2 shows the S parameters of a two-port network.

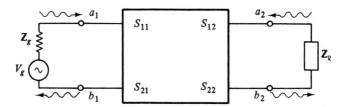

Figure 4-1-2 Two-port network

The S-parameter theory and its general properties are applicable to any n-port devices in the frequencies of microwave range. A microwave junction may have n ports, each of which is a lossless uniform transmission line, as shown in Fig. 4-1-3.

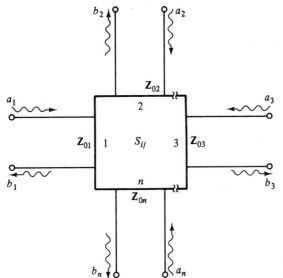

Figure 4-1-3 A microwave junction with n-ports.

As shown in Fig. 4-1-3, a_j is the incident traveling wave coming toward the junction, and b_i is the reflected traveling wave coming outward from the junction. From transmission-line theory, the incident and reflected waves are related by

$$b_i = \sum_{j}^{n} S_{ij} a_j \qquad \text{for } i = 1, 2, 3, \ldots, n \qquad (4\text{-}1\text{-}10)$$

where $S_{ij} = \Gamma_{ii}$ is the reflection coefficient of the ith port if $i = j$ with all other ports matched

$S_{ij} = T_{ij}$ is the forward transmission coefficient of the jth port if $i > j$ with all other ports matched

$S_{ij} = T_{ij}$ is the reverse transmission coefficient of the jth port if $i < j$ with all other ports matched

In general, Eq. (4-1-10) can be written as

$$b_2 = S_{11}a_1 + S_{12}a_2 + S_{13}a_3 + \cdots + S_{1n}a_n$$
$$b_2 = S_{21}a_1 + S_{22}a_2 + S_{23}a_3 + \cdots + S_{2n}a_n \qquad (4\text{-}1\text{-}11)$$
$$\text{- -}$$
$$b_n = S_{n1}a_1 + S_{n2}a_2 + S_{n3}a_3 + \cdots + S_{nn}a_n$$

In matrix notations, boldface Roman letters are used to represent matrix quantities and Eq. (4-1-11) is then expressed as

$$\mathbf{b} = \mathbf{Sa} \qquad (4\text{-}1\text{-}12)$$

where both b and a are column matrices that are usually written as

$$\mathbf{b} = \begin{bmatrix} b_1 \\ b_2 \\ \vdots \\ b_n \end{bmatrix} \quad \text{and} \quad \mathbf{a} = \begin{bmatrix} a_1 \\ a_2 \\ \vdots \\ a_n \end{bmatrix} \qquad (4\text{-}1\text{-}13)$$

The $n \times n$ matrix $\mathbf{S}$ is called the scattering matrix, which is

$$\mathbf{S} = \begin{bmatrix} S_{11} & S_{12} & S_{13} & \cdots & S_{1n} \\ S_{21} & S_{22} & S_{23} & \cdots & S_{2n} \\ & \vdots & & & \\ S_{n1} & S_{n2} & S_{n3} & \cdots & S_{nn} \end{bmatrix} \qquad (4\text{-}1\text{-}14)$$

The coefficients S_{11}, S_{12}, . . . , S_{nn} are called the scattering parameters (S parameters) or scattering coefficients. As a corollary to the S parameters, n-port voltages are linearly related to n-port currents by the impedance matrix of the junction—that is,

$$V_i = \sum_j^n Z_{ij}I_j \qquad \text{for } i = 1, 2, 3, \ldots, n \qquad (4\text{-}1\text{-}15)$$

In matrix form, Eq. (4-1-15) can be expressed as

$$\begin{bmatrix} V_1 \\ V_2 \\ \vdots \\ V_n \end{bmatrix} = \begin{bmatrix} Z_{11} & Z_{12} & \cdots & Z_{1n} \\ Z_{21} & Z_{22} & \cdots & Z_{2n} \\ \cdots & \cdots & \cdots & \cdots \\ Z_{n1} & Z_{n2} & \cdots & Z_{nn} \end{bmatrix} \begin{bmatrix} I_1 \\ I_2 \\ \vdots \\ I_n \end{bmatrix} \qquad (4\text{-}1\text{-}16a)$$

Symbolically,

$$\mathbf{V} = \mathbf{ZI} \qquad (4\text{-}1\text{-}16b)$$

4-1-2 Properties of S Parameters

Several properties of S parameters are described below.

Property I.   Symmetry of S parameters. If a microwave junction satisfies a reciprocity condition, or if there are no vacuum tubes, transistors, and so on, at the junction, the junction is a linear passive circuit, and the S parameters are equal to their corresponding transposes. That is,

$$\mathbf{S} = \tilde{\mathbf{S}} \tag{4-1-17}$$

where $\tilde{\mathbf{S}} = S_{ji} = \mathbf{S} = S_{ij}$. $\tilde{\mathbf{S}}$ is the transpose of matrix $\mathbf{S}$.

The steady-state total voltage and current at the kth port are

$$V_k = V_k^+ + V_k^- \tag{4-1-18}$$

$$I_k = \frac{V_k^+}{Z_{0k}} - \frac{V_k^-}{Z_{0k}} \tag{4-1-19}$$

Therefore the incident and reflected voltages at the kth port are

$$V_k^+ = \frac{1}{2}(V_k + Z_{0k}I_k) \tag{4-1-20}$$

$$V_k^- = \frac{1}{2}(V_k - Z_{0k}I_k) \tag{4-1-21}$$

The average incident power (complex) of the kth port is

$$\frac{1}{2}V_kI_k^* = \frac{|V_k^+|^2}{2Z_{0k}^*} \tag{4-1-22}$$

The normalized incident and reflected voltages at the kth port can be defined as

$$a_k = \frac{V_k^+}{\sqrt{Z_{0k}}} = \frac{1}{2}\left[\frac{V_k}{\sqrt{Z_{0k}}} + \sqrt{Z_{0k}}I_k\right] \tag{4-1-23}$$

$$b_k = \frac{V_k^-}{\sqrt{Z_{0k}}} = \frac{1}{2}\left[\frac{V_k}{\sqrt{Z_{0k}}} - \sqrt{Z_{0k}}I_k\right] \tag{4-1-24}$$

If the characteristic impedance is also normalized so that $\sqrt{Z_{0k}} = 1$, then

$$V_k = a_k + b_k \tag{4-1-25}$$

$$I_k = a_k - b_k \tag{4-1-26}$$

$$a_k = \frac{1}{2}(V_k + I_k) \tag{4-1-27}$$

$$b_k = \frac{1}{2}(V_k - I_k) \tag{4-1-28}$$

Since, from Eq. (4-1-15)

$$V_k = \sum_j^n Z_{kj}I_j \qquad \text{for } k = 1, 2, 3, \ldots, n \qquad (4\text{-}1\text{-}29)$$

it follows that

$$a_k = \frac{1}{2} \sum_j^n (Z_{kj} + \delta_{kj})I_k \qquad (4\text{-}1\text{-}30)$$

$$b_k = \frac{1}{2} \sum_j^n (Z_{kj} - \delta_{kj})I_k \qquad (4\text{-}1\text{-}31)$$

where δ_{kj} is called the *kronecker delta*, which is defined as

$$\delta_{kj} = 1 \qquad \text{if } k = j \qquad (4\text{-}1\text{-}32)$$

$$\delta_{kj} = 0 \qquad \text{if } k \neq j \qquad (4\text{-}1\text{-}33)$$

In matrix notation, Eqs. (4-1-30) and (4-1-31) can be written as

$$\mathbf{a} = \frac{1}{2} (\mathbf{Z} + [\mathbf{I}])\mathbf{I} \qquad (4\text{-}1\text{-}34)$$

$$\mathbf{b} = \frac{1}{2} (\mathbf{Z} - [\mathbf{I}])\mathbf{I} \qquad (4\text{-}1\text{-}35)$$

where a and b are column matrices and $[\mathbf{I}]$ is the identity matrix. Since the impedance matrix $\mathbf{Z}$ and the identity matrix $[\mathbf{I}]$ are square matrices $(n \times n)$, the matrix $(\mathbf{Z} - [\mathbf{I}])$ is surely $n \times n$ and may have an inverse. Thus

$$\mathbf{I} = 2(\mathbf{Z} + [\mathbf{I}])^{-1}\mathbf{a} \qquad (4\text{-}1\text{-}36)$$

Therefore, Eq. (4-1-35) becomes

$$\mathbf{b} = (\mathbf{Z} - [\mathbf{I}])(\mathbf{Z} + [\mathbf{I}])^{-1}\mathbf{a} \qquad (4\text{-}1\text{-}37)$$

In comparing Eq. (4-1-37) with Eq. (4-1-12), the matrix $\mathbf{S}$ can be written as

$$\mathbf{S} = (\mathbf{Z} - [\mathbf{I}])(\mathbf{Z} + [\mathbf{I}])^{-1} \qquad (4\text{-}1\text{-}38)$$

Let the matrices $\mathbf{P}$ and $\mathbf{Q}$ be so defined that

$$\mathbf{P} = \mathbf{Z} - [\mathbf{I}] \qquad (4\text{-}1\text{-}39)$$

$$\mathbf{Q} = \mathbf{Z} + [\mathbf{I}] \qquad (4\text{-}1\text{-}40)$$

Since the impedance matrix $\mathbf{Z}$ is symmetric, the matrices $\mathbf{P}$ and $\mathbf{Q}$ are also symmetric and commutative—that is,

$$\mathbf{PQ} = \mathbf{QP} \qquad (4\text{-}1\text{-}41)$$

Multiplying both the left- and right-hand sides of **PQ** and **QP** by $\mathbf{Q}^{-1}$ yields

$$\mathbf{Q}^{-1}\mathbf{PQQ}^{-1} = \mathbf{Q}^{-1}\mathbf{QPQ}^{-1} \qquad (4\text{-}1\text{-}42)$$

Then

$$\mathbf{Q}^{-1}\mathbf{P} = \mathbf{PQ}^{-1} = \mathbf{S} \qquad (4\text{-}1\text{-}43)$$

The transpose $\tilde{\mathbf{S}}$ of **S** is

$$\tilde{\mathbf{S}} = \mathbf{Q}^{-1}\mathbf{P} = \mathbf{PQ}^{-1} = \mathbf{Q}^{-1}\mathbf{P} = \mathbf{PQ}^{-1} = \mathbf{S} \qquad (4\text{-}1\text{-}44)$$

This means that the terms S_{ij} and S_{ji} of the **S** matrix are equal, and therefore the matrix S has a symmetry.

Property 2.   Unity Property. The sum of the products of each term of any one row, or of any one column, of the matrix **S** multiplied by its complex conjugate is unity—that is,

$$\sum_{i}^{n} S_{ij}S_{ij}^{*} = 1 \qquad \text{for } j = 1, 2, 3, \ldots, n \qquad (4\text{-}1\text{-}45)$$

From the principle of the conservation of energy, if the microwave devices are lossless the power input must be equal to the power output. The incident and reflected waves are related to the incident and reflected voltages by

$$a = \frac{V^{+}}{\sqrt{Z_0}} \qquad (4\text{-}1\text{-}46)$$

$$b = \frac{V^{-}}{\sqrt{Z_0}} \qquad (4\text{-}1\text{-}47)$$

It can be seen that

$$\text{Incident power} = P_{+} = \frac{1}{2} aa^{*} = \frac{1}{2} |a|^{2} \qquad (4\text{-}1\text{-}48)$$

$$\text{Reflected power} = P_{-} = \frac{1}{2} bb^{*} = \frac{1}{2} |b|^{2} \qquad (4\text{-}1\text{-}49)$$

With no loss of generality, it is assumed that a wave of unit voltage is incident upon port 1 of an n-port junction and that no voltage waves enter any of the other ports. Hence the power input is given by

$$P_{\text{in}} = a_1 a_1^{*} = |a_1|^{2} \qquad (4\text{-}1\text{-}50)$$

which is equal to the power output leaving the ith port—that is,

$$P_{\text{in}} = a_1 a_1^{*} = P_{\text{out}} = \sum_{i}^{n} b_i b_i^{*} = b_1 b_1^{*} + b_2 b_2^{*} + \cdots + b_n b_n^{*} \qquad (4\text{-}1\text{-}51)$$

Since $b_i = S_{i1}a_1$, then

$$a_1 a_1^* = (S_{11}a_1)(S_{11}a_1)^* + (S_{21}a_1)(S_{21}a_1)^* + \cdots + (S_{n1}a_1)(S_{n1}a_1)^* \qquad (4\text{-}1\text{-}52)$$

Consequently

$$1 = S_{11}S_{11}^* + S_{21}S_{21}^* + \cdots + S_{n1}S_{n1}^* \qquad (4\text{-}1\text{-}53)$$

or

$$1 = \sum_i^n S_{ij}S_{ij}^* = \sum_i^n |S_{ij}|^2 \qquad \text{for } j = 1, 2, 3, \ldots \qquad (4\text{-}1\text{-}54)$$

Since S_{ij} is symmetric,

$$1 = \sum_j^n S_{ij}S_{ij}^* = \sum_j^n |S_{ij}|^2 \qquad \text{for } i = 1, 2, 3, \ldots \qquad (4\text{-}1\text{-}55)$$

For a lossy junction, the power dissipated at the junction is

$$P_{\text{diss}} = \frac{1}{2} \sum_{n=1}^n (a_n a_n^* - b_n b_n^*) \qquad (4\text{-}1\text{-}56)$$

It can be shown that

$$\sum_{n=1}^n a_n a_n^* = \tilde{\mathbf{a}}\mathbf{a}^* \qquad (4\text{-}1\text{-}57)$$

and

$$\sum_{n=1}^n b_n b_n^* = \tilde{\mathbf{a}} \, \mathbf{SS}^*\mathbf{a}^* \qquad (4\text{-}1\text{-}58)$$

It should be noted that the right-hand terms of Eqs. (4-1-57) and (4-1-58) are 1×1 matrices or simply numbers. Hence, the power dissipated at the junction is given by

$$P_{\text{diss}} = \frac{1}{2} \tilde{\mathbf{a}}(1 - \mathbf{SS}^*)\mathbf{a}^* \qquad (4\text{-}1\text{-}59)$$

Property 3.   Zero Property. The sum of the products of each term of any row (or column) multiplied by the complex conjugate of the corresponding terms of any other row (or column) is zero.

$$\sum_i^n S_{ik}S_{ij}^* = 0 \qquad \text{for } k \neq j, \qquad \begin{matrix} k = 1, 2, 3, \ldots, n \\ j = 1, 2, 3, \ldots, n \end{matrix} \qquad (4\text{-}1\text{-}60)$$

In general, the incident and reflected waves may exist at each of the n-posts. Then, the incident and reflected power for a lossless junction are

$$\sum_j^n a_j a_j^* = \sum_i^n b_i b_i^* \qquad (4\text{-}1\text{-}61)$$

Substitution of Eq. (4-1-10) and its complex conjugate in Eq. (4-1-60) yields

$$\sum_j^n a_j a_j^* = \sum_j^n \left(\sum_i^n S_{ij} S_{ij}^* \right) a_j a_j^* + \left[\sum_k' \sum_j \left(\sum_i S_{ik} S_{ij}^* \right) a_k a_j^* \right]$$

$$+ \left[\sum_k' \sum_j \left(\sum_i S_{ik} S_{ij}^* \right) a_k a_j^* \right]^* \qquad (4\text{-}1\text{-}62)$$

where the prime on Σ' indicates that the terms of $k = j$ are not included in the sum. The first term on the right-hand side of Eq. (4-1-61) can be simplified by the use of Eq. (4-1-54). The last two terms on the right are of the form $(A + A^*)$, which is equal to twice the real part of A. Equation (4-1-61) can be simplified as

$$0 = 2\text{Re} \sum_k' \sum_j \left(\sum_i S_{ij} S_{ij}^* \right) a_k a_j^* \qquad (4\text{-}1\text{-}63)$$

Since the factor of $a_k a_j^*$ is nonzero

$$\sum_i^n S_{ik} S_{ij}^* = 0 \qquad (4\text{-}1\text{-}64)$$

where $k \neq j$.
$$k = 1, 2, 3, \ldots, n.$$
$$j = 1, 2, 3, \ldots, n.$$

For example, if only a_1 and a_2^* exist at port 1 and port 2, respectively, with all other ports terminated in their characteristic impedance, Eq. (4-1-64) becomes

$$S_{11} S_{12}^* + S_{21} S_{22}^* + S_{31} S_{32}^* + \cdots + S_{n1} S_{n2}^* = 0 \qquad (4\text{-}1\text{-}65)$$

Since S_{ij} is symmetric, it can be shown that

$$S_{11} S_{21}^* + S_{12} S_{22}^* + S_{13} S_{23}^* + \cdots + S_{1n} S_{2n}^* = 0 \qquad (4\text{-}1\text{-}66)$$

This has proved the statement of property 3 of the S parameters.

Property 4.   Phase Shift. If any of the terminal planes (or reference planes), say the kth port, is moved away from the junction by an electric distance $\beta_k \ell_k$, each of the coefficients S_{ij} involving k will be multiplied by the factor $e^{-j\beta_k \ell_k}$.

It is apparent that a change in the specified location of the terminal planes of an arbitrary junction will affect only the phase of the scattering coefficients of the junction. In matrix notation, the new S parameter S' can be written as

$$\mathbf{S'} = \boldsymbol{\phi} \mathbf{S} \boldsymbol{\phi} \qquad (4\text{-}1\text{-}67)$$

where S is the old scattering matrix, and

$$\phi = \begin{bmatrix} \phi_{11} & 0 & 0 & \cdots & 0 \\ 0 & \phi_{22} & 0 & \cdots & 0 \\ \cdots\cdots\cdots\cdots\cdots\cdots\cdots\cdots \\ 0 & 0 & \cdots & 0 & \phi_{nn} \end{bmatrix}$$ (4-1-68)

where $\phi_{11} = \phi_{22} = \phi_{kk} = e^{-j\beta_k\ell_k}$ for $k = 1, 2, 3, \ldots, n$

It should be noted that while S_{ii} involves the product of two factors of $e^{-j\beta\ell}$, the new S' will be changed by a factor of $e^{-j2\beta\ell}$.

4-2 CAVITY RESONATORS AND QUALITY FACTOR Q

In general, a cavity resonator is a metallic enclosure that confines the electromagnetic energy. The stored electric and magnetic energies inside the cavity determine its equivalent inductance and capacitance. The energy dissipated by the finite conductivity of the cavity walls determines its equivalent resistance. In practice, the rectangular-cavity resonator, circular-cavity resonator, and reentrant-cavity resonator are commonly used as shown in Figs. 4-2-1 and 4-2-2.

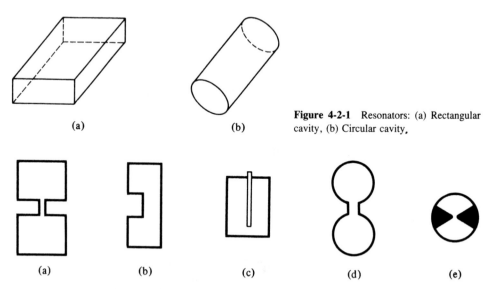

(a)   (b)

Figure 4-2-1 Resonators: (a) Rectangular cavity, (b) Circular cavity.

(a)   (b)   (c)   (d)   (e)

Figure 4-2-2 Reentrant cavities: (a) Coaxial cavity, (b) Radial cavity, (c) Tunable cavity, (d) Toroidal cavity, (e) Butterfly cavity.

Theoretically, a given resonator has an infinite number of resonant modes, and each mode corresponds to a definite resonant frequency. When the frequency of an impressed signal is equal to a resonant frequency, a maximum amplitude of the standing wave occurs, and the peak energies stored in the electric and magnetic fields are equal. The mode having the lowest resonant frequency is known as the *dominant mode*.

As the operating frequency is increased, both the inductance and capacitance of the resonant circuit must be decreased in order to maintain resonance at the operating frequency. Because the gain-bandwidth product is limited by the resonant circuit, the ordinary resonator cannot generate a large output. Several nonresonant periodic circuits or slow-wave structures (see Fig. 4-2-3) are designed for producing large gain over a wide bandwidth.

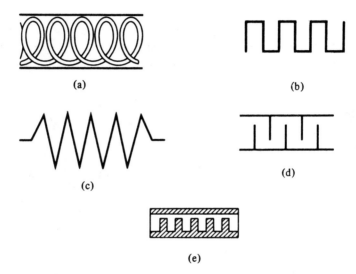

Figure 4-2-3   Slow-wave structures: (a) Helical line, (b) Folded-back line, (c) Zigzag line, (d) Interdigital line, (e) Corrugated waveguide.

4-2-1 Rectangular-Cavity Resonator

The electromagnetic field inside the cavity should satisfy Maxwell's equations subject to the boundary conditions that the electric field tangential to and the magnetic field normal to the metal walls must vanish. The geometry of a rectangular cavity is shown in Fig. 4-2-4.

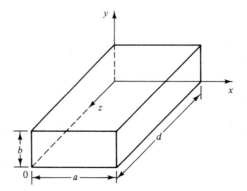

Figure 4-2-4   Coordinates of a rectangular cavity.

The wave equations in the rectangular resonator should satisfy the boundary condition of zero tangential **E** at four of the walls. It is merely necessary to choose the harmonic functions in z to satisfy this condition at the remaining two-end walls. These can be accomplished if

$$H_z = H_{0z} \cos\left(\frac{m\pi x}{a}\right) \cos\left(\frac{n\pi x}{b}\right) \sin\left(\frac{p\pi z}{d}\right) \qquad \text{TE}_{mnp} \qquad (4\text{-}2\text{-}1)$$

where $m = 0, 1, 2, 3, \ldots$ represents the number of the half-wave periodicity in the x-direction

$n = 0, 1, 2, 3, \ldots$ represents the number of the half-wave periodicity in the y-direction

$p = 1, 2, 3, 4, \ldots$ represents the number of the half-wave periodicity in the z-direction

and

$$E_z = E_{0z} \sin\left(\frac{m\pi x}{a}\right) \sin\left(\frac{n\pi x}{b}\right) \cos\left(\frac{p\pi z}{d}\right) \qquad \text{TM}_{mnp} \qquad (4\text{-}2\text{-}2)$$

where $m = 1, 2, 3, 4, \ldots$

$n = 1, 2, 3, 4, \ldots$

$p = 0, 1, 2, 3, \ldots$

The separation equation for both the TE and TM modes is given by

$$k^2 = \left(\frac{m\pi}{a}\right)^2 + \left(\frac{n\pi}{b}\right)^2 + \left(\frac{p\pi}{d}\right)^2 \qquad (4\text{-}2\text{-}3)$$

For a lossless dielectric, $k^2 = \omega^2 \mu\epsilon$; therefore, the resonant frequency is expressed by

$$f_r = \frac{1}{2\sqrt{\mu\epsilon}} \sqrt{\left(\frac{m}{a}\right)^2 + \left(\frac{n}{b}\right)^2 + \left(\frac{p}{d}\right)^2} \qquad \text{TE and TM} \qquad (4\text{-}2\text{-}4)$$

For $a > b < d$, the dominant mode is the TE_{101} mode.

In general, a straight-wire probe inserted at the position of maximum electric intensity is used to excite a desired mode; and the loop coupling placed at the position of maximum magnetic intensity is utilized to launch a specific mode. Figure 4-2-5 shows the methods of excitation for the rectangular resonator. The maximum amplitude of the standing wave occurs when the frequency of the impressed signal is equal to the resonant frequency.

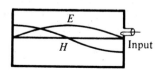
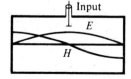

Figure 4-2-5 Methods of exciting wave modes in a resonator.

Example 4-2-1: Rectangular Resonator

An air rectangular resonator has the following dimensions:

$$
\begin{array}{lll}
\text{Width} & a = 2.286 \text{ cm} \\
\text{Height} & b = 1.016 \text{ cm} \\
\text{Length} & d = 3.429 \text{ cm}
\end{array}
$$

Compute the resonant frequency.

Solution: From Eq. (4-2-4), the resonant frequency is

$$
f_r = \frac{3 \times 10^8}{2} \sqrt{\left(\frac{1}{2.286 \times 10^{-2}}\right)^2 + 0 + \left(\frac{1}{3.429 \times 10^{-2}}\right)^2}
$$

$$
= 7.886 \text{ GHz}
$$

4-2-2 Circular-Cavity Resonator and Semicircular-Cavity Resonator

Circular-cavity resonator.   A circular-cavity resonator is a circular waveguide with two ends closed by metal wall. This is shown in Fig. 4-2-6.

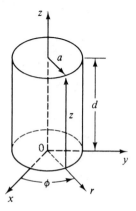

Figure 4-2-6 Coordinates of a circular resonator.

The wave function in the circular resonator should satisfy Maxwell's equations subject to the same boundary conditions as described for a rectangular-cavity resonator. It is merely necessary to choose the harmonic functions in z to satisfy the boundary conditions at the remaining two-end walls. These can be achieved if

$$H_z = H_{0z} J_n \left(\frac{X'_{np} r}{a} \right) \cos n_\phi \sin \left(\frac{q\pi z}{d} \right) \qquad \text{TE}_{npq} \qquad (4\text{-}2\text{-}5)$$

where $n = 0, 1, 2, 3, \ldots$ is the number of the periodicity in the ϕ direction
$\quad\quad p = 1, 2, 3, 4, \ldots$ is the number of zeros of the field in the radial direction
$\quad\quad q = 1, 2, 3, 4, \ldots$ is the number of the half-wave in the axial direction
$\quad\quad J_n = $ Bessel function of the first kind
$\quad\quad H_{0z} = $ the amplitude of the magnetic field

and

$$E_z = E_{0z} J_n \left(\frac{X_{np} r}{a} \right) \cos (n\phi) \cos \left(\frac{q\pi z}{d} \right) \qquad \text{TM}_{npq} \qquad (4\text{-}2\text{-}6)$$

where $n = 0, 1, 2, 3, \ldots$
$\quad\quad p = 1, 2, 3, 4, \ldots$
$\quad\quad q = 0, 1, 2, 3, \ldots$
$\quad\quad E_{0z} = $ amplitude of the electric field

The separation equation for the TE and TM modes are given, respectively, by

$$k^2 = \left(\frac{X'_{np}}{a} \right)^2 + \left(\frac{q\pi}{d} \right)^2 \qquad \text{TE mode} \qquad (4\text{-}2\text{-}7a)$$

$$k^2 = \left(\frac{X_{np}}{a} \right)^2 + \left(\frac{q\pi}{d} \right)^2 \qquad \text{TM mode} \qquad (4\text{-}2\text{-}7b)$$

Substitution of $k^2 = \omega^2 \mu\epsilon$ in Eqs. (4-2-7) yields the resonant frequencies for the TE and TM modes, respectively, as

$$f_r = \frac{1}{2\pi\sqrt{\mu\epsilon}} \sqrt{\left(\frac{X'_{np}}{a} \right)^2 + \left(\frac{q\pi}{d} \right)^2} \qquad \text{TE} \qquad (4\text{-}2\text{-}8a)$$

$$f_r = \frac{1}{2\pi\sqrt{\mu\epsilon}} \sqrt{\left(\frac{X_{np}}{a} \right)^2 + \left(\frac{q\pi}{d} \right)^2} \qquad \text{TM} \qquad (4\text{-}2\text{-}8b)$$

It is interesting to note that for $2a > d$, the TM_{010} mode is dominant; and for $d \geq 2a$, the TE_{111} mode is dominant.

Example 4-2-2: Circular-Cavity Resonator

An air circular resonator has the following dimensions:

$$\begin{array}{lll} \text{Radius of circular cavity} & a = 2.383 \text{ cm} \\ \text{Length of circular cavity} & d = 5.123 \text{ cm} \end{array}$$

(a) Determine the dominant mode.

(b) Calculate the dominant resonant frequency.

Solution:

(a) The dominent mode is TE_{111} for $2a < d$.

(b) From Eq. (4-2-8a), the resonant frequency is

$$f_r = \frac{3 \times 10^8}{2 \times 3.1416} \sqrt{\left(\frac{1.841}{2.383 \times 10^{-2}}\right)^2 + \left(\frac{1 \times 3.1416}{5.123 \times 10^{-2}}\right)^2}$$

$$= 0.47746 \times 10^8 \times 98.6356$$

$$= 4.71 \text{ GHz}$$

Semicircular-cavity resonator.   A semicircular-cavity resonator is shown in Fig. 4-2-7.

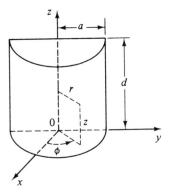

Figure 4-2-7   Semicircular resonator.

The wave function of the TE_{npq} mode in the semicircular resonator can be written as

$$H_z = H_{0z}J_n\left(\frac{X'_{np}r}{a}\right)\cos(n\phi)\sin\left(\frac{q\pi}{d}\right) \qquad \text{TE mode} \qquad (4\text{-}2\text{-}9)$$

where $n = 0, 1, 2, 3, \ldots$
$\quad\quad\;\; p = 1, 2, 3, 4, \ldots$
$\quad\quad\;\; q = 1, 2, 3, 4, \ldots$
$\quad\quad\;\; a =$ radius of the semicircular-cavity resonator
$\quad\quad\;\; d =$ length of the resonator

The wave function of the TM_{npq} in the semicircular cavity resonator can be written as

$$E_z = E_{0z}J_n\left(\frac{X_{np}r}{a}\right)\sin(n\phi)\cos\left(\frac{q\pi}{d}z\right) \qquad \text{TM mode} \qquad (4\text{-}2\text{-}10)$$

where $n = 1, 2, 3, 4, \ldots$
$\quad\quad\;\; p = 1, 2, 3, 4, \ldots$
$\quad\quad\;\; q = 0, 1, 2, 3, \ldots$

With the separation equations as given in Eqs. (4-2-7) the equations of resonant frequency for the TE and TM modes in a semicircular-cavity resonator are the same as in the circular-cavity resonator. These are repeated as follows:

$$f_r = \frac{1}{2\pi a\sqrt{\mu\epsilon}}\sqrt{(X'_{np})^2 + \left(\frac{q\pi a}{d}\right)^2} \qquad \text{TE}_{npq} \text{ mode} \qquad (4\text{-}2\text{-}11)$$

$$f_r = \frac{1}{2\pi a\sqrt{\mu\epsilon}}\sqrt{(X_{np})^2 + \left(\frac{q\pi a}{d}\right)^2} \qquad \text{TM}_{npq} \text{ mode} \qquad (4\text{-}2\text{-}12)$$

However, the values of the subscripts n, p, and q are different from that for the circular-cavity resonator. Also it must be emphasized that the TE_{111} mode is dominant if $d > a$, and the TM_{110} mode is dominant if $d < a$.

Example 4-2-2a: Semicircular Resonator

An air semicircular resonator has the following parameters:

Radius of semicircular cavity $a = 3.254$ cm
Length of semicircular cavity $d = 2.768$ cm

(a) Determine the dominant mode.
(b) Compute the resonant frequency.

Solution:

(a) The dominant mode is TM_{110} for $a > d$.
(b) From Eq. (4-2-12), the resonant frequency is

$$f_r = \frac{3 \times 10^8}{2 \times 3.1416 \times 3.254 \times 10^{-2}}\sqrt{(3.832)^2 + 0}$$
$$= 5.623 \text{ GHz}$$

4-2-3 The Q of a Cavity Resonator

The quality factor Q is a measure of the frequency selectivity of a resonant or antiresonant circuit, and it is defined as

$$Q \equiv 2\pi \frac{\text{maximum energy stored}}{\text{energy dissipated per cycle}} = \frac{\omega W}{P} \qquad (4\text{-}2\text{-}13)$$

where W = maximum stored energy
P = average power loss

At resonant frequency, the electric and magnetic energies are equal and in time quadrature. When the electric energy is maximum, the magnetic energy is zero, and vice versa. The total energy stored in the resonator is obtained by integrating, respectively, the energy density over the volume of the resonator, namely,

$$W_e = \int_v \frac{\epsilon}{2} |E|^2 dv = W_m = \int_v \frac{\mu}{2} |H|^2 dv = W \tag{4-2-14}$$

where $|E|$ and $|H|$ are the peak values of the field intensities.

The average power loss in the resonator may be evaluated by integrating the power density equation over the inner surface of the resonator, hence

$$P = \frac{R_s}{2} \int_s |H_t|^2 da \tag{4-2-15}$$

where H_t is the peak value of the tangential magnetic intensity and R_s is the surface resistance of the resonator.

Substitution of Eqs. (4-2-14) and (4-2-15) in Eq. (4-2-13) yields

$$Q = \frac{\omega\mu \int_v |H|^2 dv}{R_s \int_s |H_t|^2 da} \tag{4-2-16}$$

Since the peak value of the magnetic intensity is related to its tangential and normal components by

$$|H|^2 = |H_t|^2 + |H_n|^2$$

where H_n is the peak value of the normal magnetic intensity, the value of $|H_t|^2$ at the resonator walls is approximately twice the value of $|H|^2$ averaged over the volume. Hence, the Q of a cavity resonator as shown in Eq. (4-2-16) can be expressed approximately by

$$Q = \frac{\omega\mu \text{ (volume)}}{2R_s \text{ (surface areas)}} \tag{4-2-17}$$

An unloaded resonator may be represented by either a series or parallel resonant circuit. The resonant frequency and the unloaded Q_0 of the cavity resonator are

$$f_0 = \frac{1}{2\pi\sqrt{LC}} \tag{4-2-18}$$

and

$$Q_0 = \frac{\omega_0 L}{R} \tag{4-2-19}$$

If the cavity is coupled by means of an ideal $N:1$ transformer and a series inductance L_s to a generator having internal impedance Z_g, Fig. 4-2-8 shows the coupling circuit and its equivalent.

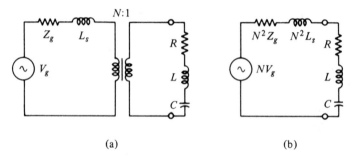

Figure 4-2-8   Cavity coupled to a generator.

The loaded Q_ℓ of the system is given by

$$Q_\ell = \frac{\omega_0 L}{R + N^2 Z_g} \qquad \text{for } |N^2 L_s| \ll |R + N^2 Z_g| \qquad (4\text{-}2\text{-}20)$$

The coupling coefficient of the system is defined as

$$K \equiv \frac{N^2 Z_g}{R} \qquad (4\text{-}2\text{-}21)$$

The loaded Q_ℓ would become

$$Q_\ell = \frac{\omega_0 L}{R(1 + K)} = \frac{Q_0}{1 + K} \qquad (4\text{-}2\text{-}22)$$

Rearrangement of Eq. (4-3-10) yields

$$\frac{1}{Q_\ell} = \frac{1}{Q_0} + \frac{1}{Q_{\text{ext}}} \qquad (4\text{-}2\text{-}23)$$

where $Q_{\text{ext}} = \dfrac{Q_0}{K} = \dfrac{\omega_0 L}{KR}$ is the external Q.

There are three cases:

1. *Critical coupling:* If the resonator is matched to the generator, then

$$K = 1 \qquad (4\text{-}2\text{-}24)$$

The loaded Q_ℓ is given by

$$Q_\ell = \frac{1}{2} Q_{\text{ext}} = \frac{1}{2} Q_0 \qquad (4\text{-}2\text{-}25)$$

2. *Overcoupling:* If $K > 1$, the cavity terminals are at a voltage maximum in the input line at resonance. The normalized impedance at the voltage maximum is the standing-wave ratio ρ. That is

$$K = \rho \tag{4-2-26}$$

The loaded Q_ℓ is given by

$$Q_\ell = \frac{Q_0}{1 + \rho} \tag{4-2-27}$$

3. *Undercoupling:* If $K < 1$, the cavity terminals are at a voltage minimum and the input terminal impedance is equal to the reciprocal of the standing-wave ratio. That is

$$K = \frac{1}{\rho} \tag{4-2-28}$$

The loaded Q_ℓ is given by

$$Q_\ell = \frac{\rho}{\rho + 1} Q_0 \tag{4-2-29}$$

The relationship of the coupling coefficient K and the standing-wave ratio ρ is shown in Fig. 4-2-9.

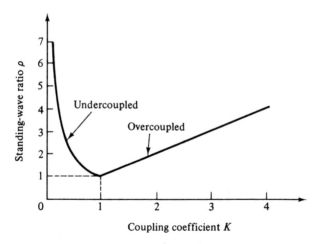

Figure 4-2-9 Coupling coefficient versus standing-wave ratio.

Example 4-2-3: Quality Factor Q

An air resonator has the following parameters:

Unloaded quality factor	$Q_0 = 3000$
Overcoupled coefficient	$K = 2$

(a) Determine the loaded Q_ℓ.

(b) Compute the external Q_{ext}.

Solution:

(a) The loaded quality factor is

$$Q_\ell = \frac{Q_0}{1 + K} = \frac{3000}{1 + 2} = 1000$$

(b) The external quality factor is

$$Q_{\text{ext}} = \frac{Q_0}{K} = \frac{3000}{2} = 1500$$

4-3 WAVEGUIDE TEES AND DIRECTIONAL COUPLERS

4-3-1 Waveguide Tees

Waveguide tees may consist of the E-plane tee, H-plane tee, magic tee, hybrid rings, corners, bends, and twists. All such waveguide components are discussed in this section.

Tee junctions.   In microwave circuits, a waveguide or coaxial-line junction with three independent ports is commonly referred to as a *tee junction*. From the S-parameter theory of a microwave junction, it is evident that a tee junction should be characterized by a matrix of third order containing nine elements, six of which should be independent. The characteristics of a three-port junction may be explained in three theorems of the tee junction. These theorems are derived from the equivalent-circuit representation of the tee junction. Their statements follow [4].

1. A short circuit may always be placed in one of the arms of a three-port junction in such a way that no power can be transferred through the other two arms.
2. If the junction is symmetrical about one of its arms, a short circuit can always be placed in that arm so that no reflections occur in power transmission between the other two arms. (That is, the arms present matched impedances.)
3. It is impossible for a general three-port junction of arbitrary symmetry to present matched impedances at all three arms.

The E-plane tee and H-plane tee are described below.

E-Plane Tee (Series Tee).   An E-plane tee is a waveguide tee in which the axis of its side arm is parallel to the E field of the main guide (see Fig. 4-3-1). If the collinear arms are symmetric about the side arm, there are two different transmission characteristics (see Fig. 4-3-2). It can be seen from Fig. 4-3-1 that if the E-plane tee is perfectly matched with the aid of screw tuners or inductive or capactive windows at the junction, the diagonal components of the scattering matrix S_{11}, S_{22}, and S_{33} are zero because there will be no reflection. When the waves are fed into the side arm (port 3), the waves appearing at port 1 and port 2 of the collinear arm will be in opposite phase and in the same magnitude. Therefore,

$$S_{13} = -S_{23} \qquad (4\text{-}3\text{-}1)$$

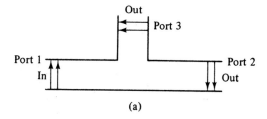

(a)

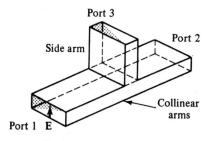

Figure 4-3-1 E-plane tee.

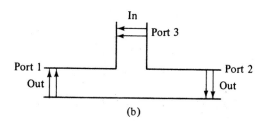

(b)

Figure 4-3-2 Two-way transmission of E-plane tee: (a) Input through main arm, (b) Input from side arm.

It should be noted that Eq. (4-3-1) does not mean that S_{13} is always positive and S_{23} is always negative. The negative sign merely means that S_{13} and S_{23} have opposite signs. For a matched junction, the S-matrix is given by

$$\mathbf{S} = \begin{bmatrix} 0 & S_{12} & S_{13} \\ S_{21} & 0 & S_{23} \\ S_{31} & S_{32} & 0 \end{bmatrix} \qquad (4\text{-}3\text{-}2)$$

From property 1 of the **S** matrix in Eq. (4-1-36), the symmetric terms in Eq. (4-3-2) are equal and they are

$$S_{12} = S_{21} \qquad S_{13} = S_{31} \qquad S_{23} = S_{32} \qquad (4\text{-}3\text{-}3)$$

From property 3 of the **S** matrix in Eq. (4-1-52), the sum of the products of each term of any column (or row) multiplied by the complex conjugate of the corresponding terms of any other column (or row) is zero and it is

$$S_{11}S_{12}^* + S_{21}S_{22}^* + S_{31}S_{32}^* = 0 \qquad (4\text{-}3\text{-}4)$$

Hence

$$S_{13}S_{23}^* = 0 \qquad (4\text{-}3\text{-}5)$$

This means that either S_{13} or S_{23}^*, or both, should be zero. However, from property 2 of the **S** matrix in Eq. (4-1-37), the sum of the products of each term of any one row (or column) multiplied by its complex conjugate is unity; that is,

$$S_{21}S_{21}^* + S_{31}S_{31}^* = 1 \tag{4-3-6}$$

$$S_{12}S_{21}^* + S_{32}S_{32}^* = 1 \tag{4-3-7}$$

$$S_{13}S_{13}^* + S_{23}S_{23}^* = 1 \tag{4-3-8}$$

Substitution of Eq. (4-3-3) in Eq. (4-3-6) results in

$$|S_{12}|^2 = 1 - |S_{13}|^2 = 1 - |S_{23}|^2 \tag{4-3-9}$$

Equations (4-3-8) and (4-3-9) are contradictory, for if $S_{13} = 0$, S_{23} is also zero, and thus Eq. (4-3-8) is false. In a similar fashion, if $S_{23} = 0$, S_{13} becomes zero, and therefore Eq. (4-3-8) is not true. This inconsistency proves the statement that a tee junction cannot be matched to the three arms. In other words, the diagonal elements of the **S** matrix of a tee junction are not all zeros.

In general, when an E-plane tee is constructed of an empty waveguide, it is poorly matched at the tee junction. Hence, $S_{ij} \neq 0$ if $i = j$. However, since the collinear arm is usually symmetrical about the side arm, $|S_{13}| = |S_{23}|$ and $S_{11} = S_{22}$. Then the **S** matrix can be simplified to

$$\mathbf{S} = \begin{bmatrix} S_{11} & S_{12} & S_{13} \\ S_{12} & S_{11} & -S_{13} \\ S_{13} & -S_{13} & S_{33} \end{bmatrix} \tag{4-3-10}$$

H-Plane Tee (Shunt Tee). An H-plane tee is a waveguide tee in which the axis of its side arm is "shunting" the E field or parallel to the H field of the main guide as shown in Fig. 4-3-3.

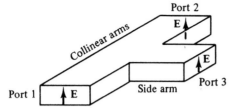

Figure 4-3-3   H-plane tee.

It can be seen that if two input waves are fed into port 1 and port 2 of the collinear arm, the output wave at port 3 will be in phase and additive. On the other hand, if the input is fed into port 3, the wave will split equally into port 1 and port 2 in phase and in the same magnitude. Therefore, the **S** matrix of the H-plane tee is similar to Eqs. (4-3-2) and (4-3-10), except that

$$S_{13} = S_{23} \tag{4-3-11}$$

Magic Tee (Hybrid Tee).   A magic tee is a combination of the *E*-plane tee and *H*-plane tee (refer to Fig. 4-3-4). The magic tee has several characteristics:

1. If two waves of equal magnitude and same phase are fed into port 1 and port 2, the output will be zero at port 3 and additive at port 4.

2. If a wave is fed into port 4 (the *H* arm), it will be divided equally between port 1 and port 2 of the collinear arms and will not appear at port 3 (the *E* arm).

3. If a wave is fed into port 3 (the *E* arm), it will produce an output of equal magnitude and opposite phase at port 1 and port 2. The output at port 4 is zero. That is, $S_{43} = S_{34} = 0$.

4. If a wave is fed into one of the collinear arms at port 1 or port 2, it will not appear in the other collinear arm at port 2, or port 1 because the *E* arm causes a phase delay while the *H* arm causes a phase advance. That is, $S_{12} = S_{21} = 0$

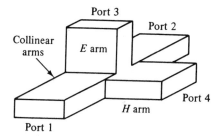

Figure 4-3-4   Magic tee.

Therefore, the **S** matrix of a magic tee can be expressed as

$$\mathbf{S} = \begin{bmatrix} 0 & 0 & S_{13} & S_{14} \\ 0 & 0 & S_{23} & S_{24} \\ S_{31} & S_{32} & 0 & 0 \\ S_{41} & S_{42} & 0 & 0 \end{bmatrix} \qquad (4\text{-}3\text{-}12)$$

The magic tee is commonly used for mixing, duplexing, and impedance measurements. For example, there are two identical radar transmitters in equipment stock. A particular application requires twice more input power to an antenna than either transmitter can deliver. A magic tee may be used to couple the two transmitters to the antenna in such a way that the transmitters do not load each other. The two transmitters should be connected to ports 3 and 4, respectively, as shown in Fig. 4-3-5. Transmitter 1, connected to port 3, causes a wave to emanate from port 1 and another to emanate from port 2; these waves are equal in magnitude but opposite in phase. Likewise, transmitter 2, connected to port 4, gives rise to a wave at port 1 and another at port 2, both equal in magnitude and in phase. At port 1, the two opposite waves cancel each other. At port 2, the two in-phase waves add together, so double output power at port 2 is obtained for the antenna as shown in Fig. 4-3-5.

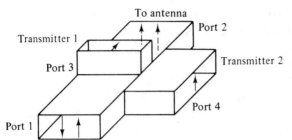

Figure 4-3-5 Magic tee coupled transmitter to antenna.

Hybrid Rings (Rat-Race Circuits). A hybrid ring consists of an annular line of proper electrical length to sustain standing waves, to which four arms are connected at proper intervals by means of series or parallel junctions. Figure 4-3-6 shows a hybrid ring with series junctions.

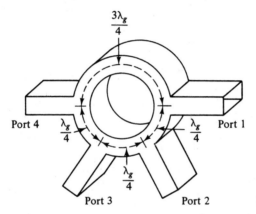

Figure 4-3-6 Hybrid ring.

The hybrid ring has similar characteristics as those of the hybrid tee. When a wave is fed into port 1, it will not appear at port 3 because the difference of phase shifts for the waves traveling in the clockwise and counterclockwise directions is 180°. Thus, the waves are cancelled at port 3. For the same reason, the waves fed into port 2 will not emerge at port 4, and so on.

The **S** matrix for an ideal hybrid ring can be expressed as

$$\mathbf{S} = \begin{bmatrix} 0 & S_{12} & 0 & S_{14} \\ S_{21} & 0 & S_{23} & 0 \\ 0 & S_{32} & 0 & S_{34} \\ S_{41} & 0 & S_{43} & 0 \end{bmatrix} \tag{4-3-13}$$

It should be noted that the phase cancellation occurs only at a designed frequency for an ideal hybrid ring. In actual hybrid rings, there are small leakage couplings, and therefore the zero elements in the matrix Eq. (4-3-13) are not quite equal to zero.

Waveguide Corners, Bends, and Twists.   The waveguide corner, bend, and twist are shown in Fig. 4-3-7. These waveguide components are commonly used to change the direction of the guide through an arbitrary angle.

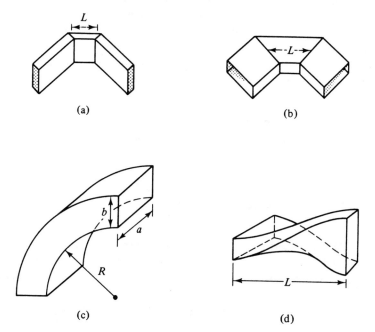

Figure 4-3-7   Waveguide corner, bend and twist: (a) E-plane corner, (b) H-plane corner, (c) Bend, (d) Continuous twist.

In order to minimize reflections from the discontinuities, it is desirable to have the mean length L between continuities equal to an odd number of quarter-wavelengths. That is,

$$L = (2n + 1) \frac{\lambda_g}{4}$$

(4-3-14)

where $n = 0, 1, 2, 3, \ldots$, and λ_g is the wavelength in the waveguide. If the mean length L is an odd number of quarter-wavelengths, the reflected waves from both ends of the waveguide section are completely cancelled. For the waveguide bend, the minimum radius of curvature for a small reflection is given by

$$R = 1.5b \qquad \text{for an } E \text{ bend}$$

(4-3-15)

and

$$R = 1.5a \qquad \text{for an } H \text{ bend}$$

(4-3-16)

where a and b are the dimensions of the waveguide bend as illustrated in Fig. 4-3-7(c).

4-3-2 Directional Couplers

A directional coupler is a four-port waveguide junction as shown in Fig. 4-3-8. It consists of a primary waveguide 1–2 and a secondary waveguide 3–4. When all ports are terminated in their characteristic impedances, there is free transmission of power, without reflection, between port 1 and port 2, and there is no transmission of power between port 1 and port 3 or between port 2 and port 4 because no coupling exists between these two pairs of ports. The degree of coupling between port 1 and port 4 and between port 2 and port 3 depends upon the structure of the coupler.

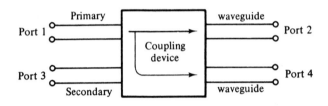

Figure 4-3-8   Directional coupler.

The characteristics of a directional coupler can be expressed in terms of its coupling factor and its directivity. Assuming the wave is propagating from port 1 to port 2 in the primary line, the coupling factor and the directivity are defined, respectively, by

$$\text{Coupling factor (dB)} = 10 \log_{10} \frac{P_1}{P_4} \tag{4-3-17}$$

$$\text{Directivity (dB)} = 10 \log_{10} \frac{P_4}{P_3} \tag{4-3-18}$$

where P_1 = power input to port 1
P_3 = power output from port 3
P_4 = power output from port 4

It should be noted that port 2, port 3, and port 4 are terminated in their characteristic impedances. The coupling factor is a measure of the ratio of power levels in the primary and secondary lines. Therefore, if the coupling factor is known, a fraction of power measured at port 4 may be used to determine the power input at port 1. This significance is very desirable for microwave power measurements because no disturbance, which may be caused by the power measurements, occurs in the primary line. The directivity is a measure of how well the forward traveling wave in the primary waveguide couples only to a specific port of the secondary waveguide. An ideal directional coupler should have infinite directivity. This means that the power at port 3 must be zero because port 2 and port 4 are perfectly matched. Actually, well-designed directional couplers have a directivity of only 30 to 35 dB.

Several types of directional couplers exist, such as a two-hole directional coupler, four-hole directional coupler, reverse coupling directional coupler (Schwinger coupler), and Bethe-hole directional coupler, as shown in Fig. 4-3-9. Only the very commonly used two-hole directional coupler is described here.

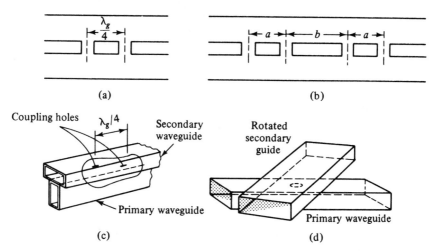

(a)

(b)

(c)

(d)

Figure 4-3-9   Different directional couplers: (a) Two-hole directional coupler, (b) Four-hole directional coupler, (c) Schwinger coupler, (d) Bethe-hole directional coupler.

Two-Hole Directional Couplers.   A two-hole directional coupler with traveling waves propagating in it is illustrated in Fig. 4-3-10. The spacing between the centers of two holes must be

$$L = (2n + 1) \frac{\lambda_g}{4} \qquad (4\text{-}3\text{-}19)$$

where n is any positive integer.

A fraction of the wave energy entered into port 1 passes through the holes and is radiated into the secondary guide as the holes act as slot antennas. The forward waves in the secondary guide are in same phase, regardless of the hole space, and are added at port 4. The backward waves in the secondary guide (waves are progressing from the right to left) are out of phase by $(2L/\lambda_g)2\pi$ radians and are canceled at port 3.

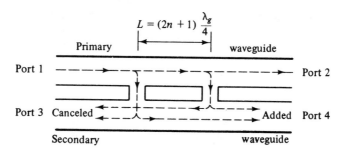

Figure 4-3-10   Two-hole directional coupler.

S *Matrix of a Directional Coupler.* In a directional coupler, all four ports are completely matched. Hence, the diagonal elements of the **S** matrix are zeros and

$$S_{11} = S_{22} = S_{33} = S_{44} = 0 \qquad (4\text{-}3\text{-}20)$$

As noted, there is no coupling between port 1 and port 3 and between port 2 and port 4. Thus,

$$S_{13} = S_{31} = S_{24} = S_{42} = 0 \qquad (4\text{-}3\text{-}21)$$

Consequently, the **S** matrix of a directional coupler becomes

$$\mathbf{S} = \begin{bmatrix} 0 & S_{12} & 0 & S_{14} \\ S_{21} & 0 & S_{23} & 0 \\ 0 & S_{32} & 0 & S_{34} \\ S_{41} & 0 & S_{43} & 0 \end{bmatrix} \qquad (4\text{-}3\text{-}22)$$

Equation (4-3-22) can be reduced further by means of the properties of the S parameters as described in Section 4-1. From Eq. (4-1-52), two relationships are given by

$$S_{12}S_{14}^* + S_{32}S_{34}^* = 0 \qquad (4\text{-}3\text{-}23)$$

and

$$S_{21}S_{23}^* + S_{41}S_{43}^* = 0 \qquad (4\text{-}3\text{-}24)$$

From Eq. (4-1-37), one may write

$$S_{12}S_{12}^* + S_{14}S_{14}^* = 1 \qquad (4\text{-}3\text{-}25)$$

Equations (4-3-23) and (4-3-24) can be also written as

$$|S_{12}|\,|S_{14}| = |S_{32}|\,|S_{34}| \qquad (4\text{-}3\text{-}26)$$

and

$$|S_{21}|\,|S_{23}| = |S_{41}|\,|S_{43}| \qquad (4\text{-}3\text{-}27)$$

Since $S_{12} = S_{21}$, $S_{14} = S_{41}$, $S_{23} = S_{32}$, and $S_{34} = S_{43}$, then

$$|S_{12}| = |S_{34}| \qquad (4\text{-}3\text{-}28)$$

$$|S_{14}| = |S_{23}| \qquad (4\text{-}3\text{-}29)$$

Let

$$S_{12} = S_{34} = p \qquad (4\text{-}3\text{-}30)$$

where p is positive and real. Then, from Eq. (4-3-24)

$$p(S_{23}^* + S_{41}) = 0 \qquad (4\text{-}3\text{-}31)$$

Let

$$S_{23} = S_{41} = jq \tag{4-3-32}$$

where q is positive and real. Then, from Eq. (4-3-25)

$$p^2 + q^2 = 1 \tag{4-3-33}$$

The **S** matrix of a directional coupler is reduced to

$$\mathbf{S} = \begin{bmatrix} 0 & p & 0 & jq \\ p & 0 & jq & 0 \\ 0 & jq & 0 & p \\ jq & 0 & p & 0 \end{bmatrix} \tag{4-3-34}$$

Example 4-3-1:   Directional Coupler

A symmetrical directional coupler with infinite directivity and a forward attenuation of 20 dB is used to monitor the power delivered to a load Z_ℓ, as shown in Fig. 4-3-11. Bolometer 1 introduces a VSWR of 2.00 on arm 4; bolometer 2 is matched to arm 3. If bolometer 1 reads 8 mW and bolometer 2 reads 2 mW, find

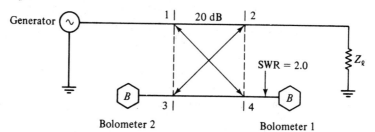

Figure 4-3-11   Power measurements by directional coupler.

(a) The amount of power dissipated in the load Z_ℓ.
(b) The VSWR on arm 2.

Solution: The wave propagation in the directional coupler is shown in Fig. 4-3-12.

(a) Power dissipation at Z_ℓ

    1. The reflection coefficient at port 4 is

$$|\Gamma| = \frac{\rho - 1}{\rho + 1} = \frac{2 - 1}{2 + 1} = \frac{1}{3}$$

    2. Since the incident power and reflected power are related by

$$P^- = P^+ |\Gamma|^2$$

where P^+ = incident power
       P^- = reflected power

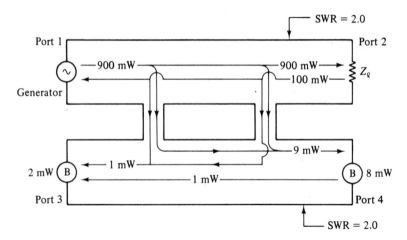

Figure 4-3-12 Wave propagation in the directional coupler.

then

$$|\Gamma| = \frac{1}{3} = \sqrt{\frac{P^-}{P^+}} = \sqrt{\frac{P^-}{8 + P^-}}$$

The incident power to port 4 is $P_4^+ = 9$ mW, and the reflected power from port 4 is $P_4^- = 1$ mW.

3. Since port 3 is matched and the bolometer at port 3 reads 2 mW, 1 mW must be radiated through the holes.
4. Since 20 dB are equivalent to a power ratio of 100 to 1, the power input at port 1 is given by

$$P_1 = 100 \, P_4^+ = 900 \text{ mW}$$

and the power reflected from the load is

$$P_2^- = 100 \, (1 \text{ mW}) = 100 \text{ mW}$$

5. The power dissipated in the load is

$$P_L = P_2^+ - P_2^- = 900 - 100 = 800 \text{ mW}$$

(b) The reflection coefficient is calculated as

$$|\Gamma| = \sqrt{\frac{P^-}{P^+}} = \sqrt{\frac{100}{900}} = \frac{1}{3}$$

then the VSWR is

$$\rho = \frac{1 + |\Gamma|}{1 - |\Gamma|} = \frac{1 + \dfrac{1}{3}}{1 - \dfrac{1}{3}} = 2.0 \qquad \text{on arm 2}$$

4-4 REENTRANT CAVITIES AND SLOW-WAVE STRUCTURES

4-4-1 Reentrant Cavities

At a frequency well below the microwave range, the cavity resonator can be represented by a lumped-constant resonant circuit. When the operating frequency is increased to several tens of megahertz, both the inductance and capacitance must be reduced to a minimum in order to maintain resonance at the operating frequency. Ultimately, the inductance is reduced to a minimum by short wire. Therefore, the reentrant cavities are designed for use in klystrons and microwave triodes. A reentrant cavity is one in which the metallic boundaries extend into the interior of the cavity. Several types of the reentrant cavities are shown in Fig. 4-2-2. One of the commonly used reentrant cavities is the coaxial cavity as shown in Fig. 4-4-1.

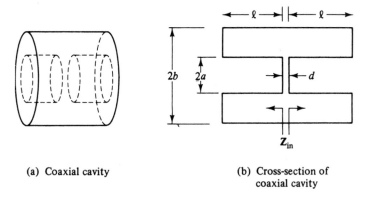

(a) Coaxial cavity

(b) Cross-section of coaxial cavity

Figure 4-4-1   Coaxial cavity and its equivalent: (a) Coaxial cavity, (b) Cross-section of coaxial cavity.

It can be seen that not only the inductance has been considerably decreased, but also the resistance losses are markedly reduced, and the shelf-shielding enclosure prevents radiation losses. It is difficult to calculate the resonant frequency of a coaxial cavity. However, an approximation can be made from the transmission-line theory. The characteristic impedance of a coaxial line is given by

$$Z_0 = \frac{1}{2\pi} \sqrt{\frac{\mu}{\epsilon}} \, \ell n \left(\frac{b}{a} \right) \qquad \text{ohms} \qquad (4\text{-}4\text{-}1)$$

The coaxial cavity is similar to a coaxial line shorted at two ends and joined at the center by a capacitance. The input impedance to each shorted coaxial line is given by

$$Z_{\text{in}} = j Z_0 \tan{(\beta \ell)} \qquad (4\text{-}4\text{-}2)$$

where ℓ is the length of the coaxial line.

Subsitution of Eq. (4-4-1) in Eq. (4-4-2) results in

$$Z_{in} = j \frac{1}{2\pi} \sqrt{\frac{\mu}{\epsilon}} \, \ell n \left(\frac{b}{a}\right) \tan(\beta \ell) \tag{4-4-3}$$

The inductance of the cavity is given by

$$L = \frac{2X_{in}}{\omega} = \frac{1}{\pi\omega} \sqrt{\frac{\mu}{\epsilon}} \, \ell n \left(\frac{b}{a}\right) \tan(\beta \ell) \tag{4-4-4}$$

The capacitance of the gap is expressed by

$$C_g = \frac{\epsilon \pi a^2}{d} \tag{4-4-5}$$

At resonance, the inductive reactance of the two shorted coaxial lines in series is equal in magnitude to the capacitive reactance of the gap. That is, $\omega L = \dfrac{1}{\omega C_g}$. Hence

$$\tan(\beta \ell) = \frac{d \mathcal{V}}{\omega a^2 \, \ell n \left(\dfrac{b}{a}\right)} \tag{4-4-6}$$

where $\mathcal{V} = \dfrac{1}{\sqrt{\mu\epsilon}}$ is the phase velocity in any medium.

    The solution to this equation gives the resonant frequency of a coaxial cavity. Since Eq. (4-4-6) contains the tangent function, it has an infinite number of solutions with larger values of frequency. Therefore, this type of reentrant cavity can support an infinite number of resonant frequencies or modes of oscillations. It can be shown that a shorted coaxial-line cavity stores more magnetic energy than electric energy. The balance of the electric stored energy appears in the gap, since at resonance the magnetic and electric stored energies are equal.

    The radial reentrant cavity as shown in Fig. 4-4-2 is also a commonly used reentrant resonator.

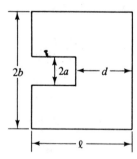

Figure 4-4-2   Radial reentrant cavity.

The inductance and capacitance [1] of a radial reentrant cavity is expressed by

$$L = \frac{\mu \ell}{2\pi} \ln \left(\frac{b}{a} \right) \tag{4-4-7}$$

$$C = \epsilon_0 \left[\frac{\pi a^2}{d} - 4a \ln \left(\frac{0.765}{\sqrt{\ell^2 + (b - a)^2}} \right) \right] \tag{4-4-8}$$

The resonant frequency [1] is given by

$$f_r = \frac{c}{2\pi \sqrt{\epsilon_r}} \left\{ a\ell \left[\frac{a}{2d} - \frac{2}{\ell} \ln \left(\frac{0.765}{\sqrt{\ell^2 + (b - a)^2}} \right) \right] \ln \left(\frac{b}{a} \right) \right\}^{-1/2} \tag{4-4-9}$$

where $c = 3 \times 10^8$ m/s is the velocity of light in vacuum.

4-4-2 Slow-Wave Structures

Slow-wave structures are special circuits that are used in the microwave tubes to reduce the wave velocity in a certain direction so that the electron beam and signal wave can interact. The phase velocity of a wave in the ordinary waveguides is greater than the velocity of light in vacuum. In the operation of traveling-wave and magnetron-type devices, the electron beam must keep in step with the microwave signal. Since the electron beam can be accelerated only to velocities that are about a fraction of the velocity of light, a slow-wave structure must be incorporated in the microwave devices so that the phase velocity of the microwave signal can keep pace with that of the electron beam for effective interactions. Several types of slow-wave structures are shown in Fig. 4-2-3. The commonly used slow-wave structure is a helical coil with a concentric conducting cylinder as shown in Fig. 4-4-3.

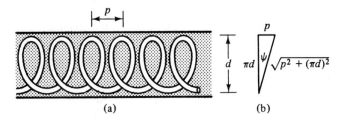

(a)   (b)

Figure 4-4-3   Helical slow-wave structure: (a) Helical coil, (b) One turn of a helix.

It can be shown that the ratio of the phase velicity $\mathcal{V}_p$ along the pitch to the phase velocity along the coil are given by

$$\frac{\mathcal{V}_p}{c} = \frac{p}{\sqrt{p^2 + (\pi d)^2}} = \sin \psi \tag{4-4-10}$$

where $c = 3 \times 10^8$ m/s is the velocity of light in free space
    p = helix pitch
    d = diameter of the helix
    ψ = pitch angle

In general, the helical coil may be within a dielectric-filled cylinder. The phase velocity in the axial direction is expressed as

$$\mathcal{V}_{p\epsilon} = \frac{p}{\sqrt{\mu\epsilon[p^2 + (\pi d)^2]}} \tag{4-4-11}$$

However, if the dielectric constant is too large, the slow-wave structure may introduce sufficient loss to the microwave devices, and reduce their efficiency. For a very-small-pitch angle, the phase velocity along the coil in free space is approximately represented by

$$\mathcal{V}_p \simeq -\frac{pc}{\pi d} = \frac{\omega}{\beta} \tag{4-4-12}$$

Figure 4-4-4 shows the $\omega - \beta$ (or Brillouin) diagram for a helical slow-wave structure.

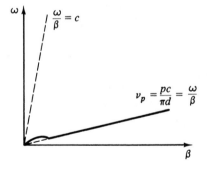

Figure 4-4-4 ω-β diagram for a helix structure.

    The helix $\omega - \beta$ diagram is very useful for designing a helix slow-wave structure. Once β is found, $\mathcal{V}_p$ can be computed from Eq. (4-4-12) for a given dimension of the helix. Furthermore, the group velocity of the wave is merely the slope of the curve as given by

$$\mathcal{V}_{gr} = \frac{\partial \omega}{\partial \beta} \tag{4-4-13}$$

For a circuit to be a slow-wave structure, the circuit must have a property of periodicity in the axial direction. The phase velocity of some of the spatial harmonics in the axial direction obtained by Fourier analysis of the waveguide field may be smaller than the velocity of light. In the helical slow-wave structure, a translation back or forth through a distance of one pitch length results in identically the same structure again. Thus, the period of a helical slow-wave structure is its pitch.

Example 4-4-2: Helical Slow-Wave Structure

An air helical coil has the following parameters:

Diameter of helical coil $d = 5$ cm
Pitch angle $\psi = 5$ degrees

(a) Determine the helix pitch.

(b) Calculate the phase velocity in the axial direction.

Solution:

(a) The helix pitch is

$$p = \pi d \tan 5°$$
$$= 3.1416 \times 5 \times 10^{-2} \times 0.08749 = 1.374 \times 10^{-2} \text{ m}$$

(b) The phase velocity is

$$\mathcal{V}_{ph} = \frac{3 \times 10^8 \times 1.374 \times 10^{-2}}{\sqrt{(1.374 \times 10^{-2})^2 + (3.1416 \times 5 \times 10^{-2})^2}}$$
$$= 2.6 \times 10^7 \text{ m/s}$$

In general, the field of the slow-wave structure must be distributed according to Floquet's theorem for periodic boundaries. Floquet's periodicity theorem states that:

The steady-state solutions for the electromagnetic fields of a single propagating mode in a periodic structure have the property that fields in adjacent cells are related by a complex constant.

Mathematically, the theorem may be stated as

$$E(x,y,z - L) = E(x,y,z)e^{j\beta_0 L}$$

$$(4\text{-}4\text{-}14)$$

where $E(x,y,z)$ is a periodic function of z with period L.

β_0 is the phase constant in the axial direction. Therefore, in a slow-wave structure, β_0 is the phase constant of average electron velocity.

It is postulated that the solution to Maxwell's equations in a periodic structure can be written in the following form

$$E(x,y,z) = f(x,y,z)e^{-j\beta_0 z}$$ $(4\text{-}4\text{-}15)$

where $f(x,y,z)$ is a periodic function of z with period L that is the period of the slow-wave structure.

For a periodic structure, Eq. (4-4-15) can be rewritten with z replaced by $z - L$

$$E(x,y,z - L) = f(x,y,z - L)e^{-j\beta_0(z-L)}$$ $(4\text{-}4\text{-}16)$

Since $f(x,y,z - L)$ is a periodic function with period L, then

$$f(x,y,z - L) = f(x,y,z) \tag{4-4-17}$$

Substitution of Eq. (4-4-17) in Eq. (4-4-16) results in

$$E(x,y,z - L) = f(x,y,z)e^{-j\beta_0 z} \, e^{j\beta_0 L} \tag{4-4-18}$$

Substitution of Eq. (4-4-15) in Eq. (4-4-17) gives

$$E(x,y,z - L) = E(x,y,z)e^{j\beta_0 L} \tag{4-4-19}$$

This expression is the mathematical statement of Floquet's theorem, Eq. (4-4-14). Therefore, Eq. (4-4-15) does indeed satisfy Floquet's theorem.

From the theory of Fourier series, any function that is periodic, single-valued, finite, and continuous may be represented by a Fourier series. Hence, the field distribution function $E(x,y,z)$ may be expanded into a Fourier series of fundamental period L as

$$E(x,y,z) = \sum_{n=-\infty}^{\infty} E_n(x,y)e^{-j(2\pi n/L)z} \, e^{-j\beta_0 z} = \sum_{n=-\infty}^{\infty} E_n(x,y)e^{-j\beta_n z} \tag{4-4-20}$$

where

$$E_n(x,y) = \frac{1}{L} \int_0^L E(x,y,z)e^{j(2\pi n/L)z} \, dz \tag{4-4-21}$$

are the amplitudes of n harmonics.

$$\beta_n = \beta_0 + \frac{2\pi n}{L} \tag{4-4-22}$$

is the phase constant of the nth modes, where $n = -\infty, \ldots, -2, -1, 0, 1, 2, 3, \ldots, \infty$.

The quantities $E_n(x,y)e^{-j\beta_n z}$ are known as spatial harmonics by analogy with time-domain Fourier series. The question is whether Eq. (4-4-20) can satisfy the electric wave equation, Eq. (2-4-20). Substitution of Eq. (4-4-20) into the wave equation results in

$$\nabla^2 \left[\sum_{n=-\infty}^{\infty} E_n(x,y)e^{-j\beta_n z} \right] - \gamma^2 \left[\sum_{n=-\infty}^{\infty} E_n(x,y)e^{-j\beta_n z} \right] = 0 \tag{4-4-23}$$

Since the wave equation is linear, Eq. (4-4-23) can be rewritten as

$$\sum_{n=-\infty}^{\infty} \left[\nabla^2 E_n(x,y)e^{-j\beta_n z} - \gamma^2 E_n(x,y)e^{-j\beta_n z} \right] = 0 \tag{4-4-24}$$

It is evident from the above equation that if each spatial harmonic is itself a solution of the wave equation for each value of n, the summation of space harmonics also satisfies the wave equation of Eq. (4-4-23). This means that only the complete solution of Eq. (4-4-23) can satisfy the boundary conditions of a periodic structure.

Furthermore, Eq. (4-4-20) shows that the field in a periodic structure can be expanded as an infinite series of waves, all at the same frequency but with different phase velocities $\mathcal{V}_{pn}$. That is,

$$\mathcal{V}_{pn} = \frac{\omega}{\beta_n} = \frac{\omega}{\beta_0 + \dfrac{2\pi n}{L}} \tag{4-4-25}$$

The group velocity $\mathcal{V}_{gr}$, defined by $\mathcal{V}_{gr} = \dfrac{\partial \omega}{\partial \beta}$, is then given as

$$\mathcal{V}_{gr} = \left[\frac{d\left(\beta_0 + \dfrac{2\pi n}{L}\right)}{d\omega} \right]^{-1} = \frac{\partial \omega}{\partial \beta_0} \tag{4-4-26}$$

which is independent of n.

It is important to note that the phase velocity $\mathcal{V}_{pn}$ in the axial direction decreases for higher values of positive n and β_0. Therefore, it appears possible for a microwave of suitable n to have a phase velocity less than the velocity of light. It follows that interactions between the electron beam and microwave signal are possible, so the amplification of active microwave devices can be achieved.

Figure 4-4-5 shows the ω-β (or Brillouin) diagram for a helix with several spatial harmonics. This ω-β diagram demonstrates some important properties which need more explanations. First, the second quadrant of the ω-β diagram indicates the negative phase velocity which corresponds to the negative n. This means that the electron beam moves in the positive z direction, while the beam velocity coincides with the negative spatial harmonic's phase velocity. This type of tube is called the backward-wave oscillator. Second, the shaded areas are the forbidden regions for propagation. This is due to the fact that if the axial phase velocity of any spatial harmonic exceeds the velocity of light, the structure radiates energy. This property has been verified by experiments [2].

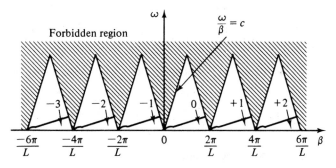

Figure 4-4-5 ω-β diagram of spatial harmonics for helical structure.

4-5 CIRCULATORS AND ISOLATORS

Both microwave circulators and microwave isolators are nonreciprocal transmission devices that use the property of Faraday rotation in the ferrite material. In order to understand the principles of operation of circulators and isolators it is appropriate to describe the behavior of ferrites in the nonreciprocal phase shifter.

A nonreciprocal phase shifter consists of a thin slab of ferrite placed in a rectangular waveguide at a point where the dc magnetic field of the incident wave mode is circularly polarized. Ferrite is a family of $MO:Fe_2O_3$, where M is a divalent metal iron such as Mn, Mg, Ni, Zn. When a piece of ferrite is affected by a dc magnetic field, the ferrite exhibits a phenomenon of Faraday rotation. This is because the ferrite is nonlinear material and its permeability is an asymmetrical tensor [3] as expressed by

$$\mathbf{B} = \hat{\mu}\mathbf{H} \tag{4-5-1}$$

where $\hat{\mu} = \mu_0(1 + \hat{\chi}_m)$ and $\tag{4-5-2}$

$$\hat{\chi}_m = \begin{bmatrix} \chi_m & jk & 0 \\ jk & \chi_m & 0 \\ 0 & 0 & 0 \end{bmatrix} \tag{4-5-3}$$

which is the tensor magnetic susceptibility. χ is the diagonal susceptibility and k is the off-diagonal susceptibility.

When a dc magnetic field is applied to a ferrite, the unpaired electrons in the ferrite material tend to line up with the dc field due to their magnetic dipole moment. However, the nonreciprocal precession of unpaired electrons in the ferrite causes their relative permeabilities (μ_r^+, μ_r^-) unequal and the wave in the ferrite is then circularly polarized. The propagation constant for a linearly polarized wave inside the ferrite can be expressed as [3]

$$\gamma\pm = j\omega\sqrt{\epsilon\mu_0(\mu + k)} \tag{4-5-4}$$

where
$$\mu = 1 + \hat{\chi}_m \tag{4-5-5}$$

$$\mu_r^+ = \mu + k \tag{4-5-6}$$

$$\mu_r^- = \mu - k \tag{4-5-7}$$

The relative permeability μ_r changes with the applied dc magnetic field as given by

$$\mu_r^\pm = 1 + \frac{\gamma_e M_e}{|\gamma_e|H_{dc} \mp \omega} \tag{4-5-8}$$

where γ_e = gyromagnetic ratio of an electron
M_e = saturation magnetization
ω = angular frequency of microwave field
H_{dc} = dc magnetic field

μ_r^+ = relative permeability in the clockwise direction (right or positive circular polarization)

μ_r^- = relative permeability in the counterclockwise direction (left or negative circular polarization)

It can be seen from Eq. (4-5-8) that if $\omega = |\gamma_e|H_{dc}$, μ_r^+ is infinite. This phenomenon is called the *gyromagnetic resonance of the ferrite*. A graph of μ_r is plotted as a function of H_{dc} for longitudinal propagation in Fig. 4-5-1.

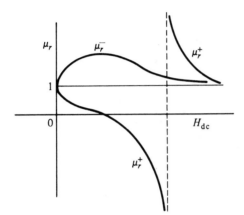

Figure 4-5-1   Curves of μ_r versus H_{dc} for axial propagation.

If μ_r^+ is much larger than μ_r^-, $(\mu_r^+ \gg \mu_r^-)$, the wave in the ferrite is rotated to the clockwise direction. Consequently, the propagation phase constant β^+ in the forward direction is different from the propagation phase constant β^- in the backward direction. By choosing the length of the ferrite slab and the dc magnetic field so that

$$\omega = (\beta^+ - \beta^-)\ell = \frac{\pi}{2} \tag{4-5-9}$$

a differential phase shift of 90° for the two directions of propagation can be obtained.

Microwave Circulators.   A microwave circulator is a multiport waveguide junction in which the wave can flow only from the nth port to the $(n + 1)$th port in one direction as shown in Fig. 4-5-2. Although there is no restriction on the number of ports, the four-port microwave circulator is most commonly used. One type of four-port microwave circulator is a combination of two 3-dB side-hole directional couplers and a rectangular waveguide with two nonreciprocal phase shifters as shown in Fig. 4-5-3.

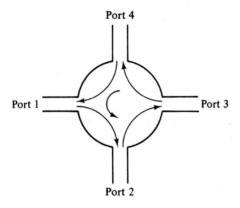

Figure 4-5-2 The symbol of a circulator.

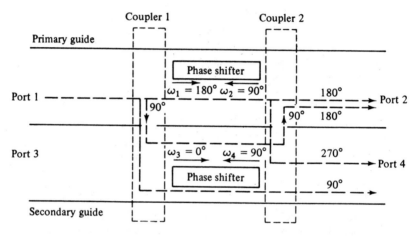

Figure 4-5-3 A four-port circulator.

The principle of operation of a typical microwave circulator can be analyzed with the aid of Fig. 4-5-3. Each of the two 3-dB couplers in the circulator introduces a phase shift of 90°, and each of the two phase shifters produces a certain amount of phase change in certain direction as indicated. When a wave is incident to port 1, the wave is split into two components by the coupler 1. The wave in the primary guide arrives at port 2 with a relative phase change of 180°. The second wave propagates through the two couplers and the secondary guide and arrives at port 2 with a relative phase shift of 180°. Since the two waves arrived at port 2 are in phase, so the power transmission is obtained from port 1 to port 2. However, the wave propagates through the primary guide, phase shifter, and the coupler 2, and arrives at port 4 with a phase change of 270°. The wave travels through the coupler 1 and the secondary guide, and it arrives at port 4 with a phase shift of 90°. Since the two waves arrived at port 4 are out of phase by 180° so the power transmission from port 1 to port 4 is zero. In general, the differential propagation constants in the two directions of propagation in a waveguide containing ferrite phase shifters should be

$$\omega_1 - \omega_3 = (2m + 1)\pi \qquad \text{rad/s} \qquad (4\text{-}5\text{-}10)$$

$$\omega_2 - \omega_4 = 2n\pi \qquad \text{rad/s} \qquad (4\text{-}5\text{-}11)$$

where m and n are any integers including zeros.

A similar analysis shows that a wave incident in port 2 emerges at port 3, and so on. As a result, the sequence of power flow is designated as $1\rightarrow2\rightarrow3\rightarrow4\rightarrow1$.

Many types of microwave circulators are constructed in use today. However, their principles of operation remain the same. Figure 4-5-4 shows a four-port circulator constructed of two magic tee's and a phase shifter. The phase shifter produces a phase shift of 180°. The explanation of how this circulator works left to the reader in the problem section.

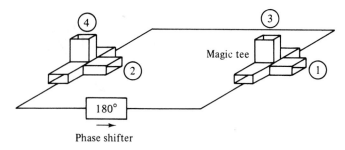

Figure 4-5-4   A four-port circulator.

A perfectly matched, lossless, and nonreciprocal four-port circulator has a **S** matrix of the form

$$\mathbf{S} = \begin{bmatrix} 0 & S_{12} & S_{13} & S_{14} \\ S_{21} & 0 & S_{23} & S_{24} \\ S_{31} & S_{32} & 0 & S_{34} \\ S_{41} & S_{42} & S_{43} & 0 \end{bmatrix} \qquad (4\text{-}5\text{-}12)$$

By use of the properties of S parameters as described in Section 4-1, the **S** matrix in Eq. (4-5-12) can be simplified to be

$$\mathbf{S} = \begin{bmatrix} 0 & 0 & 0 & 1 \\ 1 & 0 & 0 & 0 \\ 0 & 1 & 0 & 0 \\ 0 & 0 & 1 & 0 \end{bmatrix} \qquad (4\text{-}5\text{-}13)$$

Microwave isolator.   An isolator is a nonreciprocal transmission device that is used to isolate one component from reflections of other components in the transmission line. An ideal isolator completely absorbs the power for propagation in one direction and provides lossless transmission in the opposite direction. So the isolator is usually called uniline. Isolators are commonly used to improve the frequency stability of microwave generators, such as klystrons and magnetrons, in which the reflection from the load affects the generating frequency. In such cases, the isolator placed between the generator and load prevents the reflected power from the unmatched load from returning to the generator. As a result, the isolator maintains the frequency stability of the generator.

There are many ways to construct the isolators. An isolator can be made by terminating ports 3 and 4 of a four-port circulator with matched loads. On the other hand, isolators can be made by inserting a ferrite rod along the axis of a rectangular waveguide as shown in Fig. 4-5-5.

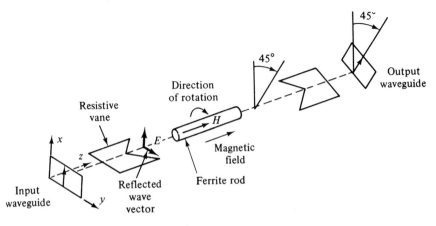

Figure 4-5-5 Faraday-rotation isolator.

The isolator shown in Fig. 4-5-5 is Faraday-rotation isolator. The principle of the Faraday-rotation isolator can be explained as follows [4]. The input resistive card is in the y-z plane, and the output resistive card is displaced 45° with respect to the input card. The dc magnetic field, which is applied longitudinally to the ferrite rod, rotates the wave plane of polarization by 45°. The degrees of rotation depend on the length and diameter of the rod and upon the applied dc magnetic field. An input TE_{10} dominant mode is incident to the left end of the isolator. Since the TE_{10} mode wave is perpendicular to the input resistive card, the wave passes through the ferrite rod without attenuation. The wave in the ferrite rod section is rotated clockwise by 45° and is normal to the output resistive card. As the result of rotation, the wave arrives at the output end without attenuation at all. On the contrary, a reflected wave from the output end is similarly rotated clockwise 45° by the ferrite rod. However, since the reflected wave is in parallel with the input resistive card, the wave is thereby absorbed by the input card. Typical performance of isolators is about 1-dB insertion loss in forward transmission and about 20- to 30-dB isolation in reverse attenuation.

REFERENCES

[1] Fujisawa, K., General treatment of klystron resonant cavities. *IRE Trans.* **MTT-6,** pp. 344–58, October 1958.

[2] Beck, A. H. W., *Space-Charge Waves.* Pergamon Press, New York, 1958.

[3] Soohoo, R. F., *Theory and Applications of Ferrites.* Prentice-Hall, Inc., Englewood Cliffs, N.J., 1960.

[4] Montgomery, C. G., et al., *Principles of Microwave Circuits.* McGraw-Hill Book Company, New York, 1948.

PROBLEMS

4-1. A coaxial resonator is constructed of a section of coaxial line and is open-circuited at both ends. The length of the resonator is 5 cm long and is filled with dielectric of $\epsilon_r = 9$. The inner conductor has a radius of 1 cm and the outer conductor has a radius of 2.5 cm.

(a) Find the resonant frequency of the resonator.

(b) Determine the resonant frequency of the same resonator with one end open and one end shorted.

4-2. An air-filled cylindrical waveguide has a radius of 3 cm and is used as a resonator for TE_{01} mode at 10 GHz by placing two perfectly conducting plates at its two ends. Determine the minimum distance between the two end plates.

4-3. A four-port circulator is constructed of two magic Tees and one phase shifter as shown in Fig. 4-5-4. The phase shifter produces a phase shift of 180°. Explain how this circulator works.

4-4. A coaxial resonator is constructed of a section of coaxial line 6 cm long and is short-circuited at both ends. The cylindrical cavity has an inner radius of 1.5 cm and an outer radius of 3.5 cm. The line is dielectric-filled with $\epsilon_r = 2.25$.

(a) Determine the resonant frequency of the cavity for TEM_{001}.

(b) Calculate the quality Q of the cavity.

4-5. A rectangular-cavity resonator has dimensions of $a = 2$ cm, $b = 5$ cm, and $d = 15$ cm. Compute

(a) the resonant frequency of the dominant mode for an air-filled cavity.

(b) the resonant frequency of the dominant mode for a dielectric-filled cavity of $\epsilon_r = 2.56$.

4-6. An undercoupled resonant cavity is connected to a lossless transmission line as shown in Fig. P4-6. The directional coupler is assumed to be ideal and matched on all arms. The unloaded Q of the cavity is 1,000 and the VSWR at resonance is 2.5.

(a) Calculate the loaded Q_ℓ of the cavity.

(b) Find the reading of bolometer 2 if bolometer 1 reads 4 mW.

(c) Compute the power dissipated in the cavity.

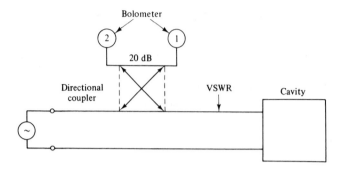

Figure p4-6   (For Problem 4-6).

4-7. A microwave transmission system consists of a generator, an overcoupled cavity, two ideal but not identical dual directional couplers with matched bolometers, and a load Z_ℓ. The lossless transmission line has a characteristic impedance Z_0. The readings of the four bolometers (1, 2, 3, and 4) are 2 mW, 4 mW, 0 mW, and 1 mW, respectively. The system is shown in Fig. P4-7.

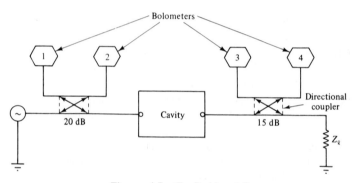

Figure p4-7   (For Problem 4-7).

(a) Find the load impedance Z_ℓ in terms of Z_0.
(b) Calculate the power dissipated by Z_ℓ.
(c) Compute the power dissipated in the cavity.
(d) Determine the VSWR on the input transmission line.
(e) Find the ratio of Q_ℓ/Q_0 for the cavity.

4-8. A symmetrical directional coupler has an infinite directivity and a forward attenuation of 20 dB. The coupler is used to monitor the power delivered to a load Z_ℓ as shown in Fig. P4-8. Bolometer 4 introduces a VSWR of 2.0 on arm 4, bolometer 3 is matched to arm 3. If bolometer 4 reads 9 mW and bolometer 3 reads 3 mW,
(a) find the amount of power dissipated in the load Z_ℓ,
(b) determine the VSWR on arm 2.

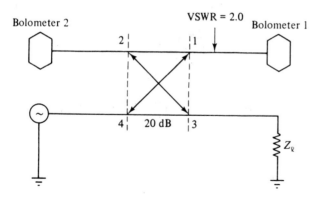

Bolometer 2

VSWR = 2.0

Bolometer 1

Z_ℓ

Figure p4-8 (For Problem 4-8).

4-9. A semicircular-cavity resonator has a length of 5 cm and a radius of 2.5 cm.
   (a) Calculate the resonant frequency for the dominant mode if the cavity is air-filled.
   (b) Repeat part (a) if the cavity is loaded by a dielectric with a relative constant of 9.

4-10. The impedance matrix of a certain lumped-element network is given by

$$[z_{ij}] = \begin{bmatrix} 4 & 2 \\ 2 & 4 \end{bmatrix}$$

Determine the scattering matrix by using S-parameter theory and indicate the values of the components.

$$[S_{ki}] = \begin{bmatrix} S_{11} & S_{12} \\ S_{21} & S_{22} \end{bmatrix}$$

4-11. A hybrid waveguide is constructed of two identical rectangular waveguides across each other at the center and works as a four-port device. Write a general scattering matrix and then simplify it as much as possible by inspection of geometrical symmetry and using the known phases of the electric waves.

4-12. A helix slow-wave structure has a pitch P of 2 mm and a diameter of 4 cm. Calculate the wave velocity in the axial direction of the helix.

CHAPTER 5

Microwave Linear-Beam Amplifiers

5-0 INTRODUCTION

Linear-beam tubes started with the Heil oscillators [1] in 1935 and the Varian brothers' klystron amplifiers [1] in 1939. The work was advanced by the space-charge-wave propagation theory of Hahn and Ramo [1] in 1939 and continued with the invention of the helix-type traveling-wave tube (TWT) by R. Kompfner in 1944 [2]. From the early 1950s, the low-power output of linear-beam tubes were made progress toward achieving high power levels, first rivaling and finally surpassing magnetrons that were the early sources of microwave high power. Subsequently, several additional devices were developed, two of which have demonstrated lasting importance. They are the extended interaction klystron [3] and the Twystron hybrid amplifier [4].

A linear-beam tube is the one which uses a magnetic field whose axis coincides with that of the electron beam to hold the beam together as it travels the length of the tube. The O-type tubes derive their name from the French TPO (*tubes à propagation des ondes*) or the original type of tube. In these tubes the electrons receive the potential energy from the dc beam voltage before they arrive in the microwave interaction region, and this energy is converted into the kinetic energy of the electrons. In the microwave interaction region the electrons are either accelerated or decelerated by the microwave field and then bunched as they drift down the tube. The bunched electrons, in turn, induce current in the output structure. The electrons then give up their kinetic energy to the microwave fields and are collected by the collector.

In this chapter, several conventional gridded vacuum tubes such as triode, tetrode, pentode, two-cavity klystron amplifier, and klystrode are studied in detail.

5-1 GRIDDED TUBES: TRIODES, TETRODES, PENTODES

The introduction of a third element between the cathode and plate of the diode by DeForest in 1907 was the start of the extensive developments involving vacuum tubes. This type of vacuum tube was called the triode. Subsequently, the tetrode and the pentode were developed. In the following sections, these three gridded tubes will be described in some detail.

5-1-1 Triodes

The triode consists of a cathode surrounded by a control grid, which is surrounded by a plate as shown in Fig. 5-1-1. The grid is usually a mesh of fine wires supported quite close to the cathode. The plate is spaced several times as far away. The entire structure is supported in an evacuated glass or metal envelope with leads to the electrodes coming out through glass on the bottom of the tube.

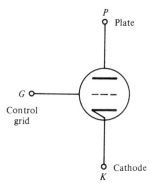

Figure 5-1-1 Schematic diagram of a triode.

The function of the grid is to control the flow of electrons from the cathode to the plate without itself drawing any electrons because its voltage is negative. As a result, a small voltage on the grid is capable of producing a large voltage drop in the plate circuit and the tube is operating on voltage control mode without power required. The term "electric valve" for a vacuum tube is particularly expressive because the grid has an electric valve function. Actually, vacuum tubes do not really amplify power, and the grid controls the flow of power from the plate power supply.

Current-voltage characteristics.   The voltage applied to the grid of a triode is usually several volts negative relative to the cathode, whereas the plate is usually maintained several hundred volts positive with respect to the cathode. Clearly, the electric field resulting from the potential of the grid tends to maintain a large space-charge cloud, whereas the field of the plate tends to reduce the space charge. This would seem to imply that a proportionality should exist between the relative effectiveness of the grid and plate potentials on the space charge. Therefore, the space-charge current can be expressed as

$$I_s = I_p + I_g = K \left(V_g + \frac{V_p}{\mu} \right)^{3/2} \tag{5-1-1}$$

where I_s = space current
  I_p = plate current
  I_g = grid current
  K = tube perveance
  V_p = plate voltage
  V_g = grid voltage
  μ = amplification factor

When the grid voltage is negative, the entire space current flows to the plate. Even when the grid is positive, the fraction of the space current going to the grid is very small, so that Eq. (5-1-1) is a reasonably good approximation for the plate current under all conditions and it can be simplified as

$$I_p = K \left(V_g + \frac{V_p}{\mu} \right)^{3/2} \tag{5-1-2}$$

Figure 5-1-2 shows the plate current versus plate voltage with grid voltage as parameters.

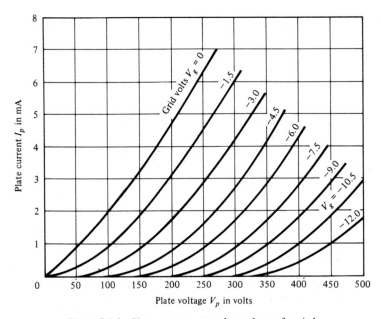

Figure 5-1-2   Plate current versus plate voltage of a triode.

Perveance K.   The perveance of an electron gun or cathode is independent of actual size of emitting structure and depends only on the electrode shape of the electron gun. Its equation is expressed by

$$K = I/V^{3/2} \qquad (5\text{-}1\text{-}3)$$

If the cathode produces 2 mA at 25 V its perveance is 16×10^{-6} A/V$^{3/2}$. The factor 10^{-6} is usually omitted; and it can be said to have a perveance of 16.

Amplification factor μ.   The amplification factor of a triode or multigrid tube is the ratio of the plate voltage increment over the grid voltage increment at a constant plate current as given by

$$\mu = -\frac{\partial I_p}{\partial V_g}\bigg/\frac{\partial I_p}{\partial V_p} = -\frac{dV_p}{dV_g}\bigg|_{I_p = \text{constant}} \qquad (5\text{-}1\text{-}4)$$

The factor is a measure of the voltage amplifying capability of the tube and is a dimensionless constant. Its values run from 2 to 200 in ordinary triodes.

Mutual conductance g_m.   The mutual conductance (or transconductance) of a tube is the rate of change of plate current over control-grid voltage at a constant plate voltage. Mathematically this is given by

$$g_m = \frac{\partial I_p}{\partial V_g} = \frac{dI_p}{dV_g}\bigg|_{V_p = \text{constant}} \qquad (5\text{-}1\text{-}5)$$

Plate resistance r_p.   The plate resistance of a tube is the reciprocal of the rate of change of plate current with plate voltage at a constant grid voltage and it is given by

$$r_p = \frac{1}{\partial I_p/\partial V_p} = \frac{\partial V_p}{\partial I_p} = \frac{dV_p}{dI_p}\bigg|_{V_g = \text{constant}} \qquad (5\text{-}1\text{-}6)$$

The plate resistance of a tube is the ac resistance of the plate circuit to a small alternating voltage superimposed upon the direct voltage. The typical values of plate resistance for a triode may vary from 1 to 50 kΩ.

Relation between tube constants.   If the change in plate current is held to zero, the product of the mutual conductance and the plate resistance is equal to the amplification factor. This is

$$g_m r_p = -\frac{dV_p}{dV_g}\bigg|_{I_p = \text{constant}} = \mu \qquad (5\text{-}1\text{-}7)$$

Equivalent circuits.   If the triode parameters μ, g_m, and r_p are reasonably constant in some region of operation, the tube behaves linearly over this range. Two linear equivalent circuits, one involving a voltage source and the other a current source, are shown in Fig. 5-1-3.

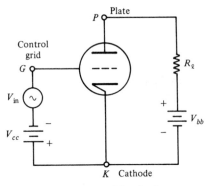

(a) Triode amplifier circuit

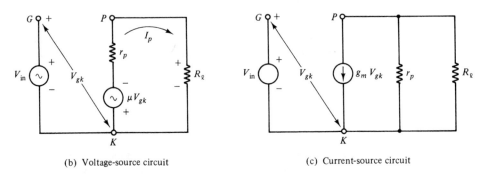

(b) Voltage-source circuit   (c) Current-source circuit

Figure 5-1-3   Equivalent circuits of a triode amplifier.

Example 5-1-1: Triode Amplifier

Calculate the output voltage and the voltage gain of a small-signal triode amplifier.

Solution:

(a) From the voltage-source equivalent circuit as shown in Fig. 5-1-3(b), we have

$$I_p(r_p + R_\ell) + \mu V_{gk} = 0$$

$$V_{gk} = V_{in}$$

Then the plate current is

$$I_p = -\frac{\mu V_{in}}{r_p + R_\ell}$$

and the output voltage is

$$V_0 = I_p R_\ell = -\frac{\mu V_{\text{in}} R_\ell}{r_p + R_\ell} = -\frac{\mu V_{\text{in}}}{1 + r_p/R_\ell}$$

(b) The voltage gain or amplification of the triode amplifier is

$$A = \frac{V_0}{V_{\text{in}}} = -\frac{\mu}{r_p + R_\ell}$$

5-1-2 Tetrodes

The tetrode is a four-electrode tube. The four electrodes are composed of the cathode, the control grid, the screen grid, and the plate as shown in Fig. 5-1-4.

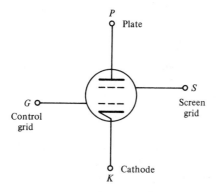

Figure 5-1-4   Schematic diagram of a tetrode.

There are two types of tetrode: screen-grid tube and beam-power tube. The screen-grid tube has a fine control grid surrounding the cathode, which in turn is surrounded by a coarser screen grid. The screen grid is, in turn, surrounded by a plate and connected to a fixed direct voltage with a large capacitor to ground in order to eliminate alternating voltage. As a result, the function of the screen grid is to shield the control grid electrostatically from the plate. The tetrode is a predominant choice for voltage amplification at the low microwave frequencies. The beam-power tube is a special tetrode with aligned control and screen grids as shown in Fig. 5-1-ba.

Current-voltage characteristics.   The plate current of a tetrode can be written as

$$I_p = K\left(V_g + \frac{V_s}{\mu_s} + \frac{V_p}{\mu_p}\right)^{3/2} \tag{5-1-8}$$

where K = perveance
    V_g = control-grid voltage
    V_s = screen-grid voltage

$$\mu_s = - \left. \frac{dV_s}{dV_g} \right|_{I_p \,=\, \text{constant}}$$

is the amplification factor of the screen grid

$$\mu_p = - \left. \frac{dV_p}{dV_g} \right|_{I_p \,=\, \text{constant}}$$

is the plate amplification factor

$$\mu_p \gg \mu_s$$

Alternatively, the plate current of a tetrode can be expressed as

$$I_p = g_m V_g + \frac{V_p}{r_p} \qquad (5\text{-}1\text{-}9)$$

where $\mu = g_m r_p$. The plate amplification factor can be determined approximately by calculating a triode amplification factor, considering the control grid as the cathode, the screen grid as the control grid, and the plate as the plate, and then multiplying the amplification factor by the screen-grid amplification factor. The product of the screen-grid amplification factor and the triode amplification factor is equal to the plate amplification factor.

Equivalent circuits.   The basic equivalent circuits of a tetrode amplifier are essentially that of the triode, even though a screen grid exists in the tetrode. The screen grid is connected to the cathode through a dc voltage that is usually less than, or at most equal to, the plate voltage. The circuit diagram of a simple tetrode amplifier with its equivalent circuits are shown in Fig. 5-1-5.

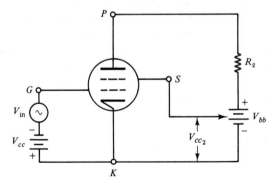

(a) Tetrode amplifier circuit

Figure 5-1-5   Circuits of a tetrode amplifier,

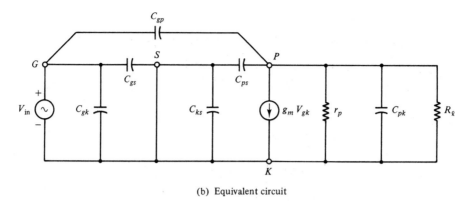

(b) Equivalent circuit

Figure 5-1-5 (*Continued*)   Circuits of a tetrode amplifier.

In practice, the screen grid is at ac ground voltage. Thus, as indicated in the equivalent circuit, the grounded screen grid effectively shorts out C_{ks} and puts C_{gk} and C_{gs} in parallel. Let this parallel combination be denoted C_1, and then

$$C_1 = C_{gk} + C_{gs} \tag{5-1-10}$$

The capacitor C_{ps} now appears from plate to ground and effectively in parallel with C_{pk}. Let this parallel combination be denoted C_2 and then

$$C_2 = C_{pk} + C_{ps} \tag{5-1-11}$$

The capacitance between the plate and the control grid C_{gp} is very small and it is assumed to be an open circuit. Owing to the shielding action of the screen grid, the capacitance C_{pk} is much smaller than the capacitance C_{ps}. Then

$$C_2 \doteq C_{ps} \tag{5-1-12}$$

As a result, the equivalent circuit of a tetrode amplifier can be simplified to Fig. 5-1-6.

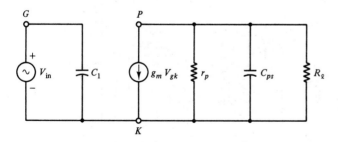

Figure 5-1-6 Ideal equivalent circuit of a tetrode amplifier.

Example 5-1-2: Tetrode Amplifier

Calculate the input admittance, the output voltage, and the amplification for a tetrode amplifier.

Solution:

(a) From the equivalent circuit in Fig. 5-1-6, we have

$$Y_{in} = j\omega C_1 = j\omega(C_{ps} + C_{pk})$$

(b) The output voltage is

$$V_0 = -g_m V_{gk} Z$$

where $\quad Z = \dfrac{1}{Y} = \dfrac{1}{1/r_p + j\omega C_{ps} + 1/R_\ell}$

(c) The amplification is

$$A = \frac{V_0}{V_{gk}} = -g_m Z$$

$$= -\frac{g_m r_p R_\ell}{r_p + R_\ell + j\omega C_{ps} r_p R_\ell}$$

Beam-power tube.   The beam-power tube is another class of tetrode. The tube has a large screen-plate spacing so that space charge of the electron beam depresses the potential between the screen and plate. As a result, secondary electrons from one electrode are prevented from reaching the other and the screen current is very small ranging from 5 to 8 percent of the plate current. The electron flow between the grid wires toward the plate is a dense beam. Figure 5-1-6(a) shows the schematic view of a beam-power tube.

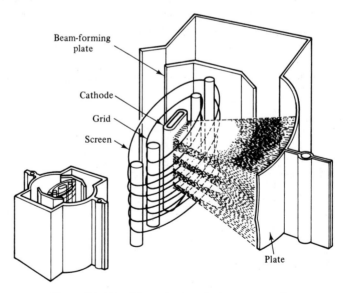

Beam-forming plate

Cathode

Grid

Screen

Plate

Figure 5-1-6(a)   Schematic view of a beam-power tube.

It is noticed that each screen-grid wire is directly opposite a control-grid wire. This construction greatly reduces the interception of the electron beam by the screen grid, since the negative bias on the control grid causes a "shadowing" of the screen-grid wires from the beam. In beam-power tubes, an additional electrode, called a "beam-forming electrode," is located near the edge of the electron beam and maintained at cathode voltage. This electrode also helps to depress the potential between the screen grid and plate and prevent secondary electrons emitted from the plate from reaching the screen grid. The approximate potential profile is shown in Fig. 5-1-6(b).

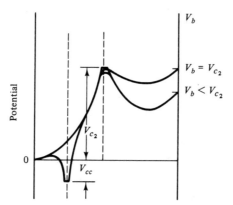

Figure 5-1-6(b)   Potential profile of a beam-power tube.

Thus, because of the beam formation, which serves to keep the screen current small, and because of the variable suppressor action, which serves to suppress secondary electron emission from the screen and the plate, the ideal power-tube characteristic is closely approximated as shown in Fig. 5-1-6(c).

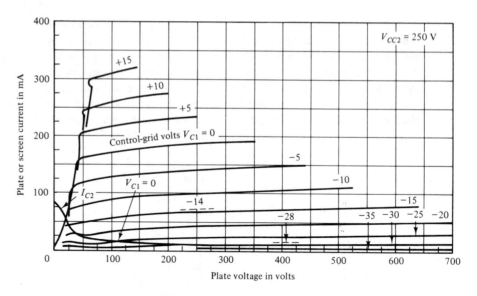

Figure 5-1-6(c) Current-voltage characteristic of a beam-power tube.

5-1-3 Pentodes

The pentode is a five-electrode tube. The five electrodes, in order, are cathode, control grid, screen grid, suppressor grid, and plate as shown in Fig. 5-1-7.

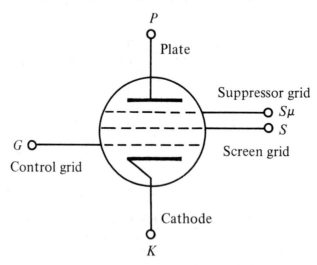

Figure 5-1-7 Schematic diagram of a pentode.

The suppressor grid is placed between the screen grid and the plate, and operated at cathode potential. As a result, the suppressor grid is able to suppress the secondary electron emission from the plate. The suppressor grid may prevent some of the primary electrons that pass through the screen from reaching the plate when the plate voltage is low. This effect arises from the deflections that may be given to some of the primary electrons if they approach close to a screen wire. Thus, if an electron is deflected by this action, its velocity in the plate direction will be reduced and the field of the suppressor grid may repel it back toward the screen. Therefore, the shape of the suppressor grid in some pentodes has been so dimensioned that the effects of secondary emission are just suppressed or are only admitted to a slight extent at the low plate voltage.

Equivalent circuit.   When a pentode is used in a circuit as a voltage amplifier, the pentode is connected exactly like a tetrode with the addition that the suppressor grid is connected to the cathode. Figure 5-1-8 shows the circuit of a pentode amplifier with its equivalent circuit.

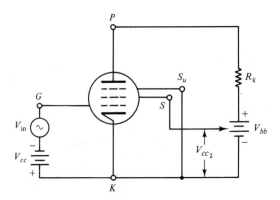

(a) Pentode amplifier circuit

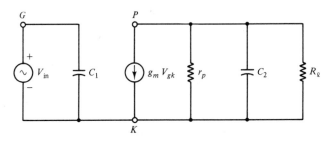

(b) Equivalent circuit

Figure 5-1-8 Circuits of a pentode amplifier.

The capacitance in a pentode circuit can be expressed as

$$C_1 = C_{gk} + C_{gs} \tag{5-1-13}$$

$$C_2 = C_{pk} + C_{ps} + C_{psu} \tag{5-1-14}$$

where C_{psu} is the capacitance between the plate and the suppressor grid. The output voltage is given by

$$V_0 = -g_m V_{gk} Z \qquad (5\text{-}1\text{-}15)$$

where Z is the combined parallel impedance in the output circuit. The amplification can be written as

$$A = \frac{V_0}{V_{in}} = -g_m Z \qquad (5\text{-}1\text{-}16)$$

where $V_{in} = V_{gk}$.

As a result, the voltage gain of a pentode amplifier is equal to the product of g_m and the output impedance Z. If the plate resistance r_p and the reactance of the output capacitor are each larger than the load resistance, as is usually the case, the voltage gain becomes

$$A \doteq -g_m R_\ell \qquad (5\text{-}1\text{-}17)$$

Example 5-1-3: Pentode Amplifier

A 6AU6 pentode has the following parameters:

Control-grid voltage	V_g	$= -70$ V
Operating frequency	f	$= 10$ MHz
Plate resistance	r_p	$= 6$ kΩ
Load resistance	R_ℓ	$= 100$ Ω
Mutual conductance	g_m	$= 3 \times 10^{-2}$ mho
Signal voltage	V_{in}	$= 90$ V
Capacitances	C_{gk}	$= 5$ $\mu\mu$F
	C_{gs}	$= 4$ $\mu\mu$F
	C_{pk}	$= 1$ $\mu\mu$F
	C_{ps}	$= 8$ $\mu\mu$F
	C_{psu}	$= 5$ $\mu\mu$F

Determine:

(a) The output impedance
(b) The output voltage
(c) The voltage gain

Solution:

(a) The output impedance

$$C_2 = 1 + 8 + 5 = 15 \ \mu\mu\text{F}$$

$$X = \frac{1}{\omega C_2} = \frac{1}{2\pi \times 10^7 \times 14 \times 10^{-12}} = 1137 \ \Omega$$

$$Z \doteq R_\ell = 100 \ \Omega$$

(b) The output voltage is

$$V_0 = -g_m V_{in} Z = -3 \times 10^{-2} \times 90 \times 100 = -270 \ V$$

(c) The voltage gain is

$$A = -g_m Z = -3 \times 10^{-2} \times 100 = -3.0$$

5-1-4 Limitations of Gridded Tubes

Conventional vacuum triodes, tetrodes, and pentodes are less useful signal sources at the frequencies above 1 GHz because of the lead-inductance and interelectrode-capacitance effects, the transit-angle effects, and the gain-bandwidth product limitations. These three effects are analyzed in detail in this section.

Lead-inductance and interelectrode-capacitance effects. At the frequencies above 1 GHz, the conventional vacuum tubes are impaired by parasitic-circuit reactances because the circuit capacitances between tube electrodes and circuit inductance of the lead wire are too large for a microwave resonant circuit. Furthermore, as the frequency is increased up to the microwave range, the real part of the input admittance may be large enough to overload the input circuit seriously and thereby reduce the operating efficiency of the tube. In order to gain a better understanding of these effects, the triode circuit shown in Fig. 5-1-9 should be studied carefully.

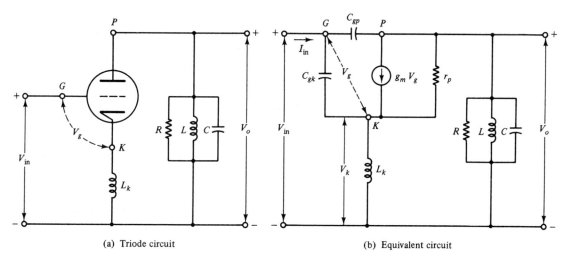

(a) Triode circuit   (b) Equivalent circuit

Figure 5-1-9 Triode circuit (a) and its equivalent (b).

Figure 5-1-9(b) shows the equivalent circuit of a triode circuit under the assumption that the interelectrode capacitances and cathode inductance are the only parasitic elements.

Since $C_{gp} \ll C_{gk}$ and $\omega L_k \ll \dfrac{1}{\omega C_{gk}}$, the input voltage V_{in} can be written

$$V_{in} = V_g + V_k = V_g + j\omega L_k g_m V_g \qquad (5\text{-}1\text{-}18)$$

and the input current is given by

$$I_{in} = j\omega C_{gk} V_g \qquad (5\text{-}1\text{-}19)$$

Substitution of Eq. (5-1-19) in Eq. (5-1-18) yields

$$V_{in} = \frac{I_{in}(1 + j\omega L_k g_m)}{j\omega C_{gk}} \qquad (5\text{-}1\text{-}20)$$

The input admittance of the tube is approximately

$$Y_{in} = \frac{I_{in}}{V_{in}} = \frac{j\omega C_{gk}}{1 + j\omega L_k g_m} = \omega^2 L_k C_{gk} g_m + j\omega C_{gk} \qquad (5\text{-}1\text{-}21)$$

in which $\omega L_k g_m \ll 1$ has been replaced. The inequality is almost always true, since the cathode lead is usually short and is quite large in diameter, and the transconductance g_m is generally much less than 1 mmho.

The input impedance at very high frequencies is given by

$$Z_{in} = \frac{1}{\omega^2 L_k C_{gk} g_m} - j\frac{1}{\omega^3 L_k^2 C_{gk} g_m^2} \qquad (5\text{-}1\text{-}22)$$

The real part of the impedance is inversely proportional to the square of the frequency, and the imaginary part is inversely proportional to the third order of the frequency. When the frequencies are above 1 GHz, the real part of impedance becomes small enough to nearly short the signal source. Consequently, the output power is decreased rapidly. Similarly, the input admittance of a pentode circuit is expressed by

$$Y_{in} = \omega^2 L_k C_{gk} g_m + j\omega(C_{gk} + C_{gs}) \qquad (5\text{-}1\text{-}23)$$

where C_{gs} is the capacitance between the grid and screen, and its input impedance is given by

$$Z_{in} = \frac{1}{\omega^2 L_k C_{gk} g_m} - j\frac{C_{gk} + C_{gs}}{\omega^3 L_k^2 C_{gk}^2 g_m^2} \qquad (5\text{-}1\text{-}24)$$

There are several ways to minimize the inductance and capacitance effects, such as a reduction in lead length and electrode area. This minimization, however, also limits the power-handling capacity.

Example 5-1-1: Input Impedance of a Triode

A 6J5 triode has the following parameters:

$$L_k = 6 \ \mu H \qquad g_m = 2.5 \times 10^{-3} \text{ mho}$$
$$C_{gk} = 3.4 \text{ pF} \qquad f = 10 \text{ GHz}$$

Compute the input impedance.

Solution: From Eq. (5-1-22), the input impedance is

$$Z_{in} = \frac{1}{(2\pi \times 10^{10})^2 (6 \times 10^{-6})(3.4 \times 10^{-12})(2.5 \times 10^{-3})}$$
$$- j \frac{1}{(2\pi \times 10^{10})^3 (6 \times 10^{-6})^2 (3.4 \times 10^{-12})(2.5 \times 10^{-3})}$$
$$= 4.97 \times 10^{-3} - j\,5.269 \times 10^{-6}$$
$$= 4.98 \times 10^{-3} \angle 0° \quad \text{(looks like a short circuit)}$$

Transit-angle effects.   Another limitation in the application of the conventional tubes at microwave frequencies is the electron transit angle between electrodes. The electron transit angle is defined as

$$\theta_g \equiv \omega \tau_g = \frac{\omega d}{\mathcal{V}_0} \tag{5-1-25}$$

where $\tau_g = \dfrac{d}{\mathcal{V}_0}$ is the transit time

$d =$ separation between cathode and grid

$\mathcal{V}_0 = 0.593 \times 10^6 \sqrt{V_0}$ is the velocity of electron

$V_0 =$ dc voltage

When the frequencies are below microwave range, the transit angle is negligible. At the range of microwave frequencies, however, the transit time (or angle) is large compared to the period of the microwave signal, and the potential between the cathode and grid may alternate from 10 to 100 times during the electron transit. The grid potential during the negative half-cycle thus removes energy that was given to the electron during the positive half-cycle. Consequently, the electrons may oscillate back and forth in the cathode-grid space or return to the cathode. The overall result of transit-angle effects is to reduce the operating efficiency of the vacuum tube. The degenerate effect becomes more serious when the frequencies are well above 1 GHz. Once the electrons pass the grid, they are quickly accelerated to the anode by the high-plate voltage.

When the frequency is below 1 GHz, the output delay is negligible in comparison with the phase of the grid voltage. This means that the transadmittance is a real large quantity, which is the usual transconductance g_m. At the microwave frequencies, the transit angle is not negligible, and the transadmittance becomes a complex number with a relatively small magnitude. This situation indicates that the output is decreased.

From the preceding analysis, it is evident that the transit-angle effect can be minimized by first accelerating the electron beam by a very high dc voltage and then velocity-modulating it. This is indeed the principal operation of such microwave tubes as klystrons and magnetrons.

Example 5-1-2: Transit Angle in a Triode

A 6J5 triode has the following parameters:

Separation between cathode and grid	d	$= 1$ cm
dc voltage	V_0	$= 150$ V
Operating frequency	f	$= 3$ GHz

(a) Compute the dc electron velocity.

(b) Calculate the transit angle.

Solution:

(a) The electron velocity is

$$\mathcal{V}_0 = 0.593 \times 10^6 \sqrt{150}$$
$$= 7.26 \times 10^6 \text{ m/s}$$

(b) The transit angle is

$$\theta_g = \frac{2\pi \times 3 \times 10^9 \times 1 \times 10^{-2}}{7.26 \times 10^6}$$
$$= 341.67 \text{ rad}$$
$$= 19576.20°$$

Gain-bandwidth product limitation. In ordinary vacuum tubes the maximum gain is generally achieved by resonating the output circuit as shown in Fig. 5-1-10.

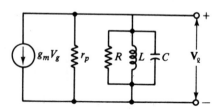

Figure 5-1-10 Output-tuned circuit of a pentode.

In Fig. 5-1-10, it has been taken that $r_p \gg \omega L_k$. The load voltage is given by

$$V_\ell = \frac{g_m V_g}{G + j\left(\omega C - \dfrac{1}{\omega L}\right)} \tag{5-1-26}$$

where $G = \dfrac{1}{r_p} + \dfrac{1}{R_\ell}$

r_p = plate resistance
R_ℓ = load resistance
L = inductive tuning element
C = capacitive tuning element

The resonant frequency is expressed by

$$f_r = \frac{1}{2\pi\sqrt{LC}}$$

(5-1-27)

and the maximum voltage gain $A_{\max}$ at resonance is given by

$$A_{\max} = \frac{g_m}{G}$$

(5-1-28)

Since the bandwidth is measured at the half-power point, the denominator of Eq. (5-1-26) must be related by

$$G = \omega C - \frac{1}{\omega L}$$

(5-1-29)

The roots of this quadratic equation are given by

$$\omega_1 = \frac{G}{2C} - \sqrt{\left(\frac{G}{2C}\right)^2 + \frac{1}{LC}}$$

(5-1-30)

$$\omega_2 = \frac{G}{2C} + \sqrt{\left(\frac{G}{2C}\right)^2 + \frac{1}{LC}}$$

(5-1-31)

Then the bandwidth can be expressed by

$$BW = \omega_2 - \omega_1 = \frac{G}{C} \qquad \text{for } \left(\frac{G}{2C}\right)^2 \gg \frac{1}{LC}$$

(5-1-32)

Hence the gain-bandwidth product of the circuit of Fig. 5-1-10 is

$$A_m(BW) = \frac{g_m}{C}$$

(5-1-33)

It is important to note that the gain-bandwidth product is independent of frequency. For a given tube, a higher gain can be achieved only at the expense of a narrower bandwidth. This restriction is applicable to a resonant circuit only. In the microwave devices either reentrant cavities or slow-wave structures are used to obtain a possible overall high gain over a broad bandwidth. Figure 5-1-11 shows the state of the art for U.S. high-power gridded tubes.

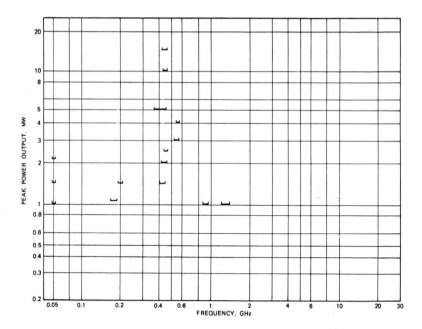

Figure 5-1-11   State of the art for U.S. high-power gridded tubes.

5-2 KLYSTRON AMPLIFIER

The two-cavity klystron is a widely used microwave amplifier operated by the principles of velocity and current modulation. All electrons injected from the cathode arrive at the first cavity with uniform velocity. Those electrons passing the first cavity gap at zeros of the gap voltage (or signal voltage) pass through with unchanged velocity; those passing through the positive half-cycles of the gap voltage undergo an increase in velocity; those passing through the negative swings of the gap voltage undergo a decrease in velocity. As a result of these actions, the electrons gradually bunch together as they travel down the drift space. The variation of velocity of the electrons in the drift space is known as *velocity modulation*. The density of the electrons in the second cavity gap varies cyclically with time. The electron beam contains an ac component and is said to be current-modulated. The maximum bunching should occur approximately at midway between the second cavity grids during its retarding phase; thus, the kinetic energy is transferred from the electrons to the field of the second cavity. The electrons then emerge from the second cavity with reduced velocity and finally terminate at the collector. The characteristics of a two-cavity klystron amplifier are as follows:

1. Efficiency: About 40%

2. Power output: Average power (CW power) is up to 500 kW and pulsed power is up to 30 MW at 10 GHz

3. Power gain: About 30 dB

The schematic diagram of a two-cavity klystron amplifier is shown in Fig. 5-2-1.

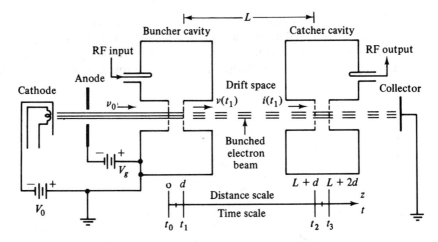

Figure 5-2-1   Two-cavity klystron amplifier.

The cavity close to the cathode is known as the *buncher cavity* or *input cavity*, which velocity-modulates the electron beam. The other cavity is called the *catcher cavity* or *output cavity*, which catches energy from the bunched electron beam. The beam then passes through the catcher cavity and is terminated at the collector. The quantitative analysis of a two-cavity klystron can be described in the next three sections under the following assumptions:

1. The electron beam is assumed to have uniform density in the cross section of the beam.
2. Space-charge effects are negligible.
3. The magnitude of the microwave signal input is assumed to be much less than the dc accelerating voltage.

5-3 VELOCITY MODULATION

When the electrons are first accelerated by the high dc voltage V_0 before entering the buncher grids, their velocity is uniform as given by

$$\mathcal{V}_0 = \sqrt{\frac{2eV_0}{m}} = 0.593 \times 10^6 \sqrt{V_0} \qquad \text{m/s} \qquad (5\text{-}3\text{-}1)$$

In Eq. (5-3-1) it is assumed that electrons leave the cathode with zero velocity. When a microwave signal is applied to the input terminal, the gap voltage between the buncher grids appears as

$$V_s = V_1 \sin \omega t \qquad (5\text{-}3\text{-}2)$$

where V_1 is the amplitude of the signal and $V_1 \ll V_0$ is assumed.

In order to find the modulated velocity in the buncher cavity in terms of either the entering time t_0 or the exiting time t_1 and the gap transit angle θ_g as shown in Fig. 5-2-1 it is necessary to determine the average microwave voltage in the buncher gap as indicated in Fig. 5-3-1.

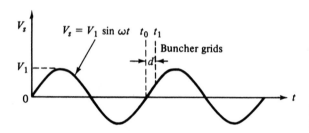

Figure 5-3-1 Signal voltage in the buncher gap.

Since $V_1 \ll V_0$, the average transit time through the buncher gap distance d is

$$\tau \cong \frac{d}{\mathcal{V}_0} = t_1 - t_0 \qquad (5\text{-}3\text{-}3)$$

The average gap transit angle can be expressed as

$$\theta_g = \omega\tau = \omega(t_1 - t_0) = \frac{\omega d}{\mathcal{V}_0} \qquad (5\text{-}3\text{-}4)$$

The average microwave voltage in the buncher gap can be found in the following way:

$$\langle V_s \rangle = \frac{1}{\tau}\int_{t_0}^{t_1} V_1 \sin(\omega t)\, dt = -\frac{V_1}{\omega\tau}[\cos(\omega t_1) - \cos(\omega t_0)]$$

$$= \frac{V_1}{\omega\tau}\left[\cos(\omega t_0) - \cos\omega\left(t_0 + \frac{d}{\mathcal{V}_0}\right)\right] \qquad (5\text{-}3\text{-}5)$$

Let

$$\omega t_0 + \frac{\omega d}{2\mathcal{V}_0} = \omega t_0 + \frac{\theta_g}{2} = A$$

and

$$\frac{\omega d}{2\mathcal{V}_0} = \frac{\theta_g}{2} = B$$

Then using the trigonometric identity that $\cos(A - B) - \cos(A + B) = 2 \sin A \sin B$, Eq. (5-3-5) becomes

$$\langle V_s \rangle = V_1 \frac{\sin\left(\dfrac{\omega d}{2\mathcal{V}_0}\right)}{\dfrac{\omega d}{2\mathcal{V}_0}} \sin\left[\omega\left(t_0 + \frac{d}{2\mathcal{V}_0}\right)\right]$$

$$= V_1 \frac{\sin(\theta_g/2)}{\theta_g/2} \sin\left(\omega t_0 + \theta_g/2\right)$$

(5-3-6)

It is defined

$$\beta_i \equiv \frac{\sin\left(\dfrac{\omega d}{2\mathcal{V}_0}\right)}{\dfrac{\omega d}{2\mathcal{V}_0}} = \frac{\sin\left(\dfrac{\theta_g}{2}\right)}{\dfrac{\theta_g}{2}}$$

(5-3-7)

β_i is known as the *beam-coupling coefficient* of the input cavity gap as shown in Fig. 5-3-2.

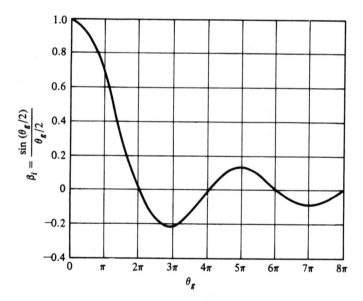

Figure 5-3-2 Beam-coupling coefficient versus gap transit angle.

It can be seen that increasing the gap transit angle θ_g decreases the coupling between the electron beam and the buncher cavity; that is, the velocity modulation of the beam for a given microwave signal is decreased. Immediately after velocity modulation, the exit velocity from the buncher gap is given by

$$
\mathcal{V}(t_1) = \sqrt{\frac{2e}{m} \left[V_0 + \beta_i V_1 \sin\left(\omega t_0 + \frac{\theta_g}{2} \right) \right]}
$$
$$
= \sqrt{\frac{2e}{m} V_0 \left[1 + \frac{\beta_i V_1}{V_0} \sin\left(\omega t_0 + \frac{\theta_g}{2} \right) \right]}
$$

(5-3-8)

where the factor $\dfrac{\beta_i V_1}{V_0}$ is called the *depth of velocity modulation*. By means of binomial expansion under the assumption of

$$
\beta_i V_1 \ll V_0
$$

(5-3-9)

Equation (5-3-8) becomes

$$
\mathcal{V}(t_1) = \mathcal{V}_0 \left[1 + \frac{\beta_i V_1}{2V_0} \sin\left(\omega t_0 + \frac{\theta_g}{2} \right) \right]
$$

(5-3-10)

Equation (5-3-10) is the equation of velocity modulation. Alternately, the equation of velocity modulation can be given by

$$
\mathcal{V}(t_1) = \mathcal{V}_0 \left[1 + \frac{\beta_i V_1}{2V_0} \sin\left(\omega t_1 - \frac{\theta_g}{2} \right) \right]
$$

(5-3-11)

Example 5-3-1: Klystron Amplifier

A two-cavity klystron amplifier has the following parameters:

Operating frequency	$f = 8$ GHz
Signal voltage	$V_1 = 2$ V (rms)
Beam voltage	$V_0 = 40$ kV
Beam current	$I_0 = 10$ A
Gap distance in either cavity	$d = 1$ mm
Length between two cavities	$L = 5$ cm

Determine:

(a) The dc electron velocity

(b) The transit time through the gap

(c) The transit angle through the gap

(d) The beam coupling coefficient

(e) The electron velocity just out from the buncher cavity

Solution:

(a) The dc electron velocity is

$$\mathcal{V}_0 = 0.593 \times 10^6\sqrt{40 \times 10^3} = 1.186 \times 10^8 \text{ m/s}$$

(b) The transit time is

$$\tau = \frac{10^{-3}}{1.186 \times 10^8} = 8.43 \text{ ps}$$

(c) The dc transit angle is

$$\theta_g = \frac{\omega d}{\mathcal{V}_0} = 2\pi \times 8 \times 10^9 \times 8.43 \times 10^{-12}$$

$$= 0.4237 \text{ rad} = 24.30°$$

(d) The beam coupling coefficient is

$$\beta_i = \frac{\sin{(\theta_g/2)}}{\theta_g/2} = \frac{\sin{(24.30°/2)}}{0.424/2} = 0.99$$

(e) The electron velocity just out from the buncher cavity at the time of $(t - \tau)$ is

$$\mathcal{V}(t_1) = 1.186 \times 10^8 \left[1 + \frac{0.99 \times 2}{2 \times 40 \times 10^3} \right.$$

$$\left. \times \sin\left(2\pi \times 8 \times 10^9 \times 8.43 \times 10^{-12} - \frac{0.424}{2} \right) \right]$$

$$= 1.186 \times 10^8 \text{ m/s} + 0.624 \times 10^3 \text{ m/s}$$

$$= \text{dc electron velocity plus RF electron velocity fluctuation}$$

5-4 BUNCHING PROCESS

Once the electrons leave the buncher cavity, they drift with a velocity given by Eq. (5-3-10) or (5-3-11) along in the field-free space between the two cavities. The effect of velocity modulation produces bunching of the electron beam or the current modulation. The electrons that pass the buncher at $V_s = 0$ travel through with unchanged velocity $\mathcal{V}_0$ and become the bunching center. Those electrons that pass the buncher cavity during the positive half-cycles of the microwave input voltage V_s travel faster than the electrons that passed the gap when $V_s = 0$. Those electrons pass the buncher cavity during the negative half-cycles of the voltage V_s travel slower than the electrons that passed the gap when $V_s = 0$. At a distance of ΔL along the beam from the buncher cavity, the beam electrons have drifted into dense clusters where the electron beam changes a half wavelength. Figure 5-4-1 shows the trajectories of minimum, zero, and maximum electron acceleration.

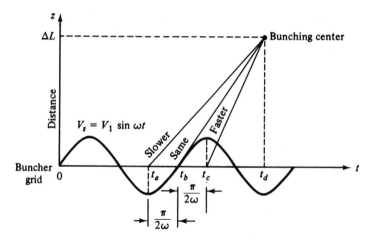

Figure 5-4-1 Bunching distance.

The distance from the buncher grid to the location of dense electron bunching for the electron at t_b is

$$\Delta L = \mathcal{V}_0(t_d - t_b) \tag{5-4-1}$$

Similarly, the distances for the electrons at t_a and t_c are

$$\Delta L = \mathcal{V}_{min}(t_d - t_a) = \mathcal{V}_{min}\left(t_d - t_b + \frac{\pi}{2\omega}\right) \tag{5-4-2}$$

and

$$\Delta L = \mathcal{V}_{max}(t_d - t_c) = \mathcal{V}_{max}\left(t_d - t_b - \frac{\pi}{2\omega}\right) \tag{5-4-3}$$

From Eq. (5-3-10) or (5-3-11), the minimum and maximum velocities are

$$\mathcal{V}_{min} = \mathcal{V}_0\left(1 - \frac{\beta_i V_1}{2V_0}\right) \tag{5-4-4}$$

and

$$\mathcal{V}_{max} = \mathcal{V}_0\left(1 + \frac{\beta_i V_1}{2V_0}\right) \tag{5-4-5}$$

Substitution of Eqs. (5-4-4) and (5-4-5) in Eqs. (5-4-2) and (5-4-3), respectively, yields the distance

$$\Delta L = \mathcal{V}_0(t_d - t_b) + \left[\mathcal{V}_0\frac{\pi}{2\omega} - \mathcal{V}_0\frac{\beta_i V_1}{2V_0}(t_d - t_b) - \mathcal{V}_0\frac{\beta_i V_1}{2V_0}\frac{\pi}{2\omega}\right] \tag{5-4-7}$$

and

$$\Delta L = \mathcal{V}_0(t_d - t_b) + \left[-\mathcal{V}_0 \frac{\pi}{2\omega} + \mathcal{V}_0 \frac{\beta_i V_1}{2V_0}(t_d - t_b) + \mathcal{V}_0 \frac{\beta_i V_1}{2V_0} \frac{\pi}{2\omega} \right] \quad (5\text{-}4\text{-}8)$$

The necessary condition for those electrons at t_a, t_b, and t_c to meet at the same distance ΔL is

$$\mathcal{V}_0 \frac{\pi}{2\omega} - \mathcal{V}_0 \frac{\beta_i V_1}{2V_0}(t_d - t_b) - \mathcal{V}_0 \frac{\beta_i V_1}{2V_0} \frac{\pi}{2\omega} = 0 \quad (5\text{-}4\text{-}9)$$

and

$$-\mathcal{V}_0 \frac{\pi}{2\omega} + \mathcal{V}_0 \frac{\beta_i V_1}{2V_0}(t_d - t_b) + \mathcal{V}_0 \frac{\beta_i V_1}{2V_0} \frac{\pi}{2\omega} = 0 \quad (5\text{-}4\text{-}10)$$

Consequently,

$$t_d - t_b \simeq \frac{\pi V_0}{\omega \beta_i V_1} \quad (5\text{-}4\text{-}11)$$

and

$$\Delta L = \mathcal{V}_0 \frac{\pi V_0}{\omega \beta_i V_1} \quad (5\text{-}4\text{-}12)$$

It should be noted that the mutual repulsion force of space charge is neglected, but the qualitative results are similar to the preceding representation when the effects of repulsion are included. Furthermore, the distance given by Eq. (5-4-12) is not the one for a maximum degree of bunching. Figure 5-4-2 shows the distance-time plot or Applegate diagram.

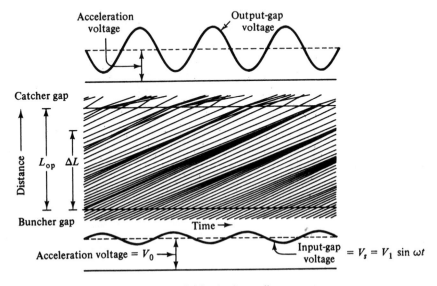

Figure 5-4-2 Applegate diagram.

What should the spacing be between the buncher and catcher cavities in order to achieve a maximum degree of bunching? Since the drift region is field free, the transit time for an electron to travel a distance of L as shown in Fig. 5-2-1 is given by

$$T = t_2 - t_1 = \frac{L}{\mathcal{V}(t_1)} = T_0 \left[1 - \frac{\beta_i V_1}{2V_0} \sin \left(\omega t_1 - \frac{\theta_g}{2} \right) \right] \qquad (5\text{-}4\text{-}13)$$

where the binomial expansion of $(1 + x)^{-1}$ for $|x| \ll 1$ has been replaced and $T_0 = \frac{L}{\mathcal{V}_0}$ is the dc transit time between cavities.

In terms of radians the preceding expression can be written as

$$\omega T = \omega t_2 - \omega t_1 = \theta_0 - X \sin \left(\omega t_1 - \frac{\theta_g}{2} \right) \qquad (5\text{-}4\text{-}14)$$

where $\theta_0 = \frac{\omega L}{\mathcal{V}_0} = 2\pi N$ is the dc transit angle between cavities $\qquad (5\text{-}4\text{-}14a)$

N = number of electron transit cycles in the drift space

$X \equiv \frac{\beta_i V_1}{2V_0} \theta_0$ is defined as the *bunching parameter* of a klystron $\qquad (5\text{-}4\text{-}14b)$

Example 5-4-1: Two-cavity Klystron

A two-cavity klystron has the following parameters:

Beam voltage	$V_0 = 45$ kV
Beam current	$I_0 = 10$ A
Length between two cavities	$L = 4$ cm
Gap distance in either cavity	$d = 1$ mm
Operating frequency	$f = 9$ GHz
Signal voltage	$V_1 = 3$ V(rms)

Calculate:

  (a) The dc electron velocity
  (b) The dc transit angle between cavities
  (c) The dc transit angle across the gap
  (d) The beam coupling coefficient
  (e) The bunching parameter

Solution:

  (a) The dc electron velocity is

$$\mathcal{V}_0 = 0.593 \times 10^6 \sqrt{45 \times 10^3} = 1.258 \times 10^8 \text{ m/s}$$

(b) The dc transit angle between cavities is

$$\theta_0 = \frac{2 \times 3.1416 \times 9 \times 10^9 \times 4 \times 10^{-2}}{1.258 \times 10^8} = 17.98 \text{ rads} = 1030.25°$$

(c) The dc transit angle across the gap is

$$\theta_g = \frac{2 \times 3.1416 \times 9 \times 10^9 \times 1 \times 10^{-3}}{1.258 \times 10^8} = 0.2264 \text{ rad} = 12.98°$$

(d) The beam coupling coefficient is

$$\beta_i = \frac{\sin (12.98/2)}{0.2264/2} = 0.998$$

(e) The bunching parameter is

$$X = \frac{\beta_i V_1}{2V_0} \theta_0 = \frac{0.998 \times 3}{2 \times 45 \times 10^3} \times 17.98 = 5.98 \times 10^{-4}$$

At the buncher gap a charge dQ_0 passing through at a time interval dt_0 is given by

$$dQ_0 = I_0 dt_0 \tag{5-4-15}$$

where I_0 is the dc current. From the principle of conversation of charges this same amount of charge dQ_0 also passes the catcher at a later time interval dt_2. Hence,

$$I_0|dt_0| = i_2|dt_2| \tag{5-4-16}$$

where the absolute value signs are necessary because a negative value of the time ratio would indicate a negative current. i_2 is the current at the catcher gap. Rewriting Eq. (5-4-13) in terms of Eq. (5-3-10) yields

$$t_2 = t_0 + \tau + T_0 \left[1 - \frac{\beta_i V_1}{2V_0} \sin \left(\omega t_0 + \frac{\theta_g}{2} \right) \right] \tag{5-4-17}$$

Alternatively,

$$\omega t_2 - \left(\theta_0 + \frac{\theta_g}{2} \right) = \left(\omega t_0 + \frac{\theta_g}{2} \right) - X \sin \left(\omega t_0 + \frac{\theta_g}{2} \right) \tag{5-4-18}$$

where $\left(\omega t_0 + \dfrac{\theta_g}{2} \right)$ = buncher cavity departure angle

$\omega t_2 - \left(\theta_0 + \dfrac{\theta_g}{2} \right)$ = catcher cavity arrival angle

Figure 5-4-3 shows the curves for the catcher cavity arrival angle as a function of the buncher cavity departure angle in terms of the bunching parameter X.

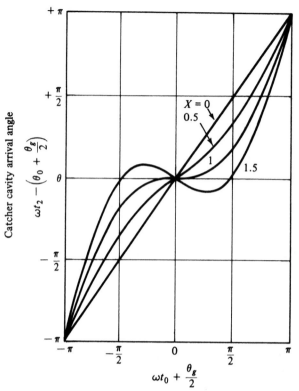

Figure 5-4-3   Catcher arrival angle versus buncher departure angle.

Differentiation of Eq. (5-4-17) with respect to t_0 results in

$$dt_2 = dt_0 \left[1 - X \cos \left(\omega t_0 + \frac{\theta_g}{2} \right) \right] \tag{5-4-19}$$

The current arriving at the catcher cavity is then given as

$$i_2(t_0) = \frac{I_0}{1 - X \cos \left(\omega t_0 + \dfrac{\theta_g}{2} \right)} \tag{5-4-20}$$

In terms of t_2 the current is

$$i_2(t_2) = \frac{I_0}{1 - X \cos \left(\omega t_2 - \theta_0 - \dfrac{\theta_g}{2} \right)} \tag{5-4-21}$$

In Eq. (5-4-21) the relationship of $t_2 = t_0 + \tau + T_0$ is used—namely, $\omega t_2 = \omega t_0 + \omega \tau + \omega T_0 = \omega t_0 + \theta_g + \theta_0$.

     Figure 5-4-4 shows curves of the beam current $i_2(t_2)$ as a function of the catcher arrival angle in terms of the bunching parameter X.

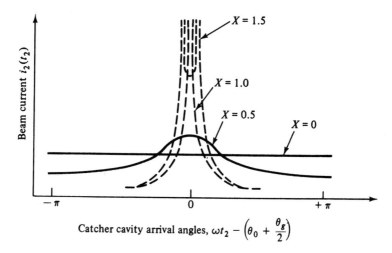

Figure 5-4-4   Beam current i_2 versus catcher cavity arrival angle.

The beam current at the catcher cavity is a periodic waveform of period $2\pi/\omega$ about dc electrons. Therefore the current i_2 can be expanded in a Fourier series. Thus,

$$i_2 = a_0 + \sum_{n=1}^{\infty} [a_n \cos(n\omega t_2) + b_n \sin(n\omega t_2)] \tag{5-4-22}$$

where n is an integer excluding zero. The series coefficients a_0, a_n, and b_n in Eq. (5-4-22) are given by the integrals.

$$a_0 = \frac{1}{2\pi} \int_{-\pi}^{\pi} i_2 d(\pi t_2) \tag{5-4-23}$$

$$a_n = \frac{1}{\pi} \int_{-\pi}^{\pi} i_2 \cos(n\omega t_2) d(\omega t_2) \tag{5-4-24}$$

$$b_n = \frac{1}{\pi} \int_{-\pi}^{\pi} i_2 \sin(n\omega t_2) d(\omega t_2) \tag{5-4-25}$$

Substitution of Eqs. (5-4-16) and (5-4-17) in Eq. (5-4-23) yields

$$a_0 = \frac{1}{2\pi} \int_{-\pi}^{\pi} I_0 d(\omega t_0) = I_0 \tag{5-4-26}$$

$$a_n = \frac{1}{\pi} \int_{-\pi}^{\pi} I_0 \cos\left[(n\omega t_0 + n\theta_g + n\theta_0) - nX \sin\left(\omega t_0 + \frac{\theta_g}{2}\right) \right] d(\omega t_0) \tag{5-4-27}$$

$$b_n = \frac{1}{\pi} \int_{-\pi}^{\pi} I_0 \sin \left[(n\omega t_0 + n\theta_g + n\theta_0) - nX \sin \left(\omega t_0 + \frac{\theta_g}{2} \right) \right] d(\omega t_0) \qquad (5\text{-}4\text{-}28)$$

By using the trigonometrical functions

$$\cos (A \pm B) = \cos A \cos B \mp \sin A \sin B$$

and

$$\sin (A \pm B) = \sin A \cos B \pm \cos A \sin B$$

the two integrals as shown in Eqs. (5-4-27) and (5-4-28) involve cosines and sines of a sine function. Each term of the integrand contains an infinite number of terms of Bessel functions. They are

$$\cos \left[nX \sin \left(\omega t_0 + \frac{\theta_g}{2} \right) \right] = J_0(nX)$$
$$+ 2 \left[J_2(nX) \cos 2 \left(\omega t_0 + \frac{\theta_g}{2} \right) \right] \qquad (5\text{-}4\text{-}29)$$
$$+ 2 \left[J_4(nX) \cos 4 \left(\omega t_0 + \frac{\theta_g}{2} \right) \right]$$
$$+ \cdots$$

and

$$\sin \left[nX \sin \left(\omega t_0 + \frac{\theta_g}{2} \right) \right] = 2 \left[J_1(nX) \sin \left(\omega t_0 + \frac{\theta_g}{2} \right) \right]$$
$$+ 2 \left[J_3(nX) \sin 3 \left(\omega t_0 + \frac{\theta_g}{2} \right) \right] \qquad (5\text{-}4\text{-}30)$$
$$+ \cdots$$

If these series are substituted into the integrands of Eqs. (5-4-27) and (5-4-28), respectively, the integrals are readily evaluated term by term and the series coefficients are

$$a_n = 2I_0 J_n(nX) \cos (n\theta_g + n\theta_0) \qquad (5\text{-}4\text{-}31)$$
$$b_n = 2I_0 J_n(nX) \sin (n\theta_g + n\theta_0) \qquad (5\text{-}4\text{-}32)$$

where $J_n(nX)$ is the nth-order Bessel function of the first kind as shown in Fig. 5-4-5.

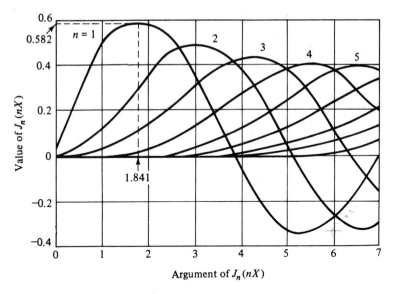

Figure 5-4-5   Bessel functions $J_n(nX)$.

Substitution of Eqs. (5-4-26), (5-4-31), and (5-4-32) in Eq. (5-4-22) yields the beam current i_2 as

$$i_2 = I_0 + \sum_{n=1}^{\infty} 2I_0 J_n(nX) \cos [n\omega(t_2 - \tau - T_0)] \qquad (5\text{-}4\text{-}33)$$

The fundamental component of the beam current at the catcher cavity has a magnitude

$$I_f = 2I_0 J_1(X) \qquad (5\text{-}4\text{-}34)$$

This fundamental component I_f has its maximum amplitude at

$$X = 1.841 \qquad (5\text{-}4\text{-}35)$$

The optimum distance L at which the maximum fundamental component of current occurs is computed from Eqs. (5-4-14a, 5-4-14b) and (5-4-35) as

$$L_{\text{optimum}} = \frac{3.682 \mathcal{V}_0 V_0}{\omega \beta_i V_1} \qquad (5\text{-}4\text{-}36)$$

It is interesting to note that the distance given by Eq. (5-4-12) is approximately 15% less than the result of Eq. (5-4-36). The discrepancy is due in part to the approximations made in deriving Eq. (5-4-12) and to the fact that the maximum fundamental component of current will not coincide with the maximum electron density along the beam because the harmonic components exist in the beam.

Example 5-4-2: Optimum Distance between Cavities

A two-cavity klystron has the following parameters:

Beam voltage	$V_0 = 10$ kV
Beam current	$I_0 = 8$ A
Beam coupling coefficient	$\beta_i = 1$
Operating frequency	$f = 40$ GHz
Signal voltage	$V_1 = 30$ V(rms)

Determine:

(a) The maximum fundamental current

(b) The dc electron velocity

(c) The optimum distance

Solution:

(a) The maximum fundamental current is

$$I_f = 2I_0 J_1(1.841) = 2 \times 8 \times 0.582 = 9.312 \text{ A}$$

(b) The dc electron velocity is

$$\mathcal{V}_0 = 0.593 \times 10^6 \sqrt{10 \times 10^3} = 0.593 \times 10^8 \text{ m/s}$$

(c) The optimum distance is

$$L = \frac{3.682 \times 0.593 \times 10^8 \times 10^4}{2\pi \times 40 \times 10^9 \times 1 \times 30}$$

$$= 0.2896 \text{ m} = 28.96 \text{ cm}$$

5-5 OUTPUT POWER AND EFFICIENCY

The maximum bunching should occur approximately midway between the catcher grids. The phase of the catcher gap voltage must be maintained in such a way that the bunched electrons, as they pass through the grids, encounter a retarding phase. When the bunched electron beam passes through a retarding phase, its kinetic energy is transferred to the field of the catcher cavity. When the electrons emerge from the catcher grids, they have reduced velocity and are finally collected by the collector. The electron bunching process in the catcher cavity is called the current modulation.

The induced current in the catcher cavity.   Since the current induced by the electron beam in the walls of the catcher cavity is directly proportional to the amplitude of the microwave input voltage V_1, the fundamental component of the induced microwave current in the catcher is given by

$$i_{2ind} = \beta_0 i_2 = \beta_0 2I_0 J_1(X) \cos [\omega(t_2 - \tau - T_0)] \qquad (5\text{-}5\text{-}1)$$

where β_0 is the beam coupling coefficient of the catcher gap. If the buncher and catcher cavities are identical, then $\beta_i = \beta_0$. The fundamental component of current induced in the catcher cavity then has a magnitude

$$I_{2ind} = \beta_0 I_2 = \beta_0 2I_0 J_1(X) \tag{5-5-2}$$

Figure 5-5-1 shows an output equivalent circuit, in which R_{sho} represents the wall resistance of catcher cavity, R_B indicates the beam loading resistance, R_L stands for the external load resistance, and R_{sh} is the effective shunt resistance.

The output power delivered to the catcher cavity and the load is given as

$$P_{out} = \frac{(\beta_0 I_2)^2}{2} R_{sh} = \frac{\beta_0 I_2 V_2}{2} \tag{5-5-3}$$

where R_{sh} is the total equivalent shunt resistance of the catcher circuit, including the load, and V_2 is the fundamental component of the catcher gap voltage.

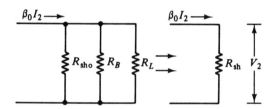

Figure 5-5-1 Output equivalent circuit.

Example 5-5-1: Output Power of a Klystron

A CW two-cavity klystron has the following parameters:

Beam voltage	$V_0 = 50$ kV
Beam current	$I_0 = 15$ A
Beam coupling coefficient	$\beta_i = \beta_0 = 1$
Separation between cavities	$L = 60$ cm
Operating frequency	$f = 4$ GHz
Signal voltage	$V_1 = 30$ V(rms)
Total shunt resistance of the catcher cavity	$R_{sh} = 100$ kΩ

Compute:

(a) The dc electron velocity

(b) The dc transit angle between cavities

(c) The bunching parameter

(d) The induced fundamental current in the catcher cavity

(e) The output power

Solution:

(a) The dc electron velocity is

$$V_0 = 0.593 \times 10^6 \sqrt{50 \times 10^3} = 1.326 \times 10^8 \text{ m/s}$$

(b) The dc transit angle between cavities is

$$\theta_0 = \frac{2\pi \times 4 \times 10^9 \times 60 \times 10^{-2}}{1.326 \times 10^8} = 113.70 \text{ rads}$$

(c) The bunching parameter is

$$X = \frac{1 \times 30 \times 113.70}{2 \times 50 \times 10^3} = 0.0341$$

$$J_1(0.0341) = 0.02$$

(d) The induced fundamental current in the catcher cavity is

$$I_{2ind} = \beta_0 I_2 = 1 \times 2 \times 15 \times 0.02 = 0.60 \text{ A}$$

(e) The output power is

$$P_{out} = \frac{(\beta_0 I_2)^2}{2} R_{sh} = \frac{(0.6)^2}{2} \times 100 \times 10^3$$

$$= 18.00 \text{ kW}$$

Efficiency of klystron.   The electronic efficiency of a klystron amplifier is defined as the ratio of the output power to the input power,

$$\text{Efficiency} \equiv \frac{P_{out}}{P_{in}} = \frac{\beta_0 I_2 V_2}{2 I_0 V_0} \tag{5-5-4}$$

in which the power losses to the beam loading and cavity walls are not included.

If the coupling is perfect, $\beta_0 = 1$, the maximum beam current approaches $I_{2max} = 2I_0(0.582)$, and the voltage V_2 is equal to V_0; then the maximum electronic efficiency is about 58%. In practice, the electronic efficiency of a klystron amplifier is in the range of 15 to 30%. Since the efficiency is a function of the catcher gap transit angle θ_g, Fig. 5-5-2 shows the maximum efficiency of a klystron as a function of catcher transit angle.

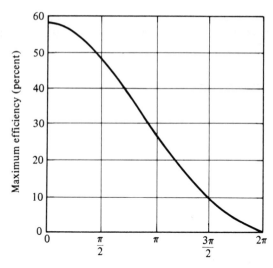

Figure 5-5-2   Maximum efficiency of klystron versus catcher transit angle.

Example 5-5-2: Klystron Efficiency

Calculate the efficiency of the klystron used in Example 5-5-1.

Solution: The efficiency is

$$\eta = \frac{P_{\text{out}}}{P_{\text{in}}} = \frac{18.00 \times 10^3}{15 \times 50 \times 10^3} = 2.4\%$$

Mutual conductance of a klystron amplifier.   The equivalent mutual conductance of the klystron amplifier can be defined as the ratio of the induced output current to input voltage. That is,

$$|G_m| \equiv \frac{i_{2ind}}{V_1} = \frac{2\beta_0 I_0 J_1(X)}{V_1} \tag{5-5-5}$$

From Eq. (5-4-14) the input voltage V_1 can be expressed in terms of the bunching parameter X as

$$V_1 = \frac{2V_0}{\beta_0 \theta_0} X \tag{5-5-6}$$

In Eq. (5-5-5) it is assumed that $\beta_0 = \beta_i$. Substitution of Eq. (5-5-6) in Eq. (5-5-5) yields the normalized mutual conductance as

$$\frac{|G_m|}{G_0} = \beta_0^2 \theta_0 \frac{J_1(X)}{X} \tag{5-5-7}$$

where $G_0 = I_0/V_0$ is the dc beam conductance. The mutual conductance is not a constant but decreases as the bunching parameter X increases. Figure 5-5-3 shows the curves of transductance as a function of X.

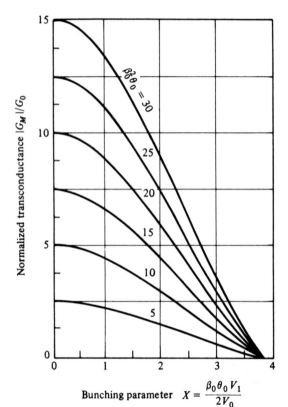

Bunching parameter $X = \dfrac{\beta_0 \theta_0 V_1}{2 V_0}$

Figure 5-5-3 Normalized transconductance G_m versus bunch ng parameter.

It can be seen from the curves that, for a small signal, the normalized transconductance is maximum—that is,

$$\frac{|G_m|}{G_0} = \frac{\beta_0^2 \theta_0}{2} \qquad (5\text{-}5\text{-}8)$$

For a maximum output at $X = 1.841$, the normalized mutual conductance is

$$\frac{|G_m|}{G_0} = 0.316 \, \beta_0^2 \theta_0 \qquad (5\text{-}5\text{-}9)$$

The voltage gain of a klystron amplifier is defined as

$$A_v \equiv \left| \frac{V_2}{V_1} \right| = \frac{\beta_0 I_2 R_{sh}}{V_1} = \frac{\beta_0^2 \theta_0}{R_0} \frac{J_1(X)}{X} R_{sh} \qquad (5\text{-}5\text{-}10)$$

where $R_0 = V_0/I_0$ is the dc beam resistance. Substitution of Eqs. (5-5-2) and (5-5-6) in Eq. (5-5-10) results in

$$A_v = G_m R_{sh} \qquad (5\text{-}5\text{-}11)$$

Example 5-5-3: Voltage Amplification of a Klystron

A two-cavity klystron has the following parameters:

Beam voltage	$V_0 = 2$ kV
Beam current	$I_0 = 50$ mA
Operating frequency	$f = 3$ GHz
Separation between cavities	$L = 6$ cm
Total shunt resistance	$R_{sh} = 50$ k
Beam coupling coefficient	$\beta_i = \beta_0 = 1$
Signal voltage	$V_1 = 80$ V (rms)

Determine:

(a) The dc electron velocity
(b) The transit angle between cavities
(c) The bunching parameter
(d) The voltage amplification

Solution:

(a) The dc electron velocity is

$$\mathcal{V}_0 = 0.593 \times 10^6 \sqrt{2 \times 10^3} = 2.67 \times 10^7 \text{ m/s}$$

(b) The transit angle between cavities is

$$\theta_0 = \frac{2\pi \times 3 \times 10^9 \times 6 \times 10^{-2}}{2.65 \times 10^7} = 42.678 \text{ rads}$$

(c) The bunching parameter is

$$X = \frac{1 \times 80}{2 \times 2 \times 10^3} \times 42.678 = 0.8536$$

$$J_1(0.8536) = 0.39$$

(d) The voltage amplification is

$$A_v = \frac{|V_2|}{|V_1|} = \frac{\beta_0^2 \theta_0}{R_0} \frac{J_1(X)}{X} R_{sh}$$

$$= \frac{1 \times 42.678}{40 \times 10^3} \times \frac{0.39}{0.8536} \times 50 \times 10^3$$

$$= 24.37$$

Power required to bunch the electron beam.   As described earlier, the bunching action takes place in the buncher cavity. When the buncher gap transit angle is small, the average energy of electrons leaving the buncher cavity over a cycle is nearly equal to the energy with which they enter. However, when the buncher gap transit angle is large, the electrons that leave the buncher gap have greater average energy than that when they enter. The difference between the average exit energy and the entrance energy must be supplied by the buncher cavity to bunch the electron beam. It is difficult to calculate the power required to produce the bunching action. Feenberg did some extensive work on beam loading [5]. The ratio of the power required to produce bunching action to the dc power required to produce the electron beam is given by Feenberg as

$$\frac{P_B}{P_0} = \frac{V_1^2}{2V_0^2}\left[\frac{1}{2}\beta_i^2 - \frac{1}{2}\beta_i \cos\left(\frac{\theta_g}{2}\right)\right] = \frac{V_1^2}{2V_0^2} F(\theta_g)$$

$$\text{where } F(\theta_g) = \frac{1}{2}\left[\beta_i^2 - \beta_i \cos\left(\frac{\theta_g}{2}\right)\right]$$

(5-5-12)

Since the dc power is

$$P_0 = V_0^2 G_0 \tag{5-5-13}$$

where $G_0 = I_0/V_0$ is the equivalent electron beam conductance, and the power given by the buncher cavity to produce beam bunching is

$$P_B = \frac{V_1^2}{2} G_B \tag{5-5-14}$$

where G_B is the equivalent bunching conductance. Substitution of Eqs. (5-5-13) and (5-5-14) in Eq. (5-5-12) yields the normalized electronic conductance as

$$\frac{G_B}{G_0} = \frac{R_0}{R_B} = F(\theta_g) \tag{5-5-15}$$

Figure 5-5-4 shows the equivalent bunching resistance as a function of the buncher gap transit angle.

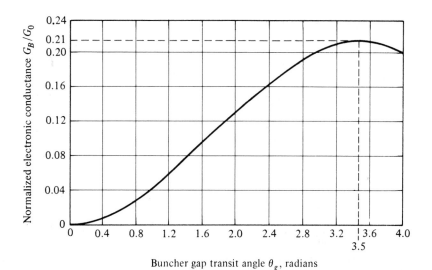

Figure 5-5-4   Normalized electronic conductance versus buncher gap transit angle.

It can be seen from Fig. 5-5-4 that there is a critical buncher gap transit angle for a minimum equivalent bunching resistance. When the transit angle θ_g is 3.5 rad, the equivalent bunching resistance is about five times the electron beam resistance. The power delivered by the electron beam to the catcher cavity can be expressed as

$$\frac{V_2^2}{2R_{sh}} = \frac{V_2^2}{2R_{sho}} + \frac{V_2^2}{2R_B} + \frac{V_2^2}{2R_L} \tag{5-5-16}$$

As a result, the effective impedance of the catcher cavity is

$$\frac{1}{R_{sh}} = \frac{1}{R_{sho}} + \frac{1}{R_B} + \frac{1}{R_L} \tag{5-5-17}$$

Finally, the loaded quality factor of the catcher cavity circuit at the resonant frequency can be written as

$$\frac{1}{Q_L} = \frac{1}{Q_0} + \frac{1}{Q_B} + \frac{1}{Q_{\text{ext}}} \tag{5-5-18}$$

where Q_L = loaded quality of the whole catcher circuit
        Q_0 = quality of the catcher walls
        Q_B = quality of the beam loading
        Q_{ext} = quality of the external load

Example 5-5-4: Klystron Amplifier

A two-cavity klystron amplifier has the following parameters:

Beam voltage	$V_0 = 1000$ V
Beam current	$I_0 = 25$ mA
dc beam resistance	$R_0 = 40$ kΩ
operating frequency	$f = 3$ GHz
Gap spacing in either cavity	$d = 1$ mm
Spacing between the two cavities	$L = 4$ cm
Effective shunt impedance,	$R_{sh} = 30$ kΩ
excluding beam loading	

(a) Find the input gap voltage to give maximum voltage V_2.

(b) Find the voltage gain, neglecting the beam loading in the output cavity.

(c) Find the efficiency of the amplifier, neglecting beam loading.

(d) Calculate the beam loading conductance and show that it was justified to neglect it in the above calculations.

Solution:

(a) For maximum V_2, $J_1(X)$ must be maximum. This means $J_1(X) = 0.582$ at $X = 1.841$. The electron velocity just leaving the cathode is

$$\mathcal{V}_0 = 0.593 \times 10^6 \sqrt{V_0} = 0.593 \times 10^6 \sqrt{10^3} = 1.88 \times 10^7 \text{ m/s}$$

The gap transit angle is

$$\theta_g = \omega \frac{d}{\mathcal{V}_0} = 2\pi(3 \times 10^9) \frac{10^{-3}}{1.88 \times 10^7} = 1 \text{ rad} = 57.3°$$

The beam coupling coefficient is

$$\beta_i = \beta_0 = \frac{\sin(\theta_g/2)}{(\theta_g/2)} = \frac{\sin(57.3°/2)}{(1/2)} = 0.96$$

The dc transit angle between the cavities is

$$\theta_0 = \omega T_0 = \omega \frac{L}{\mathcal{V}_0} = 2\pi(3 \times 10^9) \frac{4 \times 10^{-2}}{1.88 \times 10^7} = 40.10 \text{ rad}$$

The maximum input voltage V_1 is then given by

$$V_{1\max} = \frac{2V_0 X}{\beta_i \theta_0} = \frac{2(10^3)(1.841)}{(0.96)(40.1)} = 95.65 \text{ V}$$

(b) The voltage gain is found as

$$A_v = \frac{\beta_0^2 \theta_0}{R_0} \frac{J_1(X)}{X} R_{sh} = \frac{(0.96)^2(40.1)(0.582)(30 \times 10^3)}{4 \times 10^4 \times 1.841} = 8.76$$

(c) The efficiency can be found as follows

$$I_2 = 2I_0 J_1(X) = 2 \times 25 \times 10^{-3} \times 0.582 = 29.1 \times 10^{-3} \text{ A}$$

$$V_2 = \beta_0 I_2 R_{sh} = (0.96)(29.1 \times 10^{-3})(30 \times 10^3) = 838 \text{ V}$$

$$\text{Efficiency} = \frac{\beta_0 I_2 V_2}{2 I_0 V_0} = \frac{(0.96)(29.1 \times 10^{-3})(838)}{2(25 \times 10^{-3})(10^3)} = 46.8\%$$

(d) Calculate the beam loading conductance (refer to Fig. 5-5-5). The beam loading conductance G_B is

$$G_B = \frac{G_0}{2}\left[\beta_0^2 - \beta_0 \cos\frac{\theta_g}{2}\right]$$

$$= \frac{25 \times 10^{-6}}{2}[(0.96)^2 - (0.96)\cos(28.65°)] = 9.89 \times 10^{-7} \text{ mho}$$

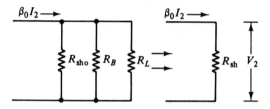

Figure 5-5-5 Equivalent circuit.

Then the beam loading resistance R_B is

$$R_B = \frac{1}{G_B} = 1.14 \times 10^6 \ \Omega$$

In comparison with R_L and R_{sho} or the effective shunt resistance R_{sh}, the beam loading resistance is like an open circuit so that it can be neglected in the above calculations.

5-6 STATE OF THE ART

Extended interaction.   The most common form of extended interaction has been attained recently by coupling two or more adjacent klystron cavities. Figure 5-6-1 shows schematically a five-section extended-interaction cavity as compared to a single-gap klystron cavity [6].

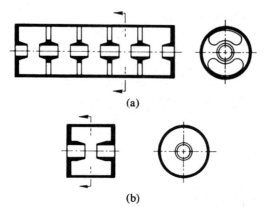

(a)

(b)

Figure 5-6-1 Comparison of a five-gap extended interaction cavity with a single-gap klystron cavity (Reprinted by permission of the IEEE, Inc.).

High efficiency and large power.   In the 1960s much effort was devoted to improving the efficiency of klystrons. For instance, a 50-kW experimental tube demonstrated 75% efficiency in the industrial heating band [7]. The VA-884D klystron is a five-cavity amplifier and its operating characteristics are listed in Table 5-6-1.

TABLE 5-6-1.   VA-884D OPERATING
CHARACTERISTICS

Frequency	5.9 to 6.45	GHz
Power output	14	kW
Gain	52	dB
Efficiency	36	%
Electronic bandwidth (1 dB)	75	MHz
Beam voltage	16.5	kV
Beam current	2.4	A

One of the better-known high-peak power klystrons is the tube developed specifically for use in the two-mile Stanford Linear Accelerator [8] at Palo Alto, California. A cutaway view of the tube is shown in Fig. 5-6-2. The operating characteristics of the tube are listed in Table 5-6-2.

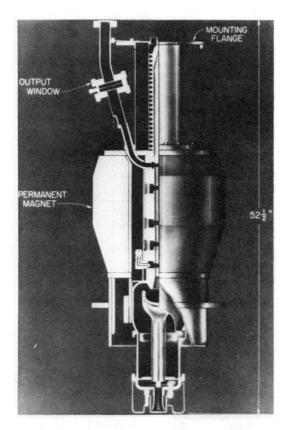

Figure 5-6-2 Cutaway of 24-MW S-band permanent magnet focused klystron (Courtesy Stanford Linear Accelerator Center).

TABLE 5-6-2.   OPERATING CHARACTERISTICS OF THE STANFORD LINEAR ACCELERATOR CENTER HIGH-POWER KLYSTRON [8]

Frequency	2.856	GHz
RF pulse width	2.5	μs
Pulse repetition rate	60 to 360	pps
Peak power output	24	MW
Beam voltage	250	kV
Beam current	250	A
Gain	50 to 55	dB
Efficiency	about 36	%
Weight of permanent (focusing) magnet	363	kg
Electronic bandwidth (1 dB)	20	MHz

The Varian CW superpower klystron amplifier VKC-8269A as shown in Fig. 5-6-3 has an output power of 500 kW (CW) at frequency of 2.114 GHz. Its power gain is 56 dB and efficiency is 50 percent. The beam voltage is 62 V (dc) and the beam current is 16.50 A (dc).

Figure 5-6-3 Photograph of Varian VKC-8269A CW super-power klystron amplifier (Courtesy of Varian Associates, Inc.).

Figure 5-6-4 demonstrates the state of the art for U.S. high-power klystron amplifiers.

Long-life improvement.   A new long-life klystron amplifier tube with a design life in excess of ten years, three times the current design life, has been developed by the Electron Dynamics Division of the Hughes Aircraft Company. Key to the long-life klystron was the development of a method of reducing the operating temperature of the tube's cathode. The cathode is made of porous tungsten impregnated with barium, calcium, and aluminum oxides. The new cathode is coated with a layer of osmium ruthenium alloy that lowers its work function, which is the temperature necessary for electrons to be emitted. This temperature reduction cuts evaporation of barium 10-fold and extends the life of the cathode.

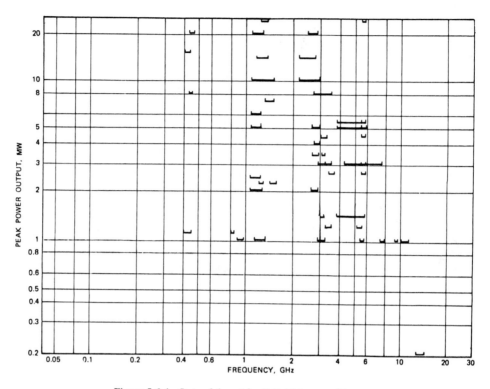

Figure 5-6-4   State of the art for U.S. high-power klystrons.

5-7 MULTICAVITY KLYSTRON AMPLIFIERS

The typical power gain of a two-cavity klystron amplifier is about 30 dB. In order to achieve higher overall gain, one way is to connect several two-cavity tubes in cascade, feeding the output of each of the tubes to the input of the following one. Besides using the multistage techniques, the tube manufacturers have designed and produced multicavity klystron to serve the high-gain requirement as shown in Fig. 5-7-1.

In a multicavity klystron each of the intermediate cavities, placed at a distance of the bunching parameter X of 1.841 away from the previous cavity, acts as a buncher with the passing electron beam inducing an enhanced RF voltage than the previous cavity, which in turn sets up an increased velocity modulation. The spacing between the consecutive cavities would therefore diminish because of the requirement of X being 1.841

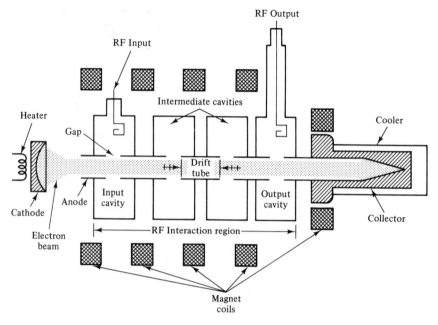

Figure 5-7-1   Schematic diagram of a four-cavity klystron amplifier (Courtesy of Varian Associates, Inc.).

and an increasing velocity modulating RF voltage as the beam progresses through the various cavities. Keeping the intercavity distance constant, an increasing beam voltage V_0 could be used in the subsequent cavities. Figure 5-7-2 shows the photograph of a four-cavity klystron amplifier.

Beam-current density.   When the two-cavity klystron amplifier was discussed in Sections 5-2 to 5-6, it was assumed that the space-charge effect is negligible and it was not considered due to the assumed small density of electrons in the beam for low-power amplifier. However, when high-power klystron tubes are analyzed the electron density of the beam is large and the forces of mutual repulsion of the electrons must be considered. When the electrons perturbate in the electron beam, the electron density consists of a dc part plus an RF perturbation due to the electron bunches. The space-charge forces within electron bunches vary with the size and shape of an electron beam. In an infinitely wide beam, the electric fields (and, thus, the space-charge forces) are constrained to act only in the axial direction. In a finite beam, the electric fields are radial as well as axial with the result that the axial component is reduced in comparison to the infinite beam. With reduced axial space-charge force, the plasma frequency is reduced and the plasma wavelength is increased.

Figure 5-7-2 Photograph of a four-cavity klystron amplifier VA-890H (Courtesy of Varian Associates, Inc.).

Mathematically, let the charge-density and velocity perturbations be simple sinusoidal variations in both time and position. They are

Charge density: $\qquad\qquad \rho = B \cos (\beta_e z) \cos (\omega_q t + \theta)$ $\qquad$ (5-7-1)

Velocity perturbation: $\qquad \mathcal{V} = -C \sin (\beta_e z) \sin (\omega_q t + \theta)$ $\qquad$ (5-7-2)

where B = constant of charge-density perturbation
$\qquad C$ = constant of velocity perturbation

$\qquad \beta_e = \dfrac{\omega}{\mathcal{V}_0}$ is the dc phase constant of the electron beam

$\qquad \omega_q = R\omega_p$ is the perturbation frequency or reduced plasma frequency
$\qquad R = \omega_q/\omega_p$ is the space-charge reduction factor and varies from 0 to 1

$\qquad \omega_p = \sqrt{\dfrac{e\rho_0}{m\epsilon_0}}$ is the plasma frequency and is a function of the electron-beam density

$\qquad \theta$ = phase angle of oscillation

The electron plasma frequency is the frequency at which the electrons will oscillate in the electron beam. This plasma frequency applies only to a beam of infinite diameter. Practical beams of finite diameter are characterized by a plasma frequency that is less than ω_p. This lower plasma frequency is called the *reduced plasma frequency* and is designated ω_q. The space-charge reduction factor R is a function of the beam radius r and the ratio n of the beam-tunnel radius to the beam radius as shown in Fig. 5-7-3 [9].

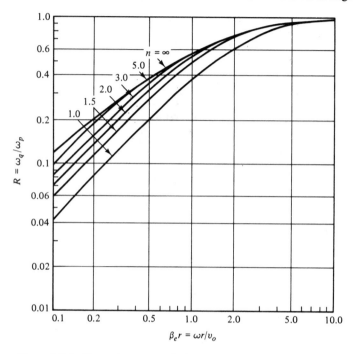

Figure 5-7-3 Plasma-frequency reduction factor R (Reprinted by permission of Pergamon Press).

As an example of the effect of the reduced space-charge forces, if $\beta_e r = \omega r/\mathcal{V}_0 = 0.85$ and $n = 2$, the reduced factor is $R = 0.5$. Therefore, $\omega_q = 2\omega_p$, which means that in a klystron the cavities would be placed twice as far apart as would be indicated by infinite-beam calculation. The total charge density and electron velocity are given by

$$\rho_{\text{tot}} = -\rho_0 + \rho \qquad (5\text{-}7\text{-}3)$$

$$\mathcal{V}_{\text{tot}} = \mathcal{V}_0 + \mathcal{V} \qquad (5\text{-}7\text{-}4)$$

where ρ_0 = dc electron charge density
       ρ = instantaneous RF charge density
       $\mathcal{V}_0$ = dc electron velocity
       $\mathcal{V}$ = instantaneous electron velocity perturbation

The total electron beam-current density can be written as

$$J_{\text{tot}} = -J_0 + J \qquad (5\text{-}7\text{-}5)$$

where J_0 = dc beam-current density
       J = instantaneous RF beam-current perturbation

The instantaneous convection beam-current density at any point in the beam is expressed as

$$J_{\text{tot}} = \rho_{\text{tot}} \, \mathcal{V}_{\text{tot}} = (-\rho_0 + \rho)(\mathcal{V}_0 + \mathcal{V})$$

$$= -\rho_0 \mathcal{V}_0 - \rho_0 \mathcal{V} + \rho \mathcal{V}_0 + \rho \mathcal{V} \qquad (5\text{-}7\text{-}6)$$

$$= -J_0 + J$$

where $J = \rho \mathcal{V}_0 - \rho_0 \mathcal{V}$ $\qquad\qquad\qquad\qquad\qquad$ (5-7-7)
$J_0 = \rho_0 \mathcal{V}_0$ is replaced
$\rho \mathcal{V}$ = very small is ignored

In accordance with the law of conservation of electric charge, the continuity equation can be written as

$$\nabla \cdot \mathbf{J} = -\frac{\partial \rho}{\partial t} \qquad\qquad (5\text{-}7\text{-}8)$$

where $J = \rho \mathcal{V}_0 - \rho_0 \mathcal{V}$ is in positive z direction only. From

$$\frac{\partial J}{\partial z} = -\omega B \sin (\beta_e z - \omega t) \cos (\omega_q t + \theta)$$
$$+ \beta_e \rho_0 C \cos (\beta_e z - \omega t) \sin (\omega_q t + \theta) \qquad (5\text{-}7\text{-}9)$$

and

$$-\frac{\partial \rho}{\partial t} = -\omega B \sin (\beta_e z - \omega t) \cos (\omega_q t + \theta)$$
$$+ \omega_q B \cos (\beta_e z - \omega t) \sin (\omega_q t + \theta) \qquad (5\text{-}7\text{-}10)$$

then we have

$$\omega_q B = \beta_e \rho_0 C \qquad\qquad (5\text{-}7\text{-}11)$$

The beam-current density and the modulated velocity are expressed as [10]

$$J = \mathcal{V}_0 B \cos (\beta_e z - \omega t) \cos (\omega_q t + \theta)$$
$$+ \frac{\omega_q}{\omega} \mathcal{V}_0 B \sin (\beta_e z - \omega t) \sin (\omega_q t + \theta) \qquad (5\text{-}7\text{-}12)$$

and

$$\mathcal{V} = -\mathcal{V}_0 \frac{\beta_i V_1}{2 V_0} \cos (\beta_q z) \sin (\beta_e z - \omega t) \qquad (5\text{-}7\text{-}13)$$

In practical microwave tubes, the ratio of ω_q / ω is much smaller than unity and the second term in Eq. (5-7-12) may be neglected in comparison with the first. Then we obtain

$$J = \mathcal{V}_0 B \cos (\beta_e z - \omega t) \cos (\omega_q t + \theta) \qquad (5\text{-}7\text{-}14)$$

Example 5-7-1: Four-cavity Klystron

A four-cavity klystron VA-828 has the following parameters:

Beam voltage	$V_0 = 14.5$ kV
Beam current	$I_0 = 1.4$ A
Operating frequency	$f = 10$ GHz
dc electron charge density	$\rho_0 = 10^{-6}$ C/m^3
RF charge density	$\rho = 10^{-8}$ C/m^3
Velocity perturbation	$\mathcal{V} = 10^5$ m/s

Compute:

(a) The dc electron velocity

(b) The dc phase constant

(c) The plasma frequency

(d) The reduced plasma frequency for $R = 0.4$

(e) The dc beam current density

(f) The instantaneous beam current density

Solution:

(a) The dc electron velocity is

$$\mathcal{V}_0 = 0.593 \times 10^6 \sqrt{14.5 \times 10^3} = 0.714 \times 10^8 \text{ m/s}$$

(b) The dc phase constant is

$$\beta_e = \frac{2\pi \times 10^{10}}{0.714 \times 10^8} = 8.80 \times 10^2 \text{ rads/m}$$

(c) The plasma frequency is

$$\omega_p = \left(1.759 \times 10^{11} \times \frac{10^{-6}}{8.854 \times 10^{-12}} \right)^{1/2} = 1.41 \times 10^8 \text{ rad/s}$$

(d) The reduced plasma frequency for $r = 0.4$ is

$$\omega_q = 0.4 \times 1.41 \times 10^8 = 0.564 \times 10^8 \text{ rad/s}$$

(e) The dc beam current density is

$$J_0 = 10^{-6} \times 0.714 \times 10^8 = 71.4 \text{ A/m}^2$$

(f) The instantaneous beam current density is

$$J = 10^{-8} \times 0.714 \times 10^8 - 10^{-6} \times 10^5 = 0.614 \text{ A/m}^2$$

The electrons leaving the input gap of a klystron amplifier has a velocity given by Eq. (5-3-10) at the exit grid as

$$\mathcal{V}(t_1) = \mathcal{V}_0 \left[1 + \frac{\beta_i V_1}{2V_0} \sin (\omega\tau) \right] \qquad (5\text{-}7\text{-}15)$$

where V_1 = magnitude of the input signal voltage

$\tau = \dfrac{d}{\mathcal{V}_0} = t_1 - t_0$ is the transit time

d = gap distance

Since the electrons under the influence of the space-charge forces exhibit simple harmonic motion, the velocity at a later time t is given by

$$\mathcal{V}_{\text{tot}} = \mathcal{V}_0 \left[1 + \frac{\beta_i V_1}{2V_0} \sin(\omega\tau) \cos(\omega_p t - \omega_p \tau) \right] \qquad (5\text{-}7\text{-}16)$$

where ω_p = plasma frequency. The current-density equation may be obtained from Eq. (5-7-14) as

$$J = -\frac{1}{2} \frac{J_0 \omega}{V_0 \omega_q} \beta_i V_1 \sin(\beta_q z) \cos(\beta_e z - \omega t) \qquad (5\text{-}7\text{-}17)$$

where $\beta_q = \dfrac{\omega_q}{\mathcal{V}_0}$ is the plasma phase constant.

Example 5-7-1A: Operation of a Four-Cavity Klystron

A four-cavity CW klystron amplifier VA-864 has the following parameters:

Beam voltage	$V_0 = 18$ kV
Beam current	$I_0 = 2.25$ A
Gap distance	$d = 1$ cm
Operating frequency	$f = 10$ GHz
Signal voltage	$V_1 = 10$ V (rms)
Beam coupling coefficient	$\beta_0 = \beta_i = 1$
dc electron beam current density	$\rho_0 = 10^{-8}$ C/m^3

Determine:

  (a) The dc electron velocity
  (b) The dc electron phase constant
  (c) The plasma frequency
  (d) The reduced plasma frequency for $r = 0.5$
  (e) The reduced plasma phase constant
  (f) The transit time across the input gap
  (g) The electron velocity leaving the input gap

Solution:

  (a) The dc electron velocity is

$$\mathcal{V}_0 = 0.593 \times 10^6 \sqrt{18 \times 10^3} = 0.796 \times 10^8 \text{ m/s}$$

(b) The dc electron phase constant is

$$\beta_e = 2\pi \times 10^{10}/(0.796 \times 10^8) = 7.89 \times 10^2 \text{ rad/m}$$

(c) The plasma frequency is

$$\omega_p = [1.759 \times 10^{11} \times 10^{-8}/(8.854 \times 10^{-12})]^{1/2} = 1.41 \times 10^7 \text{ rad/s}$$

(d) The reduced plasma frequency is

$$\omega_q = 0.5 \times 1.41 \times 10^7 = 0.705 \times 10^7 \text{ rad/s}$$

(e) The reduced plasma phase constant is

$$\beta_q = 0.705 \times 10^7/(0.796 \times 10^8) = 0.088 \text{ rad/m}$$

(f) The transit time across the gap is

$$\tau = 10^{-2}/(0.796 \times 10^8) = 0.1256 \text{ ns}$$

(g) The electron velocity leaving the input gap is

$$\mathcal{V}(t_1) = 0.796 \times 10^8[1 + 1 \times 10/(2 \times 18 \times 10^3) \sin (2\pi \times 10^{10} \times 1.256$$
$$\times 10^{-10})]$$
$$= 0.796 \times 10^8 + 2.21 \times 10^4 \text{ m/s}$$

Output current and output power of two-cavity klystron.   If the two cavities of a two-cavity klystron amplifier are assumed to be identical and are placed at the point where the RF current modulation is a maximum, the magnitude of the RF convection current at the output cavity for a two-cavity klystron can be written from Eq. (5-7-17) as

$$|i_2| = \frac{1}{2} \frac{I_0\omega}{V_0\omega_q} \beta_i|V_1| \tag{5-7-18}$$

where V_1 = magnitude of the input signal voltage. Then the magnitudes of the induced current and voltage in the output cavity are equal to

$$|I_2| = \beta_0|i_2| = \frac{1}{2} \frac{I_0\omega}{V_0\omega_q} \beta_0^2 |V_1| \tag{5-7-19}$$

and

$$|V_2| = |I_2| R_{shl} = \frac{1}{2} \frac{I_0\omega}{V_0\omega_q} \beta_0^2 |V_1| R_{shl} \tag{5-7-20}$$

where $\beta_0 = \beta_i$ is the beam coupling coefficient
R_{shl} = total shunt resistance of the output cavity in a two-cavity klystron amplifier including the external load

The output power delivered to the load in a two-cavity klystron amplifier is given by

$$P_{out} = |I_2|^2 R_{shl} = \frac{1}{4} \left(\frac{I_0 \omega}{V_0 \omega_q} \right)^2 \beta_0^4 |V_1|^2 R_{shl} \tag{5-7-21}$$

The power gain of a two-cavity klystron amplifier is then expressed by

$$\text{Power gain} = \frac{P_{out}}{P_{in}} = \frac{P_{out}}{|V_1|^2/R_{sh}} = \frac{1}{4} \left(\frac{I_0 \omega}{V_0 \omega_q} \right)^2 \beta_0^4 R_{sh} \cdot R_{shl} \tag{5-7-22}$$

where R_{sh} = total shunt resistance of the input cavity. The electronic efficiency of a two-cavity klystron amplifier is

$$\eta = \frac{P_{out}}{P_{in}} = \frac{P_{out}}{I_0 V_0} = \frac{1}{4} \left(\frac{I_0}{V_0} \right) \left(\frac{|V_1| \omega}{V_0 \omega_q} \right)^2 \beta_0^4 R_{shl} \tag{5-7-23}$$

Example 5-7-2: Characteristics of Two-Cavity Klystron

A two-cavity klystron has the following parameters:

Beam voltage	$V_0 = 20$ kV
Beam current	$I_0 = 2$ A
Operating frequency	$f = 8$ GHz
Beam coupling coefficient	$\beta_i = \beta_0 = 1$
dc electron beam charge density	$\rho_0 = 10^{-6}$ C/m³
Signal voltage	$V_1 = 10$ V (rms)
Shunt resistance of the cavity	$R_{sh} = 10$ kΩ
Total shunt resistance including load	$R = 30$ kΩ

Calculate:

(a) The plasma frequency

(b) The reduced plasma frequency for $R = 0.5$

(c) The induced current in the output cavity

(d) The induced voltage in the output cavity

(e) The output power delivered to the load

(f) The power gain

(g) The electronic efficiency

Solution:

(a) The plasma frequency is

$$\omega_p = [1.759 \times 10^{11} \times 10^{-6}/(8.854 \times 10^{-12})]^{1/2} = 1.41 \times 10^8 \text{ rad/s}$$

(b) The reduced plasma frequency is

$$\omega_q = 0.5 \times 1.41 \times 10^8 = 0.705 \times 10^8 \text{ rad/s}$$
$$\omega/\omega_q = 2\pi \times 8 \times 10^9/(0.705 \times 10^8) = 713$$

(c) The induced current in the output cavity is

$$|I_2| = \frac{1}{2} \frac{2}{20 \times 10^3} \times 713 \times 1^2 \times |10| = 0.3565 \text{ A}$$

(d) The induced voltage in the output cavity is

$$|V_2| = |I_2| R_{shl} = 0.357 \times 30 \times 10^3 = 10.71 \text{ kV}$$

(e) The output power delivered to the load is

$$P_{out} = |I_2|^2 R_{shl} = 0.357^2 \times 30 \times 10^3 = 3.82 \text{ kW}$$

(f) The power gain is

$$\text{Gain} = 1/4[2/(20 \times 10^3) \times 713]^2 \times 1^4 \times 10 \times 10^3 \times 30 \times 10^3$$
$$= 3.83 \times 10^5 = 55.8 \text{ dB}$$

(g) The electronic efficiency is

$$\eta = \frac{P_{out}}{P_{in}} = \frac{3.82 \times 10^3}{2 \times 20 \times 10^3} = 9.6\%$$

Output power of four-cavity klystron. High power may be obtained by adding additional intermediate cavities in a two-cavity klystron. Multicavity klystrons with as many as seven cavities are commercially available, although the most frequently used number of cavities is four. Each of the intermediate cavities functions in the same manner as in the two-cavity klystron amplifier.

Let us carry out a simplified analysis of the four-cavity klystron amplifier. The four cavities are assumed to be identical and they have same unloaded Q and beam coupling coefficient $\beta_i = \beta_0$. The two intermediate cavities are not externally loaded, but the input and output cavities are matched to their transmission lines. If V_1 is the magnitude of the input cavity-gap voltage, the magnitude of the RF convection current density injected into the first intermediate cavity gap is given by Eq. (5-7-17). The induced current and voltage in the first intermediate cavity are given by Eqs. (5-7-19) and (5-7-20). This gap voltage of the first intermediate cavity produces a velocity modulation on the beam in the second intermediate cavity. The RF convection current in the second intermediate cavity can be written with $|V_1|$ replaced by $|V_2|$. That is,

$$|i_3| = \frac{1}{2} \frac{I_0 \omega}{V_0 \omega_q} \beta_0 |V_2|$$

$$= \frac{1}{4} \left(\frac{I_0 \omega}{V_0 \omega_q} \right)^2 \beta_0^3 |V_1| R_{sh}$$

(5-7-24)

The output voltage of the second intermediate cavity is then given by

$$|V_3| = \beta_0 |i_3| R_{sh}$$

$$= \frac{1}{4} \left(\frac{I_0 \omega}{V_0 \omega_q} \right)^2 \beta_0^4 |V_1| R_{sh}^2$$

(5-7-25)

This voltage produces a velocity modulation again and is converted into an RF convection current at the output cavity for four-cavity klystron as

$$|i_4| = \frac{1}{2} \frac{I_0 \omega}{V_0 \omega_q} \beta_i |V_3|$$

$$= \frac{1}{8} \left(\frac{I_0 \omega}{V_0 \omega_q} \right)^3 \beta_0^5 |V_1| R_{sh}^2 \tag{5-7-26}$$

and

$$|I_4| = \beta_0 |i_4| = \frac{1}{8} \left(\frac{I_0 \omega}{V_0 \omega_q} \right)^3 \beta_0^6 |V_1| R_{sh}^2 \tag{5-7-27}$$

The output voltage is then

$$|V_4| = |I_4| R_{shl} = \frac{1}{8} \left(\frac{I_0 \omega}{V_0 \omega_q} \right)^3 \beta_0^6 |V_1| R_{sh}^2 R_{shl} \tag{5-7-28}$$

The output power from the output cavity in a four-cavity klystron amplifier can be expressed as

$$P_{out} = |I_4|^2 R_{shl} = \frac{1}{64} \left(\frac{I_0 \omega}{V_0 \omega_q} \right)^6 \beta_0^{12} |V_1|^2 R_{sh}^4 R_{shl}^2 \tag{5-7-29}$$

where R_{shl} = total shunt resistance of the output cavity including the external load.

The multicavity klystrons are often operated with their cavities stagger-tuned so as to obtain a greater bandwidth at some reduction in gain. In high-power klystrons, the cavity grids are omitted, because they would burn up due to beam interception. High-power klystron amplifiers with a power gain of 40 to 50 dB and a bandwidth of several percent are commercially available.

Example 5-7-3: Output Power of Four-Cavity Klystron

A four-cavity klystron has the following parameters:

Beam voltage	V_0 = 10 kV
Beam current	I_0 = 0.7 A
Operating frequency	f = 4 GHz
Beam coupling coefficient	$\beta_i = \beta_0 = 1$
dc electron beam charge density	ρ_0 = 5 × 10^{-5} C/m³
Signal voltage	V_1 = 2 V(rms)
Shunt resistance of cavity	R_{sh} = 10 kΩ
Total shunt resistance including load	R_{shl} = 5 kΩ

Determine:

(a) The plasma frequency

(b) The reduced plasma frequency for R = 0.6

(c) The induced current in the output cavity

(d) The induced voltage in the output cavity

(e) The output power delivered to the load

Solution:

(a) The plasma frequency is

$$\omega_p = \left[1.759 \times 10^{11} \times \frac{5 \times 10^{-5}}{8.854 \times 10^{-12}} \right]^{1/2} = 0.997 \times 10^9 \text{ rad/s}$$

(b) The reduced plasma frequency is

$$\omega_q = 0.6 \times 0.997 \times 10^9 = 0.598 \times 10^9 \text{ rad/s}$$

$$\omega/\omega_q = 2\pi \times 4 \times 10^9/(0.598 \times 10^9) = 42.03$$

(c) The induced current in the output cavity is

$$|I_4| = \frac{1}{8} \left(\frac{0.7}{10^4} \times 42.03 \right)^3 \times 1^6 \times |2| \times (10 \times 10^3)^2$$

$$= 0.6365 \text{ A}$$

(d) The induced voltage in the output cavity is

$$|V_4| = |I_4| R_{shl} = |0.6365| \times 5 \times 10^3$$
$$= 3.18 \text{ kV}$$

(e) The output power is

$$P_{out} = |I_4|^2 R_{shl} = |0.6365|^2 \times 5 \times 10^3 = 2.03 \text{ kW}$$

5-8 KLYSTRODE

In 1939, Haeff perceived that a major limitation to power output of a conventional gridded tube was the power that could be dissipated by the grids and anode themselves. He thought that nonintercepting electrodes, such as apertures, rather than wire grids, could be used if a magnetic field were deployed coaxially with the electron beam. He proposed that RF power could be removed from a bunched electron beam by passing it through a resonant cavity in which the kinetic energy of the electrons could be converted to electromagnetic energy without the necessity of collecting the electrons on the cavity walls. He designed an inductive output tube (IOT), which is now called the *klystrode,* and the tube had an output power of 100 W (CW) with 35% efficiency and 10-dB gain at 450 MHz.

5-8-1 Structure

The schematic diagram of a klystrode is shown in Fig. 5-8-1 [11]. The tube consists of a round-disc cathode, planar grid, tail pipe, and collector and is immersed in a uniform magnetic field provided by a solenoid or a permanent magnet. It can be seen that the

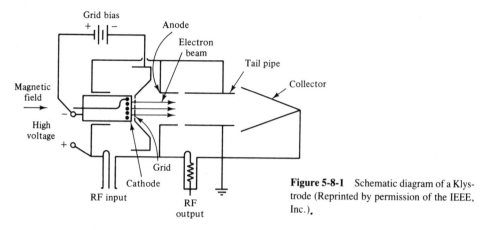

Figure 5-8-1   Schematic diagram of a Klystrode (Reprinted by permission of the IEEE, Inc.).

tube is similar to a conventional tetrode except that the output circuitry is a reentrant resonant cavity like a klystron. This is why the name *klystrode* was adopted. Figure 5-8-2 shows an illustration of a klystrode.

Figure 5-8-2   Photograph of a Klystrode (Reprinted by permission of the IEEE, Inc.).

5-8-2 Operation

In operation, if an RF voltage is applied between cathode and grid, a bunched electron beam is accelerated toward the apertured anode at a high potential and then passes through it without interception. Then the electron beam continues through a field-free region at constant velocity and passes through the output gap. In the resonant cavity the electric field decelerates the electron beam. Then the beam passes through a second field-free region, the tail pipe, with minimum interception, and finally traverses the gap between the tailpipe and collector. The tube is similar to a klystron between the anode and the collector; while between the cathode and the anode, it closely resembles the tetrode. For this reason, the name *klystrode* has been used to describe the tube. Figure 5-8-3 [11] shows the comparisons of potential, electric field (potential gradient $\mathbf{E} = -\nabla V$), and electron velocity for tetrode and klystrode.

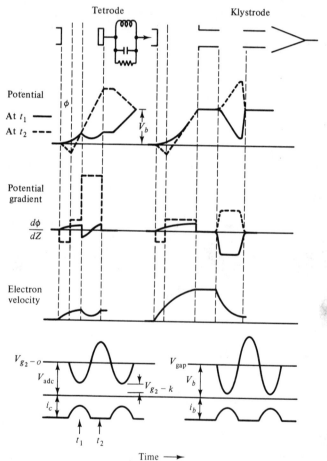

Figure 5-8-3 Comparison of potential, electric field, and electron velocity for tetrode and klystrode (Reprinted by permission of the IEEE, Inc.).

5-8-3 Characteristics

The Varian/EIMAC Division has demonstrated that the klystrode can produce an output power of 20 to 30 kW at 775 to 805 MHz. Table 5-8-1 shows the data for FM operation of a klystrode and Table 5-8-2 shows the data for class-B linear amplifier operation of a klystrode.

TABLE 5-8-1.   DATA OF FM OPERATION FOR KLYSTRODE [11]

CW operation		
Frequency	f	= 775 MHz
Beam voltage	V_b	= 30 kV
Beam current	I_b	= 1.15 A
Grid bias voltage	V_g	= 67 V
Grid current (dc)	I_g	= 0.15 A
Tailpipe current (dc)	I_{tp}	= 0.05 A
Collector current	I_c	= 1.10 A
RF power output	P_{out}	= 20 kW
RF drive power	P_{in}	= 160 W
Power gain	G	= 21 dB
Conversion efficiency	η	= 58%
Magnetic flux density	B	= 300 G
Bandwidth (-3 dB)	BW	= 2.0 MHz
Pulsed operation		
Power output	P_{out}	= 32.30 kW
Beam voltage (dc)	V_b	= 25.00 kV
Beam current (peak)	I_b	= 2.45 A
Collector current (peak)	I_c	= 2.40 A
Tailpipe current (peak)	I_{tp}	= 0.05 A
RF drive power (peak)	P_{in}	= 340.00 W
Conversion efficiency	η	= 52.7%
Power gain	G	= 19.78 dB
Grid current (peak)	I_g	= 0.40 A
Bandwidth (-3 dB)	BW	= 4.7 MHz

Reprinted by permission of the IEEE, Inc.

TABLE 5-8-2.   DATA FOR CLASS-B LINEAR OPERATION OF A KLYSTRODE [11]

Frequency	f	= 805 MHz
Beam voltage	V_b	= 30.00 kV
Beam current (peak of sync)(p.s)	I_b	= 2.70 A
Beam current (black level)(b.l)	I_b	= 2.02 A
Beam current (average picture)(a.p)	I_b	= 1.19 A
Collector-cathode voltage	V_c	= 21.00 kV
Collector current (p.s)	I_c	= 2.60 A
Collector current (b.l)	I_c	= 1.95 A
Collector current (a.p)	I_c	= 1.16 A
RF power output (p.s)	P_{out}	= 30.00 kW
RF power output (b.l)	P_{out}	= 16.90 kW
RF power output (a.p)	P_{out}	= 6.06 kW
Conversion efficiency (p.s)	η_{ps}	= 52%
Conversion efficiency (b.l)	η_{bl}	= 38%
Conversion efficiency (a.p)	η_{ap}	= 23.70%
RF drive power (p.s)	P_{in}	= 415.00 W
Power gain (p.s)	G	= 18.60 dB
Bandwidth (-3 dB)(simple cavity)	BW	= 8.45 MHz

Reprinted by permission of the IEEE, Inc.

Efficiency.   The losses due to electron transit time loading are surprisingly low in the region from the cathode to output gap, and the losses due to interception of beam current at the anode are negligible because of magnetic focusing. As a result, the conversion efficiency is very high and it can be computed by the following equation:

$$\eta = \frac{\text{RF output}}{V_b I_b + V_c I_c} \qquad (5\text{-}8\text{-}1)$$

where V_b = beam voltage
  I_b = beam current
  V_c = collector voltage
  I_c = collector current

Figure 5-8-4 shows the power-supply connections for a klystrode.

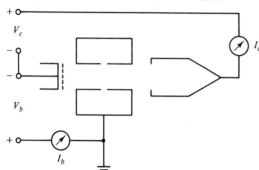

Figure 5-8-4 Power supply connections for a klystrode (Reprinted by permission of the IEEE, Inc.).

Applications.   The klystrode appears to be well suited to UHF-TV service, as an aural or visual channel amplifier. In aural FM service, the power gain, high efficiency, and expected long life are attractive. In visual service, the klystrode offers the possibility of greater overall efficiency than either klystron or tetrode can afford.

Example 5-8-1: Klystrode

A klystrode operates in a FM mode under the following parameters:

Beam voltage	V_b	= 30 kV
Beam current	I_b	= 1.15 A
Collector voltage	V_c	= 21 kV
Collector current	I_c	= 1.10 A
RF output power	P_{out}	= 30 kW

Calculate the efficiency of the klystrode.

Solution: The efficiency is

$$\eta = \frac{30 \times 10^3}{30 \times 10^3 \times 1.15 + 21 \times 10^3 \times 1.10}$$
$$= 52.10\%$$

REFERENCES

[1] WARNECKE, R. R., et al., Velocity Modulated Tubes. In *Advanced in Electrons*, vol. 3, Academic Press, New York, 1951.

[2] CHODOROW, M., and SUSSKIND, C., *Fundamentals of Microwave Electrons*. McGraw-Hill Book Company, New York, 1964.

[3] CHODOROW, M., and WESSEL-BERG, T., A High-Efficiency Klystron with Distributed Inter-action. *IRE Trans. Electron Devices*, vol. ED-8, pp. 44–55, Jan. 1961.

[4] LaRUE and RUBERT, R. R., Multi-Megawatt Hybrid TWT's at S-band and C-band. *IEEE Electron Devices Meeting*, Washington, D.C., Oct. 1964.

[5] FEENBERG, E., Notes on Velocity Modulation. *Sperry Gyroscope Laboratories Report*, pp. 5521-1043, chapter I, pp. 41–44.

[6] STAPRANS, A., et al., High-Power Linear-Beam Tubes. *Proc. IEEE*, Vol. 61, No. 3, pp. 299–300, March 1973.

[7] LIEN, E. L., High Efficiency Klystron Amplifier. In *Conv. Rec. MOGA 70* (8th Int. Conf., Amsterdam, The Netherlands, Sept. 1970).

[8] MERDINIAN, G., and LEBACQZ, J. V., High Power, Permanent Magnet Focused, S-band Klystron for Linear Accelerator Use. *Proc. 5th Int. Conf. on Hyperfrequency Tubes* (Paris, France, Sept. 1964).

[9] BECK, A. H. W., *Space Charge Waves and Slow Electromagnetic Waves*. Pergamon Press, New York, 1985. p. 106.

[10] GEWARTOWSKI, J. W., and WATSON, H. A., *Principles of Electron Tubes*. D. Van Nostrand Company, Inc., Princeton, N.J., 1965, pp. 336–38.

[11] PRIEST, DONALD H., and SHRADER, M. B., The Klystrode—An Unusual Transmitting Tube with Potential for UHF-TV. *Proc. IEEE*, vol. 70, no. 11, pp. 1318–25, Nov. 1982.

PROBLEMS

Gridded Tubes

5-1-1. A triode amplifier has the following parameters:

Amplification factor	μ	= 50
Plate resistance	r_p	= 50 kΩ
Load resistance	R_ℓ	= 10 kΩ
Input signal	V_{in}	= 10 V

Compute:

(a) The transconductance g_m

(b) The plate current I_p

(c) The output voltage V_0

(d) The voltage amplification gain

5-1-2. Derive Eq. (5-1-7) from Eq. (5-1-1).

5-1-3. A tetrode amplifier has the following parameters:

Plate amplification factor	μ_p	= 100
Screen amplification factor	μ_s	= 10
Plate resistance	r_p	= 45 kΩ
Plate voltage	V_p	= 250 V
Grid voltage	V_g	= -2.0 V

Perveance $K = 2.5 \times 10^{-6}$
Load resistance $R_\ell = 20 \text{ k}\Omega$
Source voltage $V_s = 40 \text{ V}$

Calculate:

(a) The plate current I_p
(b) The transconductance g_m
(c) The output voltage

5-1-4. A tetrode amplifier has the following parameters:

Input voltage $V_{in} = 10 \text{ V}$
Plate resistance $r_p = 50 \text{ k}$
Transconductance $g_m = 10^{-1} \text{ mhos}$
Load resistance $R_\ell = 20 \text{ k}\Omega$
Voltage across grid and cathode $V_{gk} = -10 \text{ V}$
Capacitances $C_{gk} = 100 \text{ pF}$
                                       $C_{gs} = 0.01 \text{ pF}$
                                       $C_{ks} = 0.10 \text{ pF}$
                                       $C_{ps} = 100 \text{ pF}$
                                       $C_{pk} = 0.15 \text{ pF}$
Frequency $f = 20 \text{ MHz}$

Compute:

(a) The combined capacitance C_1
(b) The combined capacitance C_2
(c) The output voltage
(d) The amplification

5-1-5. A pentode amplifier has the following parameters:

Transconductance $g_m = 10^{-2} \text{ ʊ}$
Plate resistance $r_p = 20 \text{ k}\Omega$
Load resistance $R_\ell = 50 \text{ k}\Omega$
Signal voltage $V_{in} = 50 \text{ V}$
Frequency $f = 2 \text{ MHz}$
Voltage across grid and cathode $V_{gk} = -50 \text{ V}$
Capacitances $C_{gk} = 20 \text{ pF}$
                                       $C_{gs} = 0.02 \text{ pF}$
                                       $C_{pk} = 0.20 \text{ pF}$
                                       $C_{ps} = 120 \text{ pF}$
                                       $C_{psu} = 20 \text{ pF}$

Determine:

(a) The combined parallel impedance in the output circuit
(b) The output voltage
(c) The amplification factor

5-1-6. A triode has the following parameters:

Separation between cathode and grid	$d = 1.5$ cm
dc voltage	$V_0 = 200$ V
Operating frequency	$f = 2$ GHz

(a) Compute the dc electron velocity.

(b) Calculate the transit angle.

5-1-7. A vacuum pentode tube has five grids: a cathode, a control grid, a screen grid, a suppressor grid, and an anode plate as shown in Fig. P5-1-7

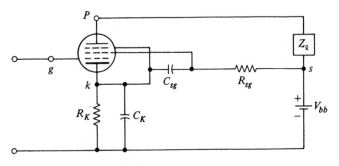

Figure P5-1-6   (Problem)

(a) Sketch the equivalent circuit.

(b) Derive an expression for the input impedance Z_{in} in terms of the angular frequency ω and the circuit parameters.

(c) Determine the transit-angle effect.

Klystrons

5-2-1. A two-cavity klystron amplifier has the following parameters:

Operating frequency	$f = 10$ GHz
Signal voltage	$V_1 = 5$ V(rms)
Beam voltage	$V_0 = 50$ kV
Beam current	$I_0 = 12$ A
Gap distance in either cavity	$d = 1$ mm
Length between two cavities	$L = 5$ cm

Compute:

(a) The dc electron velocity

(b) The transit time through the gap

(c) The transit angle through the gap

(d) The beam coupling coefficient

(e) The modulated electron velocity just out from the buncher cavity

5-2-2. A two-cavity klystron amplifier has the following parameters:

Beam voltage	$V_0 = 40$ kV
Beam current	$I_0 = 10$ A
Length between two cavities	$L = 5$ cm
Gap distance in either cavity	$d = 1$ mm
Operating frequency	$f = 12$ GHz
Signal voltage	$V_1 = 6$ V(rms)

Calculate:

(a) The dc electron beam velocity

(b) The dc transit angle between cavities

(c) The dc transit angle across the gap

(d) The beam coupling coefficient

(e) The bunching parameter

5-2-3. A two-cavity klystron amplifier has the following parameters:

Beam voltage	$V_0 = 30$ kV
Beam current	$I_0 = 8$ A
Beam coupling coefficient	$\beta_i = \beta_0 = 1$
Operating frequency	$f = 14$ GHz
Signal voltage	$V_1 = 200$ V(rms)

Determine:

(a) The maximum fundamental current

(b) The dc electron velocity

(c) The optimum distance between the two capacities

5-2-4. A CW two-cavity klystron amplifier has the following parameters:

Beam voltage	$V_0 = 45$ kV
Beam current	$I_0 = 12$ A
Beam coupling coefficient	$\beta_i = \beta_0 = 1$
Separation between cavities	$L = 50$ cm
Operating frequency	$f = 8$ GHz
Signal voltage	$V_1 = 25$ V(rms)
Total shunt resistance of catcher cavity	$R_{sh} = 90$ kΩ

Compute:

(a) The dc electron velocity

(b) The dc transit angle between cavities

(c) The bunching parameter

(d) The induced fundamental current in the catcher cavity

(e) The output power

(f) The efficiency

5-2-5. A two-cavity klystron amplifier has the following parameters:

Beam voltage	$V_0 = 20$ kV
Beam current	$I_0 = 5$ A
Operating frequency	$f = 8$ GHz
Separation between cavities	$L = 5$ cm
Total shunt resistance	$R_{sh} = 20$ kΩ
Beam coupling coefficient	$\beta_i = \beta_0 = 1$
Signal voltage	$V_1 = 50$ V(rms)

Calculate:

(a) The dc electron velocity

(b) The transit angle between cavities

(c) The bunching parameter

(d) The voltage amplification

5-2-6. A two-cavity klystron amplifier has the following parameters:

Beam voltage	$V_0 = 20$ kV
Beam current	$I_0 = 1$ A
Operating frequency	$f = 8$ GHz
Separation between cavities	$L = 5$ cm
Gap between either cavity	$d = 1$ mm
Total shunt resistance	$R_{sh} = 20$ kΩ

Compute:

(a) The input gap voltage in order to obtain maximum voltage V_2

(b) The voltage gain (neglecting the beam loading in the output cavity)

(c) The efficiency of the amplifier (neglecting beam loading)

(d) The beam loading conductance to verify that it was justified to neglect it in the above calculation

5-2-7. The parameters of a two-cavity amplifier klystron are

Beam voltage	$V_0 = 1200$ V
Beam current	$I_0 = 28$ mA
Operating frequency	$f = 8$ GHz
Gap spacing in either cavity	$d = 1$ mm
Spacing between the two cavities	$L = 4$ cm
Effective shunt resistance (excluding beam loading)	$R_{sh} = 40$ kΩ

(a) Find the input microwave voltage V_1 in order to generate a maximum output voltage V_2 (including the finite transit-time effect through the cavities).

(b) Determine the voltage gain (neglecting the beam loading in the output cavity).

(c) Calculate the efficiency of the amplifier (neglecting the beam loading).

(d) Compute the beam loading conductance and show that it is justified to neglect it in the above calculations.

5-2-8. A two-cavity amplifier klystron has the following characteristics:

Voltage gain	G = 15 dB.
Input power	P_{in} = 5 mW (milliwatts).
Total shunt impedance of the input cavity	R_{shi} = 30 kΩ.
Total shunt impedance of the output cavity	R_{sho} = 40 kΩ.
The load impedance at the output cavity	R_ℓ = 40 kΩ.

Determine:

  (a) The input voltage (RMS)

  (b) The output voltage (RMS)

  (c) The power delivered to the load in watts

5-2-9. A two-cavity amplifier klystron has the following parameters:

Beam voltage	V_0 = 900 V
Beam current	I_0 = 30 mA
Operating frequency	f = 8 GHz
Gap spacing in either cavity	d = 1 mm
Spacing between centers of cavities	L = 4 cm
Effective shunt impedance	R_{sh} = 40 kΩ

Determine:

  (a) The electron velocity

  (b) The dc transit time of electron between cavities

  (c) The input voltage for maximum output voltage

  (d) The voltage gain in dB

5-2-10. Derive Eq. (5-4-33).

Multicavity Klystrons

5-7-1. A four-cavity klystron amplifier has the following parameters:

Beam voltage	V_0 = 20 kV
Beam current	I_0 = 2 A
Operating frequency	f = 9 GHz
dc charge density	ρ_0 = 10^{-6} C/m³
RF charge density	ρ = 10^{-8} C/m³
Velocity perturbation	$\mathcal{V}$ = 10^5 m/s

Determine:

  (a) The dc electron velocity

  (b) The dc phase constant

  (c) The plasma frequency

  (d) The reduced plasma frequency for R = 0.5

  (e) The beam current density

  (f) The instantaneous beam current density

5-7-2. A four-cavity CW klystron amplifier has the following parameters:

Beam voltage	$V_0 = 30$ kV
Beam current	$I_0 = 3$ A
Gap distance	$d = 1$ cm
Operating frequency	$f = 8$ GHz
Signal voltage	$V_1 = 15$ V(rms)
Beam coupling coefficient	$\beta_i = \beta_0 = 1$
dc electron charge density	$\rho_0 = 10^{-7}$ C/m^3

Compute:

(a) The dc electron velocity
(b) The dc electron phase constant
(c) The plasma frequency
(d) The reduced plasma frequency for $R = 0.4$
(e) The reduced plasma phase constant
(f) The transit time across the input gap
(g) The modulated electron velocity leaving the input gap

5-7-3. A two-cavity klystron amplifier has the following parameters:

Beam voltage	$V_0 = 30$ kV
Beam current	$I_0 = 3$ A
Operating frequency	$f = 10$ GHz
Beam coupling coefficient	$\beta_i = \beta_0 = 1$
dc electron charge density	$\rho_0 = 10^{-7}$ C/m^3
Signal voltage	$V_1 = 15$ V(rms)
Cavity shunt resistance	$R_{sh} = 1$ kΩ
Total shunt resistance including load	$R_{sht} = 10$ kΩ

Calculate:

(a) The plasma frequency
(b) The reduced plasma frequency for $R = 0.4$
(c) The induced current in the output cavity
(d) The induced voltage in the output cavity
(e) The output power delivered to the load
(f) The power gain
(g) The electronic efficiency

5-7-4. A four-cavity klystron amplifier has the following parameters:

Beam voltage	$V_0 = 20$ kV
Beam current	$I_0 = 1.5$ A
Operating frequency	$f = 2$ GHz
Beam coupling coefficient	$\beta_i = \beta_0 = 1$
dc electron charge density	$\rho_0 = 10^{-6}$ C/m^3

Signal voltage	V_1	= 2 V(rms)
cavity shunt resistance	R_{sh}	= 2 kΩ
Total shunt resistance including load	R_{sht}	= 1 kΩ

Determine:

(a) The plasma frequency

(b) The reduced plasma frequency for $R = 0.4$

(c) The induced current in the output cavity

(d) The induced voltage in the output cavity

(e) The output power delivered to the load

(f) The electronic efficiency

Klystrodes

5-8-1. Describe the operational principles of the klystrode and its potential applications.

5-8-2. A klystrode operates in a FM mode under the following parameters:

Beam voltage	V_b	= 35 kV
Beam current	I_b	= 1.40 A
Collector voltage	V_c	= 25 kV
Collector current	I_c	= 1.50 A
RF output power	P_{ac}	= 40 kW

Compute the efficiency of the klystrode.

CHAPTER **6**

Microwave
Linear-Beam
Oscillators

6-0 INTRODUCTION

Microwave linear-beam oscillators are those tubes that generate output power even without an input signal voltage. The commonly used such oscillators are the two-cavity klystron oscillator and the reflex klystron oscillator. The basic principles governing these oscillators are investigated. Two-cavity klystron oscillator usually delivers an output power from a fraction of 1 W to 1 kW at a frequency range from 0.60 to 4 GHz. Reflex klystron oscillators are capable of supplying 0.5 W at 25 GHz.

6-1 TWO-CAVITY KLYSTRON OSCILLATOR

A two-cavity klystron is used as an oscillator by coupling the output cavity to the input cavity either directly or through a feedback line. A schematic diagram of a two-cavity klystron oscillator with a loop coupling and its equivalent circuit are shown in Fig. 6-1-1.

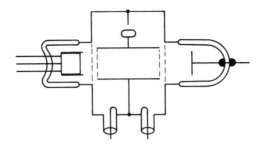

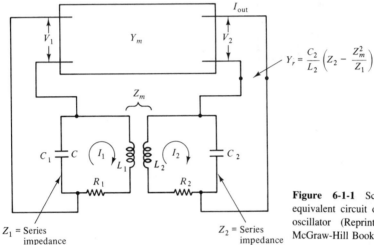

$$Y_r = \frac{C_2}{L_2}\left(Z_2 - \frac{Z_m^2}{Z_1}\right)$$

Figure 6-1-1 Schematic diagram and equivalent circuit of a two-cavity klystron oscillator (Reprinted by permission of McGraw-Hill Book Company)

Z_1 = Series impedance

Z_2 = Series impedance

The only difference between a klystron amplifier and a klystron oscillator is that the oscillator has feedback from catcher to buncher resonators. The catcher resonator is ordinarily more heavily loaded than the buncher resonator, so that its quality factor Q is lower. The satisfactory Q for oscillation must be greater than 20.

The fundamental requirement for oscillation in a two-cavity klystron oscillator is that the beam mutual admittance (or transfer admittance) of the tube must be equal to the current-voltage reduction factor or the circuit transfer admittance of the network. That is,

$$Y_m = \frac{I_{out}}{V_1} = Y_{21} \tag{6-1-1}$$

where 1 indicates the buncher resonator and 2 the catcher resonator, I_{out} = output current, and Y_{21} = circuit transfer admittance.

The output voltage is developed across a resonant admittance Y_r, which includes the effect of the coupled input cavity and has the expression

$$Y_r = \frac{C_2}{L_2}\left(Z_2 - \frac{Z_m^2}{Z_1}\right) \tag{6-1-2}$$

where C_2 = capacitance in catcher resonator
$\quad\ L_2$ = inductance in catcher resonator
$\quad\ Z_1$ = series impedance in buncher resonator
$\quad\ Z_2$ = series impedance in catcher resonator
$\quad\ Z_m$ = mutual impedance between input and output circuits

    The input voltage is related to the current flowing in a high-Q input resonant circuit by

$$V_1 = j\omega L_1 I_1 \qquad\qquad (6\text{-}1\text{-}3)$$

The circulating current in the input resonator is related to that in the output resonator by

$$I_1 = \frac{Z_m I_2}{Z_1} \qquad\qquad (6\text{-}1\text{-}4)$$

The circulating current in the output resonator is related to the output current from the tube in a high-Q circuit by

$$I_2 = \frac{I_{\text{out}}}{j\omega L_2 Y_r} \qquad\qquad (6\text{-}1\text{-}5)$$

Combination of Eqs. (6-1-2) through (6-1-5) results in

$$\frac{I_{\text{out}}}{V_1} = \frac{C_2}{L_1 Z_m} (Z_1 Z_2 - Z_m^2) \qquad\qquad (6\text{-}1\text{-}6)$$

Since $\omega L_2 = 1/(\omega C_2)$ is maintained in the output resonator, the coupling coefficient between buncher and catcher resonators is given by

$$K^2 = \frac{Z_m^2}{\omega^2 L_1 L_2} \qquad\qquad (6\text{-}1\text{-}7)$$

Substitution of Eq. (6-1-7) into Eq. (6-1-6) results in

$$I_{\text{out}} = \frac{K^2 V_1}{Z_m^3} (Z_1 Z_2 - Z_m^2) \qquad\qquad (6\text{-}1\text{-}8)$$

This ratio obtained from the circuit analysis must be equal to the ratio of output current over input voltage produced by the electron beam for oscillation—that is,

$$Y_m = \frac{K^2}{Z_m^3} (Z_1 Z_2 - Z_m^2) \qquad\qquad (6\text{-}1\text{-}9)$$

The factor of $(Z_1 Z_2 - Z_m^2)$ contributes virtually all the variation of the transfer admittance in the vicinity of resonance. The mutual impedance Z_m is relatively constant and it can be assumed to be inductive of the form

$$Z_m = j\omega L \qquad\qquad (6\text{-}1\text{-}10)$$

Substitution of Eq. (6-1-10) into Eq. (6-1-9) results in

$$Y_m = \frac{jK^2}{\omega^3 L^3} (Z_1 Z_2 + \omega^2 L^2) \tag{6-1-11}$$

If the input and output circuits have relatively high Q's, their series impedances can be expressed as

$$Z_1 = \frac{X_1}{Q_1} (1 + j2\delta Q_1) \tag{6-1-12}$$

and

$$Z_2 = \frac{X_2}{Q_2} (1 + j2\delta Q_2) \tag{6-1-13}$$

where δ = fractional deviation from resonance if $f_1 = f_2$. When Eqs. (6-1-12) and (6-1-13) substitute into Eq. (6-1-11), the beam mutual admittance becomes

$$Y_m = \frac{j}{\omega L Q_1 Q_2} [(1 + K^2 Q_1 Q_2 - 4Q_1 Q_2 \delta^2) + j2\delta(Q_1 Q_2)] \tag{6-1-14}$$

The locus of the circuit mutual admittance is a simple parabola. Figure 6-1-2 shows the loci of the beam transfer admittance Y_m and the circuit mutual admittance Y_{21} for a two-cavity klystron oscillator.

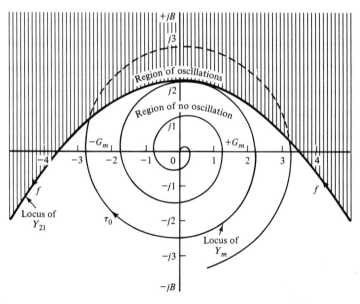

Figure 6-1-2 Loci of mutual admittances for a two-cavity klystron oscillator (Reprinted by permission of McGraw-Hill Book Company).

The two-cavity klystron oscillator will oscillate whenever the beam transadmittance locus enters the area outside the circuit-transadmittance parabola (the area outside the parabola is that in which all the possible tangents to the parabola lie). This requires that the coupling between buncher and catcher circuits be great enough, that the beam transadmittance Y_m be large enough in magnitude, and that the phase angle be correct.

For critical coupling, the coupling coefficient can be expressed by

$$K_c^2 = \frac{1}{Q_1 Q_2} \qquad (6\text{-}1\text{-}15)$$

where Q_1 = quality factor of input resonator
Q_2 = quality factor of output resonator

In general, oscillation will occur whenever the electron transit angle is in near equality with the phase angle of the circuit transfer admittance plus multiple 2π. That is,

$$\theta_e = \phi + 2\pi n \qquad (6\text{-}1\text{-}16)$$

where θ_e = electron transit angle
ϕ = phase angle of transfer admittance
n = any integer

The Varian two-cavity klystron oscillator VKX-7016 as shown in Fig. 6-1-3 has a minimum output power of 1 W at frequency range from 9.0 to 10.50 GHz. The maximum beam voltage is 5 kV (dc) and the maximum beam current is 5 mA (dc).

VRX-7016

Figure 6-1-3 Photograph of Varian two-cavity klystron oscillator VKX-7016 (Courtesy of Varian Associates, Inc.).

In general, two-cavity klystron oscillators are preferred for applications requiring higher power than that available from reflex klystrons. AM and FM noise characteristics of two-cavity tubes are superior to those of reflex klystrons. On applications requiring exceptional frequency stability, two-cavity klystron oscillators usually exhibit extremely low frequency/temperature coefficients.

Example 6-1-1: Two-cavity Klystron Oscillator

A two-resonator klystron oscillator has the following operating parameters:

Resonant frequency	f_r	$= 3$ GHz
Voltage in resonator 1	V_1	$= 1.5$ kV
Coupling coefficient	k	$= 0.02$
Mutual impedance	Z_m	$= 10 \ \Omega$
Quality factors	Q_1	$= 250$
	Q_2	$= 50$
Resonator resistances	R_1	$= 10 \ \Omega$
	R_2	$= 50 \ \Omega$
Resonator reactances	$X_{L1} = X_{L2}$	$= 2500 \ \Omega$
	$X_{c1} = X_{c2}$	$= 2500 \ \Omega$

Compute:

(a) The resonator series impedances Z_1 and Z_2

(b) The current flowing in the input resonant circuit

(c) The current flowing in the output resonant circuit

(d) The critical coupling k_c

(e) The admittance Y_r

(f) The output current I_{out}

Solution:

(a) The resonator series impedances are

$$Z_1 = R_1 + jX_{L1} - jX_{c1} = 10 \ \Omega$$

$$Z_2 = 50 \ \Omega$$

(b) The current flowing in the input resonator circuit is

$$I_1 = \frac{V_1}{jX_{L1}} = \frac{1.5 \times 10^3}{j2500} = -j0.60 \text{ A} \qquad \text{for high } Q_1$$

(c) The current flowing in the output resonant circuit is

$$I_2 = \frac{Z_1}{Z_m} I_1 = \frac{10}{10} \times (-j0.6) = -j0.60 \text{ A}$$

(d) The critical coupling is

$$k_c = \left(\frac{1}{Q_1 Q_2}\right)^{1/2} = \left(\frac{1}{2500 \times 50}\right)^{1/2} = 0.0028$$

(e) The admittance is

$$Y_r = \frac{C_2}{L_2}\left(Z_2 - \frac{Z_m^2}{Z_1}\right)$$

where

$$C_2 = \frac{1}{\omega X_{c2}} = \frac{1}{2\pi \times 3 \times 10^9 \times 2500} = 0.02 \text{ pF}$$

and

$$L_2 = \frac{X_{L2}}{\omega} = \frac{2500}{2\pi \times 3 \times 10^9} = 0.133 \ \mu\text{H}$$

Then

$$Y_r = \frac{0.02 \times 10^{-12}}{0.133 \times 10^{-6}}\left(50 - \frac{10^2}{10}\right) = +6.02 \times 10^{-6} \ \mho$$

(f) The output current is

$$\begin{aligned}
I_{\text{out}} &= I_2(jX_2)Y_r \\
&= (-j0.60)(j2500)(+6.02 \times 10^{-6}) \\
&= 9.03 \text{ mA}
\end{aligned}$$

6-2 REFLEX KLYSTRON OSCILLATOR

If a fraction of the output power in the output cavity is fed back to the input cavity and if the loop gain has a magnitude of unity with a phase shift of multiple 2π, the two-cavity klystron will oscillate. However, a two-cavity klystron oscillator is not commonly constructed because, when the oscillation frequency is varied, the resonant frequency of each cavity and the feedback path phase shift must be readjusted for a positive feedback. The reflex klystron is a single-cavity klystron that overcomes the disadvantages of the two-cavity klystron oscillator. The reflex klystron is a low-power generator of 10 to 500-mW output at a frequency range of 1 to 25 GHz. The efficiency is about 20 to 30%. It is widely used in the laboratory for microwave measurements and in the microwave receivers as local oscillators in commercial, military, and airborne Doppler radars as well as missiles. The theory of the two-cavity klystron can be applied to the analysis of reflex klystron with slight modification. A schematic diagram of the reflex klystron is shown in Fig. 6-2-1.

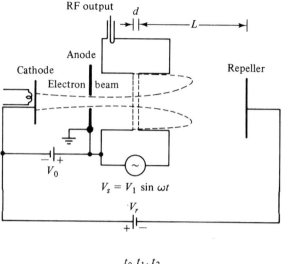

t_0 = time for electron entering cavity gap at $z = 0$

t_1 = time for same electron leaving cavity gap at $z = d$

t_2 = time for same electron returned by retarding field $z = d$ and collected on walls of cavity

Figure 6-2-1　Schematic diagram of a reflex klystron

The electron beam injected from the cathode is first velocity-modulated by the cavity-gap voltage. Some electrons accelerated by accelerating field enter the repeller space with greater velocity than those with unchanged velocity. Some electrons decelerated by the retarding field enter the repeller region with less velocity. All electrons turned around by the repeller voltage then pass through the cavity gap in bunches that occur once per cycle. On their returning journey, the bunched electrons pass through the gap during the retarding phase of the alternating field and they give up their kinetic energy to the electromagnetic energy of the field in the cavity. Oscillator output energy is then taken from the cavity. The electrons are finally collected by the walls of the cavity or other grounded metal parts of the tube. Figure 6-2-2 shows an Applegate diagram for the $1\frac{3}{4}$ mode of a reflex klystron.

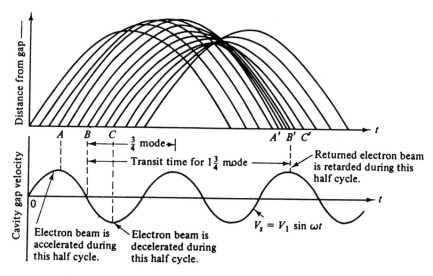

Figure 6-2-2   Applegate diagram with gap voltage for a reflex klystron

6-3 VELOCITY MODULATION

The analysis of a reflex klystron is similar to that of a two-cavity klystron. For simplicity, the effect of space-charge forces on the electron motion will be again neglected. The electron entering the cavity gap from the cathode at $z = 0$ and time t_0 is assumed to have uniform velocity

$$\mathcal{V}_0 = 0.593 \times 10^6 \sqrt{V_0} \tag{6-3-1}$$

The same electron leaves the cavity gap at $z = d$ at time t_1 with velocity

$$\mathcal{V}(t_1) = \mathcal{V}_0 \left[1 + \frac{\beta_i V_1}{2V_0} \sin \left(\omega t_1 - \frac{\theta_g}{2} \right) \right] \tag{6-3-2}$$

This expression is identical to Eq. (5-3-11) since the problems up to this point are identical with that of a two-cavity klystron amplifier. This electron is forced back to the cavity at $z = d$ and time t_2 by the retarding electric field E, which is given by

$$E = \frac{V_r + (V_0 + V_1 \sin \omega t)}{L} \tag{6-3-3}$$

where L = spacing between cavity and repeller as $L \gg d$ and d is neglected. This retarding field E is assumed to be constant in the z direction. The force equation for one electron in the repeller region is

$$M \frac{d^2 z}{dt^2} = -eE = -e \frac{(V_r + V_0)}{L} \tag{6-3-4}$$

where $\mathbf{E} = -\nabla V$ is used in the z direction only
V_r = magnitude of the repeller voltage
$|V_1 \sin \omega t| \ll (V_r + V_0)$ is assumed

Integration of Eq. (6-3-4) twice yields

$$\frac{dz}{dt} = \frac{-e(V_r + V_0)}{mL} \int_{t_1}^{t} dt = \frac{-e(V_r + V_0)}{mL} (t - t_1) + K_1 \qquad (6\text{-}3\text{-}5)$$

at $t = t_1$, $\dfrac{dz}{dt} = \mathcal{V}(t_1) = K_1$

$$z = \frac{-e(V_r + V_0)}{mL} \int_{t_1}^{t} (t - t_1)dt + \mathcal{V}(t_1) \int_{t_1}^{t} dt$$

$$z = \frac{-e(V_r + V_0)}{2mL} (t - t_1)^2 + \mathcal{V}(t_1)(t - t_1) + K_2$$

at $t = t_1$, $z = d = K_2$, then

$$z = \frac{-e(V_r + V_0)}{2mL} (t - t_1)^2 + \mathcal{V}(t_1)(t - t_1) + d \qquad (6\text{-}3\text{-}6)$$

On the assumption that the electron leaves the cavity gap at $z = d$ and time t_1 with a velocity of $\mathcal{V}(t_1)$ and returns to the gap at $z = d$ and time t_2, then, at $t = t_2$, $z = d$,

$$0 = \frac{-e(V_r + V_0)}{2mL} (t_2 - t_1)^2 + \mathcal{V}(t_1)(t_2 - t_1)$$

The round-trip transit time in the repeller region is given by

$$T' = t_2 - t_1 = \frac{2mL}{e(V_r + V_0)} \mathcal{V}(t_1) = T_0' \left[1 + \frac{\beta_i V_1}{2V_0} \sin \left(\omega t_1 - \frac{\theta_g}{2} \right) \right] \qquad (6\text{-}3\text{-}5)$$

where $T_0' = \dfrac{2mL\mathcal{V}_0}{e(V_r + V_0)}$ is the round-trip dc transit time of the center-of-the-bunch electron.

$$(6\text{-}3\text{-}5a)$$

Multiplication of Eq. (6-3-5) through by a radian frequency results in

$$\omega(t_2 - t_1) = \theta_0' + X' \sin \left(\omega t_1 - \frac{\theta_g}{2} \right) \qquad (6\text{-}3\text{-}6)$$

where $\theta_0' = \omega T_0'$ is the round-trip dc transit angle of the center-of-the-bunch electron

$$(6\text{-}3\text{-}6a)$$

$X' = \dfrac{\beta_i V_1}{2V_0} \theta_0'$ is the bunching parameter of the reflex klystron oscillator   (6-3-6b)

Example 6-3-1:   Reflex Klystron

A reflex klystron has the following parameters:

Beam voltage	$V_0 = 700\text{ V}$
Beam current	$I_0 = 34\text{ mA}$
Resonant frequency	$f_r = 12\text{ GHz}$
Repeller voltage	$V_r = 300\text{ V}$
Distance between cavity and repeller	$L = 1\text{ cm}$
Signal voltage	$V_1 = 2\text{ V (rms)}$
Beam coupling coefficient	$\beta_i = 1$

Computer:

(a) The dc electron velocity

(b) The round-trip dc transit time

(c) The round-trip dc transit angle

(d) The bunching parameter

Solution:

(a) The dc electron velocity is

$$\mathcal{V}_0 = 0.593 \times 10^6 \times (700)^{1/2} = 1.57 \times 10^7 \text{ m/s}$$

(b) The round-trip dc transit time

$$T_0' = \frac{2 \times 10^{-2} \times 1.57 \times 10^7}{1.759 \times 10^{11}\,(300 + 700)} = 1.785 \text{ ns}$$

(c) The round-trip transit angle is

$$\theta_0' = 2\pi \times 12 \times 10^9 \times 1.785 \times 10^{-9} = 134.60 \text{ rad}$$

(d) The bunching parameter is

$$X' = \frac{1 \times 2}{2 \times 700} \times 134.6 = 0.1923$$

6-4 POWER OUTPUT AND EFFICIENCY

In order for the electron beam to generate a maximum amount of energy to the oscillation, the returning electron beam must cross the cavity gap when the gap field is maximum retarding. In this way, a maximum amount of kinetic energy can be transferred from the returning electrons to the cavity walls. It can be seen from Fig. 6-2-6 that for a maximum energy transfer the round-trip transit angle, referring to the center of the bunch, must be given by

$$\omega(t_2 - t_1) = \omega T_0' = \left(n - \frac{1}{4} \right) 2\pi = N2\pi = 2\pi n - \frac{\pi}{2} \qquad (6\text{-}4\text{-}1)$$

where $V_1 \ll V_0$ is assumed

n = any positive integer for cycle number

$N = \left(n - \dfrac{1}{4} \right)$ is the number of modes

The current modulation on the electron beam as it reenters the cavity from the repeller region can be determined in the same manner as in Section 5-4 for a two-cavity klystron amplifier. It can be seen from Eqs. (5-4-14) and (6-3-6b) that the bunching parameter X' of a reflex klystron oscillator has a negative sign with respect to the bunching parameter X of a two-cavity klystron amplifier. Furthermore, the beam current injected into the cavity gap from the repeller region flows in the negative z direction. Consequently, the beam current of a reflex klystron oscillator can be written as

$$i_{2t} = -I_0 - \sum_{n=1}^{\infty} 2I_0 J_n(nX') \cos [n(\omega t_2 - \theta_0' - \theta_g)] \qquad (6\text{-}4\text{-}2)$$

The fundamental component of the current induced in the cavity by the modulated electron beam is given by

$$i_2 = -\beta_i I_2 = 2I_0 \beta_i J_1(X') \cos (\omega t_2 - \theta_0') \qquad (6\text{-}4\text{-}3)$$

in which θ_g has been neglected as a small quantity compared with θ_0'. The magnitude of the fundamental component is

$$I_2 = 2I_0 \beta_i J_1(X') \qquad (6\text{-}4\text{-}4)$$

The dc power supplied by the beam voltage V_0 is

$$P_{dc} = V_0 I_0 \qquad (6\text{-}4\text{-}5)$$

The ac power delivered to the load is given by

$$P_{ac} = \frac{V_1 I_2}{2} = V_1 I_0 \beta_i J_1(X') \qquad (6\text{-}4\text{-}6)$$

From Eqs. (6-3-6a), (6-3-6b), and (6-4-1), the ratio of V_1 over V_0 is expressed by

$$\frac{V_1}{V_0} = \frac{2X'}{\beta_i(2\pi n - \pi/2)} \qquad (6\text{-}4\text{-}7)$$

Substitution of Eq. (6-4-7) in Eq. (6-4-6) yields the power output as

$$P_{ac} = \frac{2V_0 I_0 X' J_1(X')}{2\pi n - \pi/2} \qquad (6\text{-}4\text{-}8)$$

Therefore, the electronic efficiency of a reflex klystron oscillator is defined as

$$\text{Efficiency} \equiv \frac{P_{ac}}{P_{dc}} = \frac{2X' J_1(X')}{2\pi n - \pi/2} \qquad (6\text{-}4\text{-}9)$$

The factor $X'J_1(X')$ reaches a maximum value of 1.25 at $X' = 2.408$ and $J_1(X') = 0.52$. In practice, the $n = 1$ or $n = 2$ cycles have the most power output. If the cycle $n = 2$ or the mode number $N = 1\frac{3}{4}$ mode, the maximum electronic efficiency becomes

$$\text{Efficiency}_{\text{max}} = \frac{2(2.408)\,J_1(2.408)}{2\pi(2) - \pi/2} = 22.7\% \qquad (6\text{-}4\text{-}10)$$

The maximum theoretical efficiency of a reflex klystron oscillator ranges from 20 to 30%. Figure 6-4-1 shows a curve of $X'J_1(X')$ vs. X'. For a given beam voltage V_0, the relationship between repeller voltage and cycle number n required for oscillation is found by inserting Eqs. (6-3-1) and (6-4-1) into Eq. (6-3-5), and the result is

$$\frac{V_0}{(V_r + V_0)^2} = \frac{(2\pi n - \pi/2)^2}{8\omega^2 L^2}\frac{e}{m} \qquad (6\text{-}4\text{-}11)$$

The power output can be expressed in terms of the repeller voltage V_r—that is,

$$P_{\text{ac}} = \frac{V_0 I_0 X' J_1(X')\,(V_r + V_0)}{\omega L}\sqrt{\frac{e}{2mV_0}} \qquad (6\text{-}4\text{-}12)$$

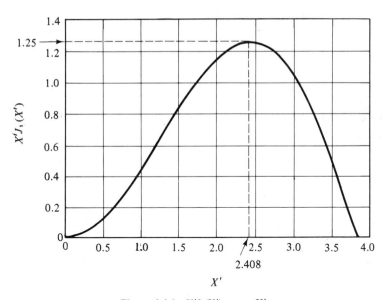

Figure 6-4-1 $X'J_1(X')$ versus X'.

It can be seen from Eq. (6-3-11) that for a given beam voltage V_0 and cycle number n or mode number N, the center repeller voltage V_r can be determined in terms of center frequency. Then the power output at the center frequency can be calculated from Eq. (6-3-12). When the frequency varies from the center frequency and the repeller voltage about the center voltage, the power output will vary accordingly like a bell shape as shown in Fig. 6-4-2.

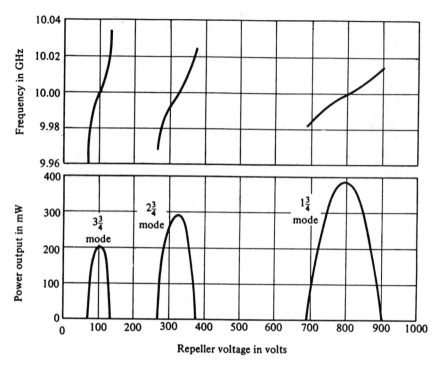

Figure 6-4-2   Power output and frequency characteristics of a reflex klystron.

Example 6-4-1:   Reflex Klystron Oscillator

A reflex klystron has the following parameters:

Beam voltage	$V_0 = 1.4$ kV
Beam current	$I_0 = 80$ mA
Repeller voltage	$V_r = 100$ V
Resonant frequency	$f_r = 7.445$ GHz
Signal voltage	$V_1 = 10$ V
Beam coupling coefficient	$\beta_i = 1$
Cycle number	$n = 2$
Distance between cavity and repeller	$L = 2$ cm

Calculate:

(a) The dc electron velocity

(b) The round-trip dc transit time

(c) The round-trip dc transit angle

(d) The beam bunching parameter

(e) The magnitude of the induced fundamental current

(f) The ac power delivered to the load

(g) The electronic efficiency

Solution:

(a) The dc electron velocity is

$$\mathcal{V}_0 = 0.593 \times 10^6 \times (1.4 \times 10^3)^{1/2} = 2.22 \times 10^7 \text{ m/s}$$

(b) The round-trip dc transit time is

$$T_0' = \frac{2 \times 2 \times 10^{-2} \times 2.22 \times 10^7}{1.759 \times 10^{11} \times (100 + 1.4 \times 10^3)} = 3.366 \text{ ns}$$

(c) The round-trip transit angle is

$$\theta_0' = 2\pi \times 7.445 \times 10^9 \times 3.366 \times 10^{-9} = 157.46 \text{ rad}$$

(d) The beam bunching parameter is

$$X' = \frac{1 \times 10}{2 \times 1400} \times 157.46 = 0.5624$$

$$J_1(0.5624) = 0.272$$

(e) The magnitude of the induced fundamental current is

$$I_2 = 2 \times 80 \times 10^{-3} \times 1 \times 0.272 = 43.52 \text{ mA}$$

(f) The ac power delivered to the load is

$$P_{ac} = \frac{2 \times 1400 \times 80 \times 10^{-3} \times 0.5624 \times 0.272}{2 \times 3.1416 \times 2 - 3.1416/2}$$

$$= 3.116 \text{ W}$$

(g) The electronic efficiency is

$$\eta = \frac{P_{ac}}{P_{dc}} = \frac{3.116}{1400 \times 0.08} = 0.0278 = 2.78 \%$$

6-5 ELECTRONIC ADMITTANCE

From Eq. (6-4-3), the induced current can be written in phasor form as

$$i_2 = 2I_0\beta_i J_1(X')\, e^{-j\theta_0'} \tag{6-5-1}$$

The voltage across the gap at time t_2 can also be written in phasor form as

$$V_2 = V_1\, e^{-j\pi/2} \tag{6-5-2}$$

The ratio of i_2 to V_2 is defined as the electronic admittance of the reflex klystron—that is,

$$Y_e \equiv \frac{I_0}{V_0}\frac{\beta_i^2\,\theta_0'}{2}\frac{2J_1(X')}{X'}\, e^{j(\pi/2 - \theta_0')} \tag{6-5-3}$$

The amplitude of the phasor admittance indicates that the electronic admittance is a function of the dc beam admittance, the dc transit angle, and the second transit of the electron beam through the cavity gap. It is evident that the electronic admittance is nonlinear since it is proportional to the factor $2J_1(X')/X'$, and X' is proportional to the signal voltage. This factor of proportionality is shown in Fig. 6-5-1. When the signal voltage goes to zero, the factor approaches unity.

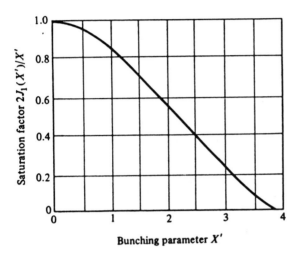

Figure 6-5-1 Reflex klystron saturation factor

The equivalent circuit of a reflex klystron is shown in Fig. 6-5-2. In this circuit, L and C are the energy storage elements of the cavity, G_c the copper losses of the cavity, G_b the beam loading conductance, and G_ℓ the load conductance.

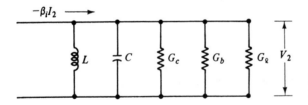

Figure 6-5-2 Equivalent circuit of a reflex klystron

The necessary condition for oscillations is that the magnitude of the negative real part of the electronic admittance as given by Eq. (6-5-3) must be not less than the total conductance of the cavity circuit. That is,

$$|-G_e| \geq G \tag{6-5-4}$$

where $G = G_c + G_b + G_\ell = 1/R_{sh}$, and R_{sh} is the effective shunt resistance.

Equation (6-5-3) can be rewritten in rectangular form.

$$Y_e = G_e + jB_e \tag{6-5-5}$$

Since the electronic admittance shown in Eq. (6-5-3) is in exponential form, its phase is $\pi/2$ when θ_0' is zero. The rectangular plot of the electron admittance Y_e is a spiral as shown in Fig. 6-5-3. Any value of θ_0' for which the spiral lies in the area to the left of line $(-G - jB)$ will yield oscillation. The value of θ_0' is then given by

$$\theta_0' = \left(n - \frac{1}{4} \right) 2\pi = N2\pi \qquad (6\text{-}5\text{-}6)$$

where N is the mode number as indicated in the plot; the phenomenon verifies the early analysis.

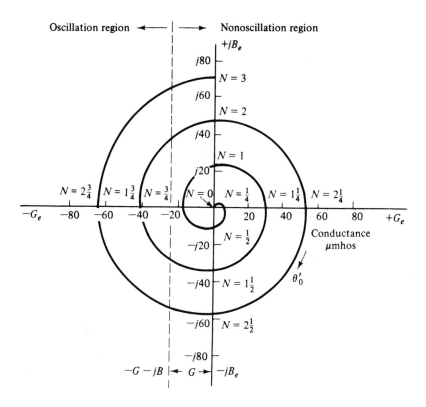

Figure 6-5-3 Electronic admittance spiral of a reflex klystron

The Varian reflex klystron oscillator VA-259C as shown in Fig. 6-5-4 has a minimum output power of 1 W at frequency from 4.4 to 8.4 GHz. The maximum beam voltage is 750 V (dc) and the maximum beam current is 80 mA dc.

Figure 6-5-4 Photograph of Varian reflex
oscillator VA-259C (Courtesy of Varian Associates, Inc.).

Example 6-5-1:   Reflex Klystron

A reflex klystron operates under the following conditions:

Beam voltage	V_0 = 600 V
Shunt resistance	R_{sh} = 15 kΩ
Resonant frequency	f_r = 9 GHz
Distance between cavity and repeller	L = 1 mm
Ratio of e and m	$\dfrac{e}{m}$ = 1.759 × 10^{11} (MKS system)

The tube is oscillating at f_r at the peak of the $n = 2$ cycles or N = $1\frac{3}{4}$ mode. Assume that
the transit time through the gap and beam loading can be neglected.

(a) Find the value of repeller voltage V_r.

(b) Find the direct current necessary to give a microwave gap voltage of 200 V.

(c) What is the electronic efficiency under this condition?

Solution:

(a) From Eq. (6-4-11),

$$\frac{V_0}{(V_r + V_0)^2} = \left(\frac{e}{m}\right)\frac{(2\pi n - \pi/2)^2}{8\omega^2 L^2} = (1.759 \times 10^{11})\frac{(2\pi 2 - \pi/2)^2}{8(2\pi \times 9 \times 10^9)^2(10^{-3})^2}$$

$$= 0.832 \times 10^{-3}$$

$$(V_r + V_0)^2 = \frac{600}{0.832 \times 10^{-3}} = 0.721 \times 10^6$$

$$V_r = 250 \text{ V}$$

(b) Assume that $\beta_0 = 1$. Since

$$V_1 = I_2 R_{sh} = 2I_0 J_1(X')R_{sh}$$

the direct current I_0 is

$$I_0 = \frac{V_1}{2J_1\,(X')R_{sh}} = \frac{200}{2(0.582)(15 \times 10^3)} = 11.45 \text{ mA}$$

(c) From Eq. (6-4-9), the electronic efficiency is

$$\text{Efficiency} = \frac{2X'J_1(X')}{2\pi n - \pi/2} = \frac{2(1.841)(0.582)}{2\pi(2) - \pi/2} = 19.5\%$$

REFERENCE

[1] SPANGENBERG, KARL R., *Vacuum Tubes,* McGraw-Hill Book Co., New York, 1948. pp. 607, 613.

PROBLEMS

Two-Cavity Klystron Oscillators

6-1-1. A two-resonator klystron oscillator has the following parameters:

Resonant frequency	f_r	= 4 GHz
Voltage in resonator 1	V_1	= 2 kV
Quality factors	Q_1	= 300
	Q_2	= 60
Coupling coefficient	k	= 0.02
Mutual impedance	Z_m	= 5 Ω
Resonator capacitive reactance	$X_{c1} = X_{c2}$	= 3000 Ω
Resonator inductive reactance	$X_{L1} = X_{L2}$	= 3000 Ω
Resonator resistances	R_1	= 10 Ω
	R_2	= 50 Ω

Compute:

(a) The resonator series impedances Z_1 and Z_2

(b) The current flowing in the input resonant circuit I_1

(c) The current flowing in the output resonant circuit I_2

(d) The critical coupling k_c

(e) The admittance Y_r

(f) The output current I_{out}

6-1-2. A two-cavity klystron oscillator has the following parameters:

Current in resonator 1	I_1	= $-j0.8$ A
Resonant frequency	f_r	= 8 GHz

Resonator capacitive reactances	$X_{c1} = X_{c2} = 2000 \ \Omega$
Resonator inductive reactances	$X_{L1} = X_{L2} = 2000 \ \Omega$
Resonator resistances	$R_1 = 10 \ \Omega$
	$R_2 = 50 \ \Omega$
Mutual impedance	$Z_m = 20 \ \Omega$

Calculate:

(a) The quality factors Q_1 and Q_2

(b) The capacitance C_2 and inductance L_2

(c) The resonant admittance Y_r

(d) The voltage in the resonator 1

(e) The current in the resonator 2

(f) The coupling coefficient k_c

(g) The output current I_{out}

Reflex Klystrons

6-3-1. A reflex klystron oscillator has the following parameters:

Beam voltage	$V_0 = 700$ V
Beam current	$I_0 = 34$ mA
Resonant frequency	$f_r = 10$ GHz
Repeller voltage	$V_r = 500$ V
Distance between cavity and repeller	$L = 2$ cm
Signal voltage	$V_1 = 5$ V(rms)
Beam coupling coefficient	$\beta_i = 1$

Compute:

(a) The dc electron velocity

(b) The round-trip dc transit time

(c) The round-trip dc transit angle

(d) The bunching parameter

6-3-2. A reflex klystron oscillator has the following parameters:

Beam voltage	$V_0 = 900$ V
Beam current	$I_0 = 50$ mA
Resonant frequency	$f_r = 20$ GHz
Repeller voltage	$V_r = 500$ V
Signal voltage	$V_1 = 10$ V
Beam coupling coefficient	$\beta_i = 1$
Cycle number	$n = 2$
Distance between cavity and repeller	$L = 2$ cm

Calculate:

(a) The dc electron velocity

(b) The round-trip dc transit time

(c) The round-trip dc transit angle

(d) The beam bunching parameter

(e) The magnitude of the induced fundamental current

(f) The ac power delivered to the load

(g) The electronic efficiency

6-3-3. A reflex klystron oscillator has the following parameters:

Beam voltage	$V_0 = 500$ V
Shunt resistance	$R_{sh} = 20$ kΩ
Resonant frequency	$f_r = 10$ GHz
Distance between cavity and repeller	$L = 1$ mm
Cycle number	$n = 2$

Calculate:

(a) The required repeller voltage V_r

(b) The dc current necessary to give a maximum gap voltage of 250 V

(c) The electronic efficiency

6-3-4. A reflex klystron operates at the peak mode of $n = 2$ with

Beam voltage	$V_0 = 300$ V
Beam current	$I_0 = 20$ mA
Signal voltage	$V_1 = 40$ V

Determine:

(a) The input power in watts

(b) The output power in watts

(c) The efficiency

6-3-5. A reflex klystron operates under the following conditions:

Beam voltage	$V_0 = 500$ V
Shunt resistance	$R_{sh} = 20$ kΩ
Resonant frequency	$f_r = 8$ GHz
Spacing between repeller and cavity	$L = 1$ mm

The tube is oscillating at f_r at the peak of the $n = 2$ cycles or N $= 1\frac{3}{4}$ mode. Assume that the transit time through the gap and the beam loading effect can be neglected.

(a) Find the value of repeller voltage V_r.

(b) Find the direct current necessary to give microwave gap voltage of 200 V.

(c) Calculate the electronic efficiency.

6-3-6. A reflex klystron operates at the peak of the $n = 2$ cycle. The dc power input is 40 mW and $V_1/V_0 = 0.278$. If 20% of the power delivered by the beam is dissipated in the cavity walls, find the power delivered to the load.

6-3-7. A reflex klystron operates at the peak of the $n = 1$ cycle or $N = \frac{3}{4}$ mode. The dc power input is 40 mW and the ratio of V_1 over V_0 is 0.278.

    (a) Determine the efficiency of the reflex klystron.

    (b) Find the total output power in milliwatts.

    (c) If 20% of the power delivered by the electron beam is dissipated in the cavity walls, find the power delivered to the load.

CHAPTER 7

Microwave Traveling-Wave Tubes

7-0 INTRODUCTION

In the previous two chapters, the fundamental principle of velocity modulation for klystrons was discussed. In this chapter, the operational principles of microwave traveling-wave tubes (TWTs) are described. In a traveling-wave tube, the electron beam and electromagnetic wave are traveling at approximately equal speed, and the energy interaction occurs between the two waves. Theoretically, the speed of the electromagnetic wave is considerably higher than the physically realizable velocity of electrons. In practice, the electromagnetic wave is passing through some kind of slow-wave structure, and its speed in the electron direction is approximately equal to the speed of the electron beam.

In general, traveling-wave tubes can be classified into two types: the O-type traveling-wave tubes and the M-type traveling-wave tubes. The O-type traveling-wave tubes are those tubes in which the electron beam is focused by a parallel dc magnetic field along the electron beam such as traveling-wave amplifiers (TWAs), backward-wave amplifiers (BWAs), and backward-wave oscillators (BWOs) investigated in this chapter. The M-type traveling-wave tubes are those tubes in which the electron beam is focused by a perpendicular dc magnetic field (or a crossed magnetic field) such as the magnetron oscillator, backward-wave crossed-field amplifier (Amplitron), and backward-wave crossed-field oscillator (Carcinotron) analyzed in Chapter 8.

7-1 HELIX TRAVELING-WAVE-TUBE AMPLIFIER

Since Kompfner invented the helix traveling-wave tube (TWT) in 1944 [1], the basic circuit of the TWT has had little change. For broadband applications, the helix TWTs are almost exclusively used, whereas for high-average-power purposes, such as radar transmitters, the coupled-cavity TWTs are commonly used.

In the previous chapters, klystrons and reflex klystrons were analyzed in some detail. Before starting to describe the TWT, it seems appropriate to compare the basic operating principles of both the TWT and the klystron. In the case of the TWT, the microwave circuit is nonresonant and the wave propagates with the same speed as the electrons in the beam. The initial effect on the beam is a small amount of velocity modulation caused by the weak electric fields associated with the traveling wave. Just as in the klystron, this velocity modulation later translates to current modulation, which then induces an RF current in the circuit, causing amplification. However, there are some major differences between the TWT and the klystron.

1. The interaction of electron beam and RF field in the TWT is continuous over the entire length of the circuit, but the interaction in the klystron occurs only at the gaps of a few resonant cavities.
2. The wave in the TWT is a propagating wave, but the wave in the klystron is not.
3. In the coupled-cavity TWT, there is a coupling effect between the cavities, whereas each cavity in the klystron operates completely independently.

A helix traveling-wave tube (TWT) consists of an electron beam and a slow-wave structure. The electron beam is focused by a constant magnetic field along the electron beam and the slow-wave structure. This is termed an O-type traveling-wave tube. The slow-wave structure is either helical type or folded-back line. The applied signal propagates around the turns of the helix and produces an electric field at the center of the helix, directed along the helix axis. The axial electric field progresses with a velocity that is very close to the velocity of light multiplied by the ratio of helix pitch to helix circumference. When the electrons enter the helix tube, an interaction takes place between the moving axial electric field and the moving electrons. On the average, the electrons transfer energy to the wave on the helix. This interaction causes the signal wave on the helix to become larger. The electrons entering the helix at zero field are not affected by the signal wave, those electrons entering the helix at the accelerating field are accelerated, and those at the retarding field are decelerated. As the electrons travel further along the helix, they bunch at the collector end. The bunching shifts the phase by $\pi/2$. Each electron in the bunch encounters a stronger retarding field. Then the microwave energy of the electrons is delivered by the electron bunch to the wave on the helix. The amplification of the signal wave is accomplished.

The characteristics of the traveling-wave tube are

Frequency range	3 GHz and higher
Bandwidth	about 0.8 GHz

Efficiency	20 to 40%
Power output	up to 10 kW average
Power gain	up to 60 dB

7-1-1 Amplification Process

The schematic diagram of a helix-type traveling-wave tube is shown in Fig. 7-1-1.

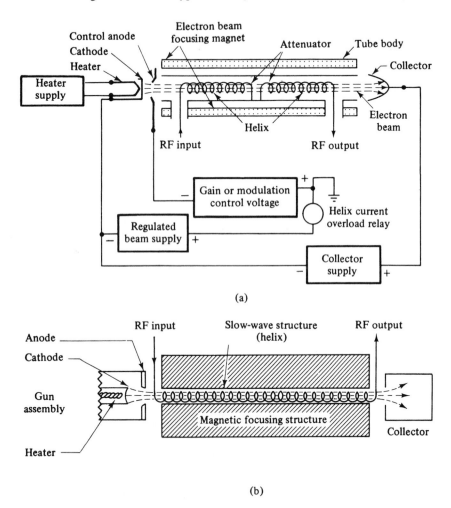

(a)

(b)

Figure 7-1-1   Diagram of helix traveling-wave tube: (a) Schematic diagram, (b) simplified circuit.

The slow-wave structure of the helix is characterized by the Brillouin diagram shown in Fig. 4-4-5. The phase shift per period of the fundamental wave on the structure is given by

$$\theta_1 = \beta_0 L \tag{7-1-1}$$

where $\beta_0 = \dfrac{\omega}{\mathcal{V}_0}$ is the phase constant of average beam velocity, and L is the period or pitch. Since the dc transit time of an electron is given by

$$T_0 = \frac{L}{\mathcal{V}_0} \tag{7-1-2}$$

the phase constant of the nth space-harmonic is

$$\beta_n = \frac{\omega}{\mathcal{V}_0} = \frac{\theta_1 + 2\pi n}{\mathcal{V}_0 T_0} = \beta_0 + \frac{2\pi n}{L} \tag{7-1-3}$$

In Eq. (7-1-3) the axial space-harmonic phase velocity is assumed in synchronism with the beam velocity for possible interactions between the electron beam and electric field— that is,

$$\mathcal{V}_{np} = \mathcal{V}_0 \tag{7-1-4}$$

Equation (7-1-3) is identical to Eq. (4-4-22). In practice, the dc velocity of the electrons is adjusted to be slightly greater than the axial velocity of the electromagnetic wave for energy transfer. When a signal voltage is coupled into the helix, the axial electric field exerts a force on the electrons as a result of the following relationships

$$\mathbf{F} = -e\mathbf{E} \qquad \text{and} \qquad \mathbf{E} = -\nabla V \tag{7-1-5}$$

The electrons entering the retarding field are decelerated, and those in the accelerating field are accelerated. They are to begin forming a bunch centered about those electrons that enter the helix during the zero field. This process is shown in Fig. 7-1-2.

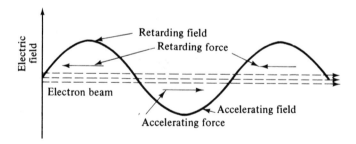

Figure 7-1-2   Interactions between electron beam and electric field.

Since the dc velocity of the electrons is slightly greater than the axial wave velocity, more electrons are in the retarding field than in the accelerating field, and a great amount of energy is transferred from the beam to the electromagnetic field. The microwave signal voltage is, in turn, amplified by the amplified field. The bunch continues to become more compact, and a larger amplification of the signal voltage occurs at the end of the helix. The magnet produces an axial magnetic field to prevent spreading of the electron beam as it travels down the tube. An attenuator placed near the center of the helix reduces all the waves traveling along the helix to nearly zero so that the reflected waves from the mismatched loads can be prevented from reaching the input and causing oscillation. The bunched electrons emerging from the attenuator induce a new electric field with the same frequency. This field, in turn, induces a new amplified microwave signal on the helix.

The motion of electrons in the helix-type traveling-wave tube may be quantitatively analyzed in terms of the axial electric field. If the traveling wave is propagating in the z direction, the z component of the electric field can be expressed as

$$E_z = E_1 \sin (\omega t - \beta_p z) \tag{7-1-6}$$

where E_1 is the magnitude of the electric field in the z direction. If $t = t_0$ at $z = 0$, the electric field is assumed maximum. Note that $\beta_p = \dfrac{\omega}{V_p}$ is the axial phase constant of the microwave, and V_p is the axial phase velocity of the wave. The equation of motion of the electron is given by

$$m \frac{dV}{dt} = -eE_1 \sin (\omega t - \beta_p z) \tag{7-1-7}$$

Assume that the velocity of the electron is

$$V = V_0 + V_e \cos (\omega_e t + \theta_e) \tag{7-1-8}$$

Then

$$\frac{dV}{dt} = -V_e \omega_e \sin (\omega_e t + \theta_e) \tag{7-1-9}$$

where V_0 = dc electron velocity
V_e = magnitude of velocity fluctuation in the velocity-modulated electron beam
ω_e = angular frequency of velocity fluctuation
θ_e = phase angle of the fluctuation

Substitution of Eq. (7-1-9) in Eq. (7-1-7) yields

$$mV_e \omega_e \sin (\omega_e t + \theta_e) = eE_1 \sin (\omega t - \beta_p z) \tag{7-1-10}$$

For interactions between the electrons and the electric field, the velocity of the velocity-modulated electron beam must be approximately equal to the dc electron velocity—that is,

$$V \simeq V_0 \tag{7-1-11}$$

Hence, the distance z traveled by the electrons is

$$z = V_0 (t - t_0) \tag{7-1-12}$$

and

$$mV_e\omega_e \sin(\omega_e t + \theta_e) = eE_1 \sin[\omega t - \beta_p V_0(t - t_0)] \qquad (7\text{-}1\text{-}13)$$

Comparison of the left-hand and the right-hand sides of Eq. (7-1-13) results in

$$V_e = \frac{eE_1}{m\omega_e}$$

$$\omega_e = \beta_p(V_p - V_0) \qquad (7\text{-}1\text{-}14)$$

and
$$\theta_e = \beta_p V_0 t_0$$

It can be seen that the magnitude of the velocity fluctuation of the electron beam is directly proportional to the magnitude of the axial electric field.

Example 7-1-1: Helical Slow-wave Structure

A continuous-wave (CW) helix traveling-wave-tube amplifier has the following parameters:

Beam voltage	V_0	$= 10 \text{ kV}$
Operating frequency	f	$= 10 \text{ GHz}$
Helical-coil diameter	d	$= 1.578 \text{ cm}$

Determine:

(a) The dc electron velocity

(b) The helical pitch for proper synchronism between the beam velocity and RF signal axial velocity

Solution:

(a) The dc electron velocity is

$$V_0 = 0.593 \times 10^6 \times (10 \times 10^3)^{1/2} = 0.593 \times 10^8 \text{ m/s}$$

(b) For proper electron interactions, the RF axial phase velocity must be approximately equal to the dc electron velocity—that is,

$$V_p = V_0 = 0.593 \times 10^8 \text{ m/s}$$

Then from Eq. (4-4-1), the helical pitch is

$$L = \pi d \left(\frac{c^2}{V_p^2} - 1\right)^{-1/2}$$

$$= 3.1416 \times 1.578 \left[\frac{(3 \times 10^8)^2}{(0.593 \times 10^8)^2} - 1\right]^{-1/2}$$

$$= 1 \text{ cm}$$

7-1-2 Convection Current

In order to determine the relations between the circuit and electron-beam quantities, the convection current induced on the electron beam by the axial electric field and the microwave axial field produced by the beam can be first developed. When the space-charge effect is considered, the electron velocity, the charge density, the current density, and the axial electric field will perturbate about their averages or dc values. Mathematically, these quantities can be expressed as

$$\mathcal{V} = \mathcal{V}_0 + \mathcal{V}_1 e^{j\omega t - \gamma z} \tag{7-1-15}$$

$$\rho = \rho_0 + \rho_1 e^{j\omega t - \gamma z} \tag{7-1-16}$$

$$J = -J_0 + J_1 e^{j\omega t - \gamma z} \tag{7-1-17}$$

$$E_z = E_1 e^{j\omega t - \gamma z} \tag{7-1-18}$$

where $\gamma = \alpha_e + j\beta_e$ is the propagation constant of the axial waves. The minus sign is attached to J_0 so that J_0 may be a positive in the negative z direction. For a small signal, the electron-beam-current density can be written as

$$J = \rho\mathcal{V} \simeq -J_0 + J_1 e^{j\omega t - \gamma z} \tag{7-1-19}$$

where $-J_0 = \rho_0\mathcal{V}_0$, $J_1 = \rho_1\mathcal{V}_0 + \rho_0\mathcal{V}_1$, and $\rho_1\mathcal{V}_1 \simeq 0$ have been replaced. If an axial electric field exists in the structure, it will perturbate the electron velocity according to the force equation. Hence the force equation can be written as

$$\frac{d\mathcal{V}}{dt} = -\frac{e}{m} E_1 e^{j\omega t - \gamma z} = \left(\frac{\partial}{\partial t} + \frac{dz}{dt}\frac{\partial}{\partial z} \right) \mathcal{V} = (j\omega - \gamma\mathcal{V}_0)\,\mathcal{V}_1 e^{j\omega t - \gamma z} \tag{7-1-20}$$

where $\dfrac{dz}{dt}$ has been replaced by $\mathcal{V}_0$. Hence,

$$\mathcal{V}_1 = \frac{-\dfrac{e}{m}}{j\omega - \gamma\mathcal{V}_0} E_1 \tag{7-1-21}$$

In accordance with the law of conservation of electric charge, the continuity equation can be written as

$$\nabla \cdot \mathbf{J} + \frac{\partial\rho}{\partial t} = (-\gamma J_1 + j\omega\rho_1)e^{j\omega t - \gamma z} = 0 \tag{7-1-22}$$

It follows that

$$\rho_1 = -\frac{j\gamma J_1}{\omega} \tag{7-1-23}$$

Substituting Eqs. (7-1-21) and (7-1-23) in the following expression

$$J_1 = \rho_1\mathcal{V}_0 + \rho_0\mathcal{V}_1 \tag{7-1-24}$$

the result becomes

$$J_1 = j\frac{\omega}{\mathcal{V}_0}\frac{e}{m}\frac{J_0}{(j\omega - \gamma\mathcal{V}_0)^2} E_1 \tag{7-1-25}$$

where $-J_0 = \rho_0 \mathcal{V}_0$ has been replaced. If the magnitude of the axial electric field is uniform over the cross-sectional area of the electron beam, the spatial ac current i will be proportional to the dc current I_0 with the same proportionality constant for J_1 and J_0. Therefore, the convection current in the electron beam is given by

$$i = j \frac{\beta_e I_0}{2V_0(j\beta_e - \gamma)^2} E_1 \tag{7-1-26}$$

where $\beta_e \equiv \dfrac{\omega}{\mathcal{V}_0}$ is defined as the phase constant of the velocity-modulated electron beam

$\mathcal{V}_0 = \sqrt{\dfrac{2e}{m}} V_0$ has been used

This equation is called the *electronic equation*, for it determines the convection current induced by the axial electric field. If the axial field and all parameters are known, the convection current can be found by means of Eq. (7-1-26).

Example 7-1-2:   Convection Current

A high-power CW helix traveling-wave amplifier has the following parameters:

Beam voltage	$V_0 = 11.4$ kV
Beam diameter	$d = 3$ mm
dc beam charge density	$\rho_0 = 2.224$ mC/m^3
Operating frequency	$f = 8$ GHz
Axial electric field	$E_1 = 38$ kV/m
Velocity perturbation	$\mathcal{V}_1 = 10^6$ m/s
Charge density perturbation	$\rho_1 = 10^{-4}$ C/m^3

Compute:

    (a) The dc electron velocity

    (b) The phase constant of the dc electron beam

    (c) The propagation constant for $\alpha_e = -50$

    (d) The dc beam current density

    (e) The beam current

    (f) The current density perturbation

    (g) The convection current

Solution:

    (a) The dc electron velocity is

$$\mathcal{V}_0 = 0.593 \times 10^6 \times (11.4 \times 10^3)^{1/2} = 0.633 \times 10^8 \text{ m/s}$$

    (b) The phase constant of the dc electron beam is

$$\beta_e = 2\pi \times 8 \times 10^9/(0.633 \times 10^8) = 794.10 \text{ rad/m}$$

(c) The propagation constant for $\alpha_e = -5$ is

$$\gamma = \alpha_e + j\beta_e = -50 + j794.10$$

(d) The dc beam current density is

$$J_0 = \rho_0 \mathcal{V}_0 = 2.224 \times 10^{-3} \times 0.633 \times 10^8 = 1.415 \times 10^5 \text{ A/m}^2$$

(e) The beam current is

$$I_0 = J_0 A_0 = 1.415 \times 10^5 \times \left(\frac{3}{2} \times 10^{-3}\right)^2 \times 3.1416 = 1 \text{ A}$$

(f) The current density perturbation is

$$J_1 = \rho_1 \mathcal{V}_0 + \rho_0 \mathcal{V}_1 = 10^{-4} \times 0.633 \times 10^8 + 2.224 \times 10^{-3} \times 10^6$$
$$= 0.8554 \times 10^4 \text{ A/m}^2$$

(g) The convection current is

$$i = j\frac{794.1 \times 1}{2 \times 11.4 \times 10^3 \times [j794.1 - (-50 + j794.1)]^2} \times 38 \times 10^3$$
$$= j0.529 \text{ A}$$

7-1-3 Axial Electric Field

The convection current in the electron beam induces an electric field in the slow-wave circuit. This induced field adds to the field already present in the circuit and causes the circuit power to increase with distance. The coupling relationship between the electron beam and the slow-wave helix is shown in Fig. 7-1-3.

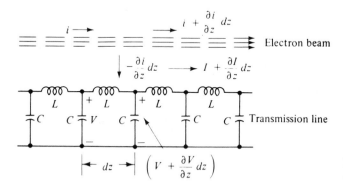

Figure 7-1-3 Electron beam coupled to equivalent circuit of a slow-wave helix.

For simplicity, the slow-wave helix is represented by a distributed lossless transmission line. The parameters are defined as follows:

$$L = \text{inductance per unit length}$$
$$C = \text{capacitance per unit length}$$

I = alternating current in transmission line
V = alternating voltage in transmission line
i = convection current

Since the transmission line is coupled to a convection-electron-beam current, a current is then induced in the line. The current flowing into the left end of the portion of the line of length dz is i, and the current flowing out of the right end of dz is $\left(i + \dfrac{\partial i}{\partial z} dz \right)$. Since the net change of current in the length dz must be zero, however, the current flowing out of the electron beam into the line must be $\left(- \dfrac{\partial i}{\partial z} dz \right)$. Application of transmission-line theory and Kirchhoff's current law to the electron beam results, after simplification, in

$$\frac{\partial I}{\partial z} = -C \frac{\partial V}{\partial t} - \frac{\partial i}{\partial z} \qquad (7\text{-}1\text{-}27)$$

Then

$$- \gamma I = -j\omega C V + \gamma i \qquad (7\text{-}1\text{-}28)$$

in which $\dfrac{\partial}{\partial z} = - \gamma$ and $\dfrac{\partial}{\partial t} = j\omega$ are replaced. From Kirchhoff's voltage law, the voltage equation, after simplification, is

$$\frac{\partial V}{\partial z} = -L \frac{\partial I}{\partial t} \qquad (7\text{-}1\text{-}29)$$

Similarly,

$$- \gamma V = -j\omega L I \qquad (7\text{-}1\text{-}30)$$

Elimination of the circuit current I from Eqs. (7-1-28) and (7-1-29) yields

$$\gamma^2 V = -V\omega^2 LC - \gamma i j\omega L \qquad (7\text{-}1\text{-}31)$$

If the convection-electron-beam current is not present, Eq. (7-1-31) reduces to a typical wave equation of a transmission line. When $i = 0$, the propagation constant is defined from Eq. (7-1-31) as

$$\gamma_0 \equiv j\omega \sqrt{LC} \qquad (7\text{-}1\text{-}32)$$

and the characteristic impedance of the line can be determined from Eqs. (7-1-28) and (7-1-30) as

$$Z_0 = \sqrt{\frac{L}{C}} \qquad (7\text{-}1\text{-}33)$$

When the electron-beam current is present, Eq. (7-1-31) can be written in terms of Eqs. (7-1-32) and (7-1-33) as

$$V = - \frac{\gamma \gamma_0 Z_0}{\gamma^2 - \gamma_0^2} i \qquad (7\text{-}1\text{-}34)$$

Since $E_z = - \dfrac{\partial V}{\partial z} = \gamma V$, the axial electric field is given by

$$E_1 = - \frac{\gamma^2 \gamma_0 Z_0}{\gamma^2 - \gamma_0^2} i \qquad (7\text{-}1\text{-}35)$$

This equation is called the *circuit equation* because it determines how the axial electric field of the slow-wave helix is affected by the spatial ac electron-beam current.

Example 7-1-3:   Characteristic Impedance of a Helical Line

The equivalent circuit of a helix traveling-wave amplifier has the following parameters:

Coupled inductance	L	= 3.75 nH
Coupled capacitance	C	= 1.5 pF
Operating frequency	f	= 9 GHz

Calculate:

(a) The propagation constant of the line

(b) The characteristic impedance of the line

Solution:

(a) The propagation constant is

$$\gamma_0 = j\omega\sqrt{LC} = j2\pi \times 9 \times 10^9 \sqrt{3.75 \times 10^{-9} \times 1.5 \times 10^{-12}}$$

$$= j4.24$$

(b) The characteristic impedance is

$$Z_0 = \sqrt{\frac{L}{C}} = \sqrt{\frac{3.75 \times 10^{-9}}{1.5 \times 10^{-12}}} = 50 \ \Omega$$

7-1-4 Wave Modes

The wave modes of a helix-type traveling-wave tube can be determined by solving the electronic and circuit equations simultaneously for the propagation constants. Each solution for the propagation constants represents a mode of traveling wave in the tube. It can be seen from Eqs. (7-1-26) and (7-1-35) that there are four distinct solutions for the propagation constants. This means there are four modes of traveling wave in the O-type traveling-wave tube. Substitution of Eq. (7-1-26) in Eq. (7-1-35) yields

$$(\gamma^2 - \gamma_0^2)(j\beta_e - \gamma)^2 = -j\frac{\gamma^2 \gamma_0 Z_0 \beta_e I_0}{2V_0} \qquad (7\text{-}1\text{-}36)$$

Equation (7-1-36) is of the fourth order in γ and thus has four roots. The exact solutions can be obtained with numerical methods or a digital computer. However, the approximate solutions may be found by equating the dc electron-beam velocity to the axial phase velocity of the traveling wave, which is equivalent to setting

$$\gamma_0 = j\beta_e \tag{7-1-37}$$

Then, Eq. (7-1-36) reduces to

$$(\gamma - j\beta_e)^3 (\gamma + j\beta_e) = 2C^3\beta_e^2\gamma^2 \tag{7-1-38}$$

where C is the traveling-wave tube gain parameter and is defined as

$$C \equiv \left(\frac{I_0 Z_0}{4V_0}\right)^{1/3} \tag{7-1-39}$$

It can be seen from Eq. (7-1-38) that there are three forward traveling waves corresponding to $e^{-j\beta_e z}$ and one backward traveling wave corresponding to $e^{+j\beta_e z}$. Let the propagation constant of the three forward traveling waves be

$$\gamma = j\beta_e - \beta_e C\delta \tag{7-1-40}$$

where it is assumed that $C\delta \ll 1$. Substitution of Eq. (7-1-40) in Eq. (7-1-38) results in

$$(-\beta_e C\delta)^3(j2\beta_e - \beta_e C\delta) = 2C^3\beta_e^2(-\beta_e^2 - 2j\beta_e^2 C\delta + \beta_e^2 C^2\delta^2) \tag{7-1-41}$$

Since $C\delta \ll 1$, Eq. (7-1-41) is reduced to

$$\delta = (-j)^{1/3} \tag{7-1-42}$$

From the theory of complex variables the three roots of $(-j)$ can be plotted in Fig. 7-1-4.

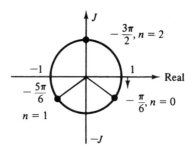

Figure 7-1-4   Three roots of $(-j)$.

Equation (7-1-42) can be written in exponential form as

$$\delta = (-j)^{1/3} = e^{-j[(\pi/2 + 2n\pi)/3]} \tag{7-1-43}$$

where $n = 0, 1, 2$. The first root δ_1 at $n = 0$ is

$$\delta_1 = e^{-j\pi/6} = \frac{\sqrt{3}}{2} - j\frac{1}{2} \tag{7-1-44}$$

The second root δ_2 at $n = 1$ is

$$\delta_2 = e^{-j5\pi/6} = -\frac{\sqrt{3}}{2} - j\frac{1}{2} \tag{7-1-45}$$

The third root δ_3 at $n = 2$ is

$$\delta_3 = e^{-j3\pi/2} = j \tag{7-1-46}$$

The fourth root δ_4 corresponding to the backward traveling wave can be obtained by setting

$$\gamma = -j\beta_e - \beta_e C \delta_4 \tag{7-1-47}$$

Similarly,

$$\delta_4 = -j\frac{C^2}{4} \tag{7-1-48}$$

Thus, the values of the four propagation constants γ are given by

$$\gamma_1 = -\beta_e C \frac{\sqrt{3}}{2} + j\beta_e \left(1 + \frac{C}{2}\right) \tag{7-1-49}$$

$$\gamma_2 = \beta_e C \frac{\sqrt{3}}{2} + j\beta_e \left(1 + \frac{C}{2}\right) \tag{7-1-50}$$

$$\gamma_3 = j\beta_e(1 - C) \tag{7-1-51}$$

$$\gamma_4 = -j\beta_e \left(1 - \frac{C^3}{4}\right) \tag{7-1-52}$$

These four propagation constants represent four different modes of wave propagation in the O-type helical traveling-wave tube. It is concluded that the wave corresponding to γ_1 is a forward wave and that its amplitude grows exponentially with distance; the wave corresponding to γ_2 is also a forward wave, but its amplitude decays exponentially with distance; the wave corresponding to γ_3 is also a forward wave, but its amplitude remains constant; and the fourth wave corresponding to γ_4 is a backward wave, and there is no change in amplitude. The growing wave propagates at a phase velocity slightly lower than the electron-beam velocity, and the energy flows from the electron beam to the wave. The decaying wave propagates the same velocity as that of the growing wave, but the energy flows from the wave to the electron beam. The constant-amplitude wave travels at a velocity slightly higher than the electron-beam velocity, but no net energy exchange occurs between the wave and the electron beam. The backward wave progresses in the negative z direction with a velocity slightly higher than the velocity of the electron beam inasmuch as the typical value of C is about 0.02.

Example 7-1-4:   Wave Modes in a TWA

A pulsed helix TWA has the following parameters:

Beam voltage	$V_0 = 10.5$ kV
Beam current	$I_0 = 2$ A
Operating frequency	$f = 10$ GHz
Characteristic impedance	$Z_0 = 5\ \Omega$

Determine:

(a) The dc electron velocity

(b) The phase constant of the electron beam

(c) The gain parameter

(d) The four propagation constants

(e) The four wave equations

Solution:

(a) The dc electron velocity is

$$\mathcal{V}_0 = 0.593 \times 10^6 \sqrt{1.5 \times 10^3} = 0.6076 \times 10^8 \text{ m/s}$$

(b) The phase constant of the electron beam is

$$\beta_e = 2\pi \times 10 \times 10^9/(0.6076 \times 10^8) = 1034 \text{ rad/m}$$

(c) The gain parameter is

$$C = \left(\frac{2 \times 5}{4 \times 10.5 \times 10^3}\right)^{1/3} = 0.062$$

(d) The four propagation constants are

$$\gamma_1 = -1034 \times 0.062 \times \frac{\sqrt{3}}{2} + j1034 \left(1 + \frac{0.062}{2}\right)$$

$$= -55.52 + j1066.05$$

$$\gamma_2 = +55.52 + j1066.05$$

$$\gamma_3 = j1034 (1 - 0.062) = j969.89$$

$$\gamma_4 = -j1034 \left(1 - \frac{0.062^3}{2}\right) = -j1034$$

(e) The four wave equations are

(1) Forward wave with increasing amplitude is

$$E_1 \exp(j\omega t - \gamma_1 z) = E_1 \exp(55.52z) \exp(j\omega t - j1066z)$$

(2) Forward wave with decreasing amplitude is

$$E_1 \exp(j\omega t - \gamma_2 z) = E_1 \exp(-55.52z) \exp(j\omega t - j1066z)$$

(3) Forward wave with constant amplitude is

$$E_1 \exp(j\omega t - \gamma_3 z) = E_1 \exp(j\omega t - j969.89z)$$

(4) Backward wave with constant amplitude is

$$E_1 \exp(j\omega t - \gamma_4 z) = E_1 \exp(j\omega t + j1034z)$$

7-1-5 Gain Consideration

For simplicity, it is assumed that the structure is perfectly matched so that there is no backward traveling wave. Such is usually the case. Even though there is a reflected wave from the output end of the tube traveling backward toward the input end, the attenuator placed around the center of the tube subdues the reflected wave to a minimum or zero level. Thus the total circuit voltage is the sum of three forward voltages corresponding to the three forward traveling waves. This is equivalent to

$$V(z) = V_1 e^{-\gamma_1 z} + V_2 e^{-\gamma_2 z} + V_3 e^{-\gamma_3 z} = \sum_{n=1}^{3} V_n e^{-\gamma_n z} \qquad (7\text{-}1\text{-}53)$$

The input current can be found from Eq. (7-1-26) as

$$i(z) = -\sum_{n=1}^{3} \frac{I_0}{2V_0 C^2} \frac{V_n}{\delta_n^2} e^{-\gamma_n z} \qquad (7\text{-}1\text{-}54)$$

in which $C\delta \ll 1$, $E_1 = \gamma V$, and $\gamma = j\beta_e(1 - C\delta)$ have been used. The input fluctuating component of velocity of the total wave may be found from Eq. (7-1-21) as

$$\mathcal{V}_1(z) = \sum_{n=1}^{3} j \frac{\mathcal{V}_0}{2V_0 C} \frac{V_n}{\delta_n} e^{-\gamma_n z} \qquad (7\text{-}1\text{-}55)$$

where $E_1 = \gamma V$, $C\delta \ll 1$, $\beta_e \mathcal{V}_0 = \omega$, and $\mathcal{V}_0 = \sqrt{\dfrac{2e}{m}} \, V_0$ have been used.

To determine the amplification of the growing wave, the input reference point is set at $z = 0$ and the output reference point is taken at $z = \ell$. It follows that at $z = 0$ the voltage, current, and velocity at the input point are given by

$$V(0) = V_1 + V_2 + V_3 \qquad (7\text{-}1\text{-}56)$$

$$i(0) = -\frac{I_0}{2V_0 C^2} \left(\frac{V_1}{\delta_1^2} + \frac{V_2}{\delta_2^2} + \frac{V_3}{\delta_3^2} \right) \qquad (7\text{-}1\text{-}57)$$

$$\mathcal{V}_1(0) = -j \frac{\mathcal{V}_0}{2V_0 C} \left(\frac{V_1}{\delta_1} + \frac{V_2}{\delta_2} + \frac{V_3}{\delta_3} \right) \qquad (7\text{-}1\text{-}58)$$

The simultaneous solution of Eqs. (7-1-56), (7-1-57), and (7-1-58) with $i(0) = 0$ and $\mathcal{V}_1(0) = 0$ is

$$V_1 = V_2 = V_3 = \frac{V(0)}{3} \tag{7-1-59}$$

Since the growing wave is increasing exponentially with distance, it dominates the total voltage along the circuit. When the length ℓ of the slow-wave structure is sufficiently large, the output voltage will be almost equal to the voltage of the growing wave. Substitution of Eqs. (7-1-49) and (7-1-59) in Eq. (7-1-53) yields the output voltage as

$$V(\ell) \simeq \frac{V(0)}{3} \exp\left(\frac{\sqrt{3}}{2} \beta_e C \ell\right) \exp\left[-j\beta_e \left(1 + \frac{C}{2}\right)\ell\right] \tag{7-1-60}$$

The factor $\beta_e \ell$ is conventionally written as $2\pi N$, where N is the circuit length in electronic wavelength—that is,

$$N = \frac{\ell}{\lambda_e} \quad \text{and} \quad \beta_e = \frac{2\pi}{\lambda_e} \tag{7-1-61}$$

The amplitude of the output voltage is then given by

$$V(\ell) = \frac{V(0)}{3} \exp(\sqrt{3}\pi NC) \tag{7-1-62}$$

The output power gain in decibels is defined as

$$A_p \equiv 10 \log \left|\frac{V(\ell)}{V(0)}\right|^2 = -9.54 + 47.3 NC \quad \text{dB} \tag{7-1-63}$$

where NC is a numerical number.

The output power gain as shown in Eq. (7-1-63) indicates an initial loss at the circuit input of 9.54 dB. This loss is due to the fact that the input voltage splits into three waves of equal magnitude and the growing wave voltage is only one-third the total input voltage. It also can be seen that the power gain is proportional to the length N in electronic wavelength of the slow-wave structure and the gain parameter C of the circuit.

The power loss can be computed from the following expression:

$$\text{Power loss} = 20 \log \left|\frac{\delta_1^2}{(\delta_1 - \delta_2)(\delta_1 - \delta_3)}\right| \tag{7-1-64}$$

Substitution of Eqs. (7-1-44), (7-1-45), (7-1-46), and (7-1-48) into Eq. (7-1-64) yields

$$\text{Power loss} = 20 \log \left|\frac{\left(\frac{\sqrt{3}}{2} - j\frac{1}{2}\right)^2}{\left[\frac{\sqrt{3}}{2} - j\frac{1}{2} - \left(-\frac{\sqrt{3}}{2} - j\frac{1}{2}\right)\right]\left[\frac{\sqrt{3}}{2} - j\frac{1}{2} - j\right]}\right| \tag{7-1-65}$$

$$= 20 \log\left(\frac{1}{3}\right) = -9.54 \text{ dB}$$

The Varian-pulsed helix-traveling-wave amplifier VTU-5191F1 as shown in Fig.

7-1-5 has a minimum peak output power of 1 kW at frequency range from 8.0 to 16.0 GHz. The power gain is 55 dB and its duty cycle is 0.025. The typical beam voltage is 11 kV dc and the beam current is 2 A dc.

Figure 7-1-5   Photograph of Varian helix-TWA VTU-5191F1 (Courtesy of Varian Associates, Inc.).

Example 7-1-5:   Power Gain of a TWA

A pulsed-helix TWA has the following parameters:

Beam voltage	$V_0 = 1.5$ kV
Beam current	$I_0 = 2$A
Length	$N = 15$ in wavelength
Characteristic impedance	$Z_0 = 20$ Ω

Compute:

(a) The gain parameter

(b) The output power gain

Solution:

(a) The gain parameter is

$$C = \left(\frac{2 \times 20}{4 \times 10.5 \times 10^3} \right)^{1/3} = 0.098$$

(b) The output power gain is

$$A_p = -9.54 + 47.3 \, NC$$
$$= -9.54 + 47.3 \times 15 \times 0.098$$
$$= 60 \text{ dB}$$

7-2 COUPLED-CAVITY TRAVELING-WAVE-TUBE AMPLIFIER

The helix traveling-wave tubes (TWTs) described in the preceding section produce at most up to several kilowatts of average power output. This section is devoted to discuss coupled-cavity traveling-wave tubes that are used for high-power applications.

7-2-1 Physical Description

The term *coupled-cavity* means that a coupling is provided by a long slot which strongly couples the magnetic component of the field in adjacent cavities in such a manner that the passband of the circuit is mainly a function of this one variable. Figure 7-2-1 shows two coupled-cavity circuits that are principally used in traveling-wave tube.

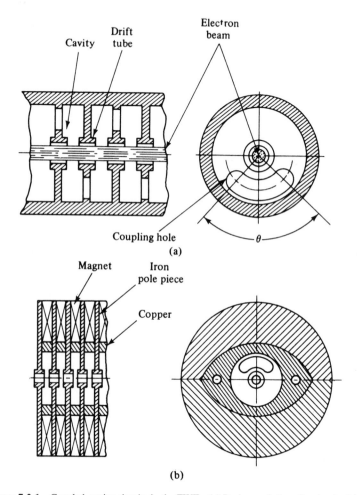

Figure 7-2-1   Coupled-cavity circuits in the TWTs: (a) Basic coupled-cavity circuit, (b) Coupled-cavity with integral periodic-permanent magnet (Reprinted by permission of the IEEE, Inc.),

There are two types of coupled-cavity circuit in the traveling-wave tubes. The first type consists of the fundamentally forward-wave circuit that is normally used for pulse applications requiring at least one-half megawatt of peak power. Such coupled-cavity circuits exhibit negative mutual inductive coupling between the cavities and operate with the fundamental space harmonic. The clover leaf [3] and centipede circuits [7], shown in Fig. 7-2-2, belong to this type.

Figure 7-2-2   Centipede and cloverleaf coupled-cavity circuit (Reprinted by permission of the IEEE, Inc.).

The second type is the first space-harmonic circuit, which has positive mutual coupling between the cavities. Such circuits operate with the first spatial harmonic and are commonly used for pulse or continuous-wave applications from one to several hundred kilowatts of power output [5]. In addition, the long-slot circuit of the positive mutual coupling-cavity circuit operates at the fundamental spatial harmonic with a higher-frequency mode. This circuit is suitable for megawatt power output. Figure 7-2-3 shows several space-harmonic coupled-cavity circuits.

Figure   7-2-3   Space-harmonic coupled-cavity circuits (Reprinted by permission of the IEEE, Inc.).

7-2-2 Principles of Operation

Any repetitive series of lumped LC elements constitute a propagating-filter–type circuit. The coupled cavities in the traveling-wave tube are usually highly over–coupled, resulting in a bandpass filter-type characteristic. When the slot angle (θ) as shown in Fig. 7-2-1(a) is larger than 180°, the passband is close to its practical limits. The drift tube is formed by the reentrant part of the cavity, just as in the case of a klystron. During the interaction of the RF field and the electron beam in the traveling-wave tube a phase change occurs between the cavities as a function of frequency. A decreasing phase characteristic is reached if the mutual inductance of the coupling slot is positive, whereas an increasing phase characteristic is obtained if the mutual inductive coupling of the slot is negative [3].

The amplification of the traveling-wave tube interaction requires that the electron beam interacts with a component of circuit field that has an increasing phase characteristic with frequency. The circuit periodicity can give rise to field components that have phase characteristics [6] as shown in Fig. 7-2-4.

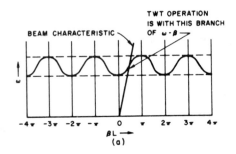

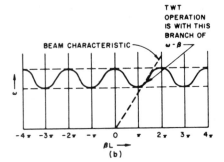

Figure 7-2-4 ω-β diagram for coupled-cavity circuits: (a) Fundamental forward-wave circuit, (b) Fundamental backward-wave circuit (Reprinted by permission of the IEEE, Inc.)

In Fig. 7-2-4, the angular frequency (ω) is plotted as a function of phase shift (βl) per cavity. The ratio of ω to β is equal to the phase velocity. For a circuit having positive mutual inductive coupling between the cavities, the electron-beam velocity is adjusted to be approximately equal to the phase velocity of the first forward-wave spatial harmonic. For the circuits with negative mutual inductive coupling, the fundamental branch component of the circuit wave is suitable for synchronism with the electron beam and is normally used by the traveling-wave tube. The coupled-cavity equivalent circuit has been developed by Curnow [6] as shown in Fig. 7-2-5.

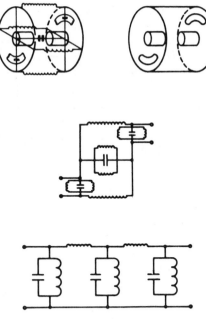

Figure 7-2-5   Equivalent circuits for a slot-coupled cavity (Reprinted by permission of the IEEE, Inc.)

In Fig. 7-2-5, inductances are used to represent current flow and capacitors to represent the electric fields of the cavities. The circuit can be evolved into a fairly simple configuration. Loss in the cavities can be approximately calculated by adding resistance in series with the circuit inductance.

7-2-3 Microwave Characteristics

When discussing the power capability of traveling-wave tubes, it is important to make a clear distinction between the average and peak power because these two figures are limited by totally different factors. The average power at a given frequency is almost always limited by thermal consideration relative to the RF propagating circuit. However, the peak RF power capability depends upon the voltage for which the tube can be designed. The beam current varies as the three-half power of the voltage, and the product of the beam current and the voltage determines the total beam power—that is,

$$I_{\text{beam}} = KV_0^{3/2} \qquad (7\text{-}2\text{-}1)$$

and

$$P_{\text{beam}} = KV_0^{5/2} \qquad (7\text{-}2\text{-}2)$$

where V_0 is the beam voltage, and K is the electron-gun perveance.

For a solid-beam electron gun with good optics, the perveance is generally considered to be between 1 to 2 × 10⁻⁶. Once the perveance is fixed, the required voltage for a given peak beam power is then uniquely determined. Figures 7-2-6 and 7-2-7 demonstrate the difference between peak and average power capability and the difference between periodic-permanent-magnet (PPM) and solenoid-focused designs [2].

Coupled-cavity traveling-wave tubes are constructed with a limited amount of gain per section of cavities to ensure stability. Each cavity section is terminated by a matched load, an input line, or an output line in order to reduce again variations with frequency. Cavity sections are cascaded to achieve higher tube gain than can be tolerated in one section of cavities. Stable gain greater than 60 dB can be obtained over about 30% bandwidth by this method.

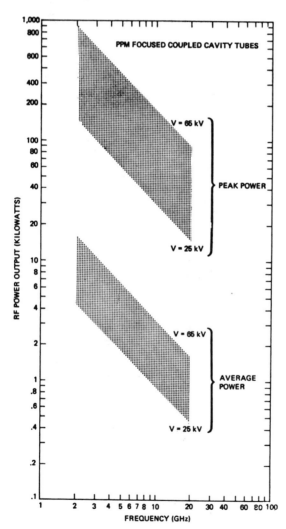

Figure 7-2-6 Peak and average power capability of typical TWTs in field use (Reprinted by permission of the IEEE, Inc.).

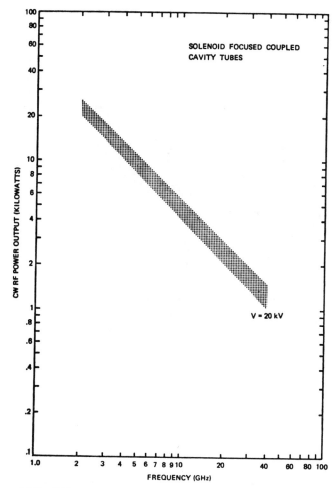

Figure 7-2-7 CW power capability of TWTs operating at nearly 20 kV (Reprinted by permission of the IEEE, Inc.).

The overall efficiency of coupled-cavity traveling-wave tubes is determined by the amount of energy converted to RF energy and the energy dissipated by the collector. Interaction efficiencies from 10 to 40% have been achieved from coupled-cavity traveling-wave tubes. Overall efficiencies of 20 to 55% have been obtained [6].

The Hughes X-band pulsed TWT amplifier 751H as shown in Fig. 7-2-8 is a typical multicoupled-cavity TWA and has a minimum peak output power of 50 kW at 8.8 GHz. Its power gain is 52 dB and its duty cycle is 0.01. The cathode voltage is -31 kV and the cathode current is 7 A. The depressed efficiency is 35%.

Example 7-2-1: Coupled-Cavity TWA

A multicoupled-cavity TWA has the following parameters:

$$
\begin{aligned}
&\text{Beam voltage} &&V_0 = 24 \text{ kV} \\
&\text{Electron-gun perveance} &&K = 1.21 \times 10^{-6} \text{ A/V}^{3/2}
\end{aligned}
$$

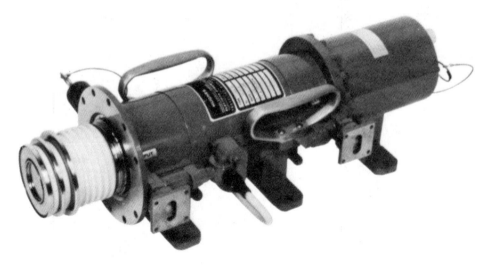

Figure 7-2-8   Photograph of Hughes TWA 751H (Courtesy of Hughes Aircraft Company, Electron Dynamics Division).

(a) Compute the beam current.

(b) Calculate the total beam power.

Solution:

(a) The beam current is

$$I_0 = KV^{3/2} = 1.21 \times 10^{-6} \times (2.4 \times 10^4)^{3/2}$$
$$= 1.21 \times 10^{-6} \times 3.72 \times 10^6$$
$$= 4.5 \text{ A}$$

(b) The total beam power is

$$P_0 = KV_0^{5/2} = 1.21 \times 10^{-6} \times (2.4 \times 10^4)^{5/2}$$
$$= 1.21 \times 10^{-6} \times 8.93 \times 10^{10}$$
$$= 108 \text{ kW}$$

7-3 GRIDDED-CONTROL TRAVELING-WAVE-TUBE AMPLIFIER

Gridded-control traveling-wave tubes (TWTs) are the most versatile devices used for amplification at microwave frequencies with high gain, high power, high efficiency, and wide bandwidth. High-power gridded-control TWTs have four main sections: electron gun for electron emission, slow-wave structure for effective beam interaction, magnetic circuit for beam focusing, and collector structure for collecting electron beams and dissipating heat energy. Specifically, the physical components of a gridded-control traveling-wave tube consists of an electron emitter, a shadow grid, a control grid, a modulating anode, a coupled-cavity circuit, a solenoid magnetic circuit, and a collector depression structure as shown in Fig. 7-3-1.

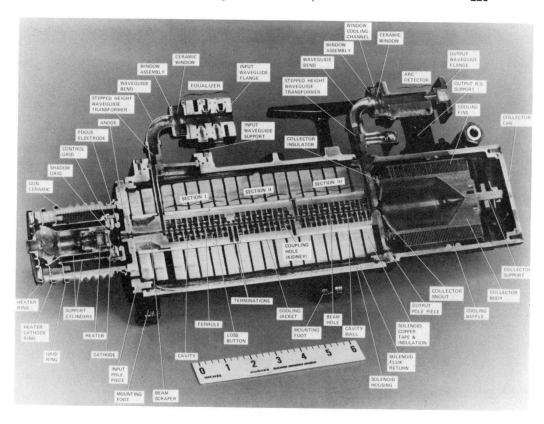

Figure 7-3-1 Cutaway of a coupled-cavity traveling-wave tube (Courtesy of Hughes Aircraft Company, Electron Dynamics Division).

   After electrons are emitted from the cathode, the electron beam has a tendency to spread out due to the electron repelling force. On the other hand, the electron beam must be small enough for effective interaction with the slow-wave circuit. Usually the diameter of the electron beam is smaller than one-tenth wavelength of the signal. The gridded-control traveling-wave high-power tube utilizes a shadow-grid technique to control the electron beam, so the device is called *gridded traveling-wave tube* (GTWT). As shown in Fig. 7-3-2, the electron emitter of a gridded traveling-wave tube has two control electrodes: one shadow grid near the cathode and one control grid slightly away from the cathode. The shadow grid, which is at cathode potential and interposed between the cathode and the control grid, suppresses electron emission from those portions of the cathode that would give rise to interception at the control grid. The control grid, which is at a positive potential, controls the electron beam. These grids can control far greater beam power than would otherwise be possible.

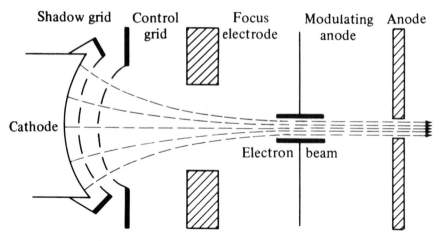

Figure 7-3-2   Electron emitter with control electrodes for gridded high-power TWT

In general, an anode modulation technique is frequently used in traveling-wave tubes to eliminate voltage pulsing through the lower unstable beam voltage and to reduce modulator power requirements for high-power pulse output. In a gridded traveling-wave tube, the modulator applies a highly regulated positive grid drive voltage with respect to the cathode to turn the electron beam on for RF amplification. An unregulated negative grid bias voltage with respect to the cathode is used to cut the electron beam off. Thus the anode modulator acts as a pulse switch for the electron beam of the gridded traveling-wave tube.

The anode of the electron gun is operated at a voltage higher than that of the slow-wave structure to prevent positive ions formed by the electron beam in the region of the slow-wave structure from draining toward the cathode and bombarding it.

7-3-1 High Efficiency and Collector Voltage Depression

After passing through the output cavity, the electron beam strikes a collector electrode. The function of the collector electrode could be performed by replacing the second grid of the output cavity with a solid piece of metal. However, a separate electrode may have two advantages. First, the collector can be made as large as desired in order to collect the electron beam at a lower density, thus minimizing localized heating. If the collector were a part of the slow-wave circuit, its size would be limited by the maximum gap capacitance consistent with good high-frequency performance. Second, by using a separate collector can reduce its potential considerably below the beam voltage in the RF interaction region, thereby reducing the power dissipated in the collector and increasing the overall efficiency of the device. Gridded traveling-wave high-power tubes have a separate collector that dissipates the electrons in the form of heat. A cooling mechanism absorbs the heat by thermal conduction to a cooler surface.

The efficiency of a gridded traveling-wave high-power tube is the ratio of the RF power output over the product of cathode voltage (beam voltage) and cathode current (beam current). It may be expressed in terms of the product of the electronic efficiency and the circuit efficiency. The electronic efficiency expresses the percentage of the dc or pulsed input power that is converted into RF power on the slow-wave structure. The circuit efficiency, on the other hand, determines the percentage of the dc input power that is delivered to the load exterior to the tube. The electron beam does not extract energy from any dc power supply unless the electrons are actually collected by an electrode connected to that power supply. If a separate power supply is connected between cathode and collector and if the cavity grids intercepts a negligible part of the electron beam, the power supply between the cathode and collector will be the only one supplying any power to the tube. For a gridded traveling-wave tube, the collector voltage is normally operated at about 40% of the cathode voltage. Thus the overall efficiency of conversion of dc to RF power is almost twice the electronic efficiency. Under this condition the tube is operating with collector voltage depression.

Example 7-3-1:   Efficiency of a TWA

A GTWT amplifier has the following parameters:

Cathode voltage	V_0	$= -36$ kV
Cathode current	I_0	$= 9$ A
Output power	P_{out}	$= 75$ kW
Collector voltage	V_c	$= -18$ kV

Determine:

(a) The electronic efficiency

(b) The overall efficiency

Solution:

(a) The electronic efficiency is

$$\eta = \frac{RFP_{out}}{V_0 I_0} = \frac{75 \times 10^3}{36 \times 10^3 \times 9}$$

$$= 23\%$$

(b) The overall efficiency is

$$\eta_{all} = \frac{75 \times 10^3}{18 \times 10^3 \times 9}$$

$$= 46\%$$

7-3-2 Normal Depression and Overdepression of Collector Voltage

Most gridded traveling-wave tubes are very sensitive to the variations of the collector depression voltages below the normal depression level since the tubes operate closer to the knee of the electron spent-beam curves. Figure 7-3-3 shows the spent-beam curves for a typical gridded traveling-wave tube [8].

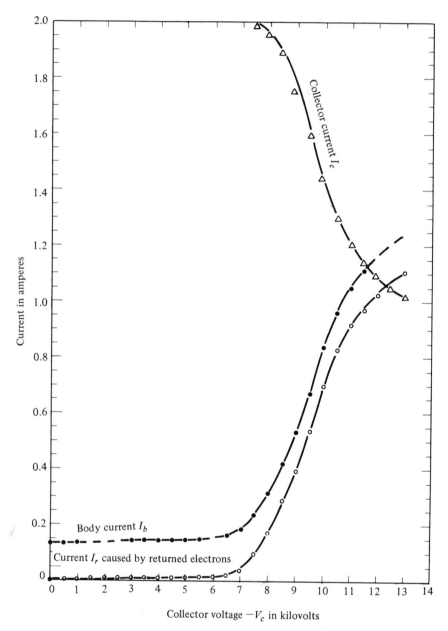

Figure 7-3-3 Spent beam curves for a typical gridded TWT.

Under normal collector depression voltage V_c at -7 kV with full saturated power output the spent beam electrons are collected by the collector and return to the cathode. Thus the collector current I_c is about 2.09 A. A small amount of electrons intercepted by the beam scraper or slow-wave circuit contributes the tube body current of about 0.178 A. Very few electrons with lower kinetic energy reverse the direction of their velocity inside the collector and fall back onto the output pole piece. These returning electrons yield a current I_r of 0.041 A, which is only a small fraction of the body current I_b. These values are shown in Fig. 7-3-4 [8].

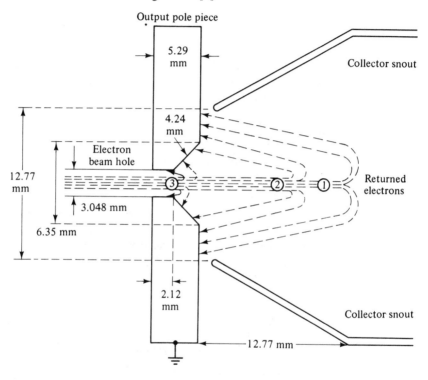

① indicates that most electrons are returned
  by low overdepression voltage
② indicates that most electrons are returned
  by high overdepression voltage
③ indicates that most electrons are returned
  by higher overdepression voltage

Figure 7-3-4   Impact probability of returned electrons by overdepressed collector voltage.

When the collector voltage is overdepressed from the normal level of -7.5 kV to the worst case of about -11.5 kV, a greater number of the spent electrons inside the collector reverse the direction of their velocity by a highly negative collector voltage and fall back onto the grounded output pole piece because the potential of the pole piece is 11.5 kV higher than the collector voltage. It can be seen from Fig. 7-3-4 that when the collector voltage is overdepressed from -7.5 to -11.5 kV, the collector current is decreased sharply from 2.01 to 1.14 A and the body current is increased rapidly from 0.237 to 1.110 A. The body current consists of two parts: one part is the current due to the electrons intercepted by the circuit or the beam scrapers; and another part is the current due to the electrons returned by the overdepressed collector voltage.

Example 7-3-2: Gridded Traveling-Wave Tube (GTWT)

A gridded traveling-wave tube is operated under overdepression of collector voltage as follows:

Overdepression voltage	V_c	$= -11$ kV
Returned current	I_r	$= 0.85$ A
Mass of heated iron pole piece	Mass	$= 250$ mg
Specific heat H of iron at 20°C	H	$= 0.108$ calories/g-°C

Determine:

(a) The number of electrons returned per second

(b) The energy in eV associated with these returning electrons in 20 ms

(c) The power in watts for the returning electrons

(d) The heat in calories associated with the returning electrons (a factor for converting joules to calories is 0.238)

(e) The temperature T in °C for the output iron pole piece [Hint: $T = 0.238 VIT/(\text{mass} \times \text{specific heat.})$]

(f) Whether the output iron pole piece is melted

Solution:

(a) The number of electrons returned per second is

$$I_r = 0.85/(1.6 \times 10^{-19}) = 5.31 \times 10^{18} \text{ electrons/s}$$

(b) The energy is

$$W = Pt = VI_r t = 11 \times 10^3 \times 5.31 \times 10^{18} \times 20 \times 10^{-3}$$
$$= 1.168 \times 10^{21} \text{ eV}$$

(c) The power is

$$P = VI_r = 11 \times 10^3 \times 0.85 = 9.35 \text{ kW}$$

(d) The created heat is

$$H(\text{heat}) = 0.238 Pt = 0.238 VI_r t$$
$$= 0.238 \times 11 \times 10^3 \times 0.85 \times 20 \times 10^{-3}$$
$$= 44.51 \text{ calories}$$

(e) The temperature is

$$T = 0.238VI_r t/(\text{mass} \times \text{specific heat})$$
$$= 44.51/(250 \times 10^{-3} \times 0.108) = 1648.52°C$$

(f) The output iron pole piece is melted because the melting point of iron is 1535°C.

7-3-3 Multistage Collector Voltage Depression Technique

The amount of power dissipated in the collector can be reduced to a minimum by making use of a collector that has multiple velocity sorting stages. A collector can have two or three stages for high-efficiency applications.

Two-stage collector voltage depression.   When the spent electron beam arrives in the collector, the kinetic energies of each electron are different. Under the normal operation at a collector voltage of about 40% of the cathode voltage, very few electrons will be returned by the negative collector voltage. Consequently, the tube body current is very small and negligible because the returned electrons are the only ones intercepted by the cavity grids and the slow-wave circuit. When the collector is more negative, however, more electrons with lower energy will reverse their direction of velocity and fall onto the output pole piece. Thus the tube body current will increase sharply. Since electrons of various energy classes exist inside the collector, two-state collector voltage depression may be utilized. Each stage is biased at a different voltage. Specifically, the main collector may be biased at 40% depression of the cathode voltage for normal operation, but the collector snout may be grounded to the output pole piece for overdepression operation. As a result, the returned electrons will be collected by the collector snout and returned to the cathode even though the collector voltage is overdepressed to be more negative. Since the collector is cooled by a cooling mechanism the overheating problem for overdepression is eased. Figure 7-3-5 shows a structure of two-stage collector voltage depression and Fig. 7-3-6 shows a basic interconnection of a gridded traveling-wave tube with its power supplies [8].

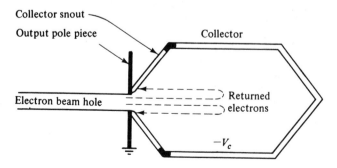

Figure 7-3-5 Diagram for two-stage collector depression

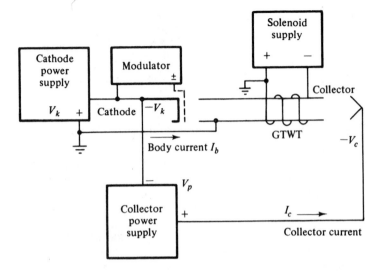

Figure 7-3-6 Basic interconnection of a gridded traveling-wave tube with its power supplies.

The Hughes X-band pulsed gridded TWA 752H as shown in Fig. 7-3-7 is a typical gridded traveling-wave amplifier. Its minimum peak output power is 100 kW at frequency from 8.4 to 9.4 GHz and its depressed efficiency is 30%. The power gain is 16 dB and the duty cycle is 0.01. The cathode voltage is -50 V, the cathode current is 15 A, and the collector voltage is -20 kV.

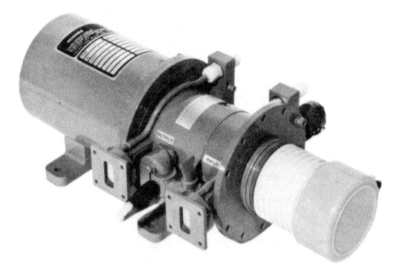

Figure 7-3-7 Photograph of Hughes gridded TWA 752H (Courtesy of Hughes Aircraft Company, Electron Dynamics Division).

Three-stage collector voltage depression.   A three-stage collector is shown in Fig. 7-3-8. The first stage is grounded, the second and the third stages are depressed below the ground potential. A collector of this design is commonly referred to as having two depressed stages. The first stage and the third stage collect those electrons with very low and very high kinetic energy, respectively. The second stage collects those electrons with intermediate kinetic energy. The number of depression stages are related to the overall TWT efficiency as follows:

> 18% efficiency with a collector at ground potential
> 41% efficiency with one depressed collector stage
> 47% efficiency with two depressed collector stages
> 50% efficiency with three depressed collector stages

For most applications, the TWT amplifiers have two depressed collector stages to provide the most logical compromise between the TWT efficiency and the complexity of the TWT and power converter.

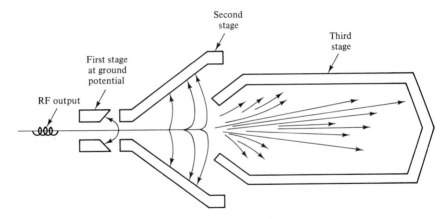

Figure 7-3-8   Three-stage collector.

7-4 VOLTAGE STABILIZATION TECHNIQUES

The cathode voltage of a gridded traveling-wave tube is negative with respect to ground so the electrons can be emitted from the cathode. In order to maintain a constant beam power for a uniform gain, the cathode voltage must be constant. In addition, the phase shift through the tube is directly related to the beam velocity; thus high resolution and low ripple are required in the cathode voltage power supply to avoid undesirable phase-shift variations. Consequently, the cathode power supply of the gridded traveling-wave high-power tube is usually regulated for better than 1% over line and load changes and is also well filtered because of the critical requirements on the cathode voltage with respect to ground. The cathode power supply provides the tube body current. Under normal operation the body current is very small in comparison with the collector current. Figure 7-4-1 shows a basic interconnection for a gridded traveling-wave tube with two regulated power supplies.

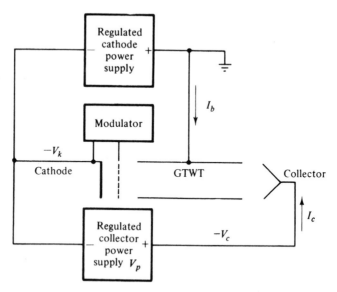

Figure 7-4-1 Interconnections for a gridded traveling-wave high-power tube.

Figure 7-4-2 illustrates a voltage-regulator circuit for a cathode power supply. The voltage regulator indicated in the circuit consists of two devices: one differential amplifier and one tetrode tube. The solid-state differential amplifier amplifies the difference between the preset reference voltage and the voltage that is one-thousandth of the output voltage. The reference voltage is adjustable to a preset level. The output voltage of the differential amplifier drives the control grid of the regulator tube between cutoff and saturation in order to nullify the difference voltage.

As shown in Fig. 7-4-1, the negative terminal of the collector power supply is connected to the cathode and the positive terminal to the collector electrode. The collector depression voltage is the difference of the two regulated supply voltages. The cathode supply provides the tube body current and the collector supply yields the collector current. The ratio of the collector current over the body current is about 10 for an operation of 40% voltage depression. Thus the power delivered to the tube by the collector supply is about four times larger than the power furnished by the cathode supply. An electrical transient may occur in the circuit when the power supplies are just being switched on or off. The reasons for an electrical transient may be due to two factors:

1. Load Changes—When the load of a generator is suddenly increased, a larger current is demanded. Since the generator cannot meet the demand instantaneously, the output voltage of the generator drops momentarily. Conversely, when the load of a generator is suddenly decreased, the output voltage drops accordingly.

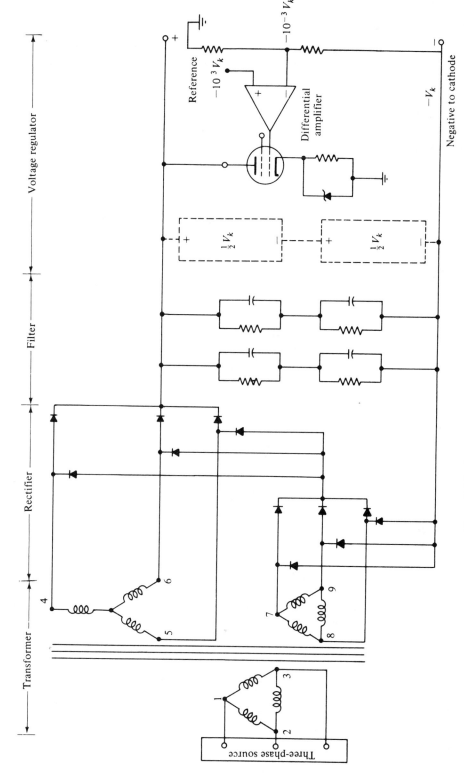

Figure 7-4-2 Voltage-regulator circuit for cathode power supply.

235

2. Switching On or Off—When the switch of a generator is just turned on or off, the armature current in the armature conductors produces armature reaction. The nature of armature reaction reduces the terminal voltage for lagging loads.

When an electrical transient is created in the circuit, the collector voltage is overdepressed. As a result, the spent electrons inside the collector reverse the direction of their velocity by the highly negative collector voltage and fall back onto the grounded output pole piece. The tube body current is sharply increased and the collector current is greatly decreased. When the returned electrons impact the output pole piece, the pole piece will be damaged by high heat. The damage of the output pole piece creates a mismatch in the interaction circuit and degrades the performance of the tube. In particular, the tube gain, efficiency, bandwidth, and the power output are affected accordingly by the circuit mismatch when the collector voltage is overdepressed below the normal depression level. Furthermore, the large body current may burn out the solid-state differential amplifier of the cathode voltage regulator and vary the electron beam of the gridded tube. If the damage is beyond the tolerance of the gridded traveling-wave high-power tube, the tube may cease to function.

In order to maintain a constant collector depression voltage, the collector voltage must be regulated. There are three possible ways to do so [8]:

1. Regulator in Series with the Collector Power Supply—In this method a voltage regulator is incorporated in series with the collector supply as shown in Fig. 7-4-3 so that the output voltage of the collector supply may be regulated at a certain level with respect to ground. Since the output voltage of the cathode supply is highly regulated at a certain level, the difference between the two regulated voltages will produce a well-regulated voltage with respect to ground at the collector electrode.

2. Regulator in Parallel with the Collector Supply—In this method a voltage regulator is inserted in parallel with the collector supply as shown in Fig. 7-4-4 so that a regulated voltage with respect to ground at normal depression may be achieved at the collector terminal.

3. Regulator Between the Cathode Voltage and the Collector Voltage—In this method the collector depression voltage is regulated with respect to the cathode voltage as shown in Fig. 7-4-5. If the collector voltage is overdepressed above the normal depression value (absolute value), the differential amplifier 2 tends to adjust the cathode voltage below its fixed level (absolute value). When the cathode voltage is dropped, the collector voltage is readjusted to its normal depression level with respect to ground.

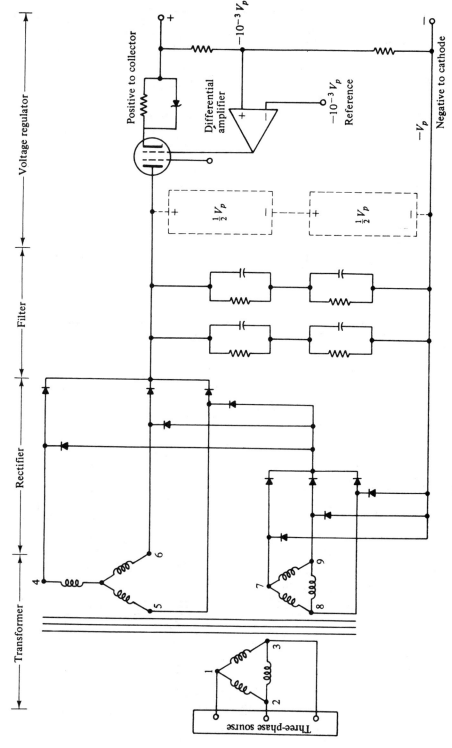

Figure 7-4-3 Regulator in series with collector supply.

237

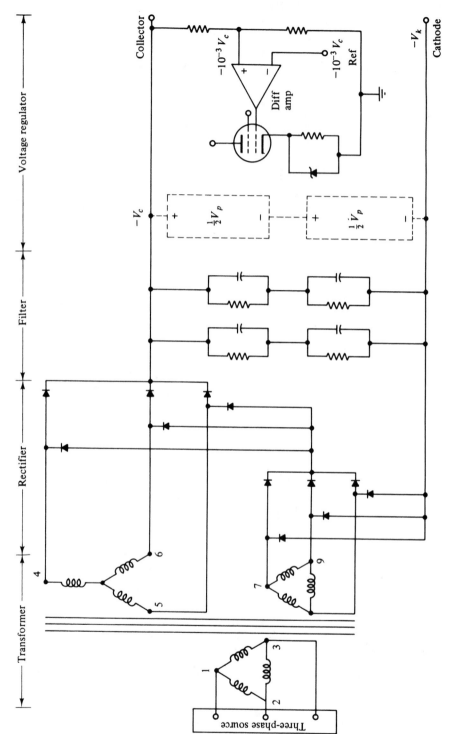

Figure 7-4-4 Regulator in parallel with collector supply.

238

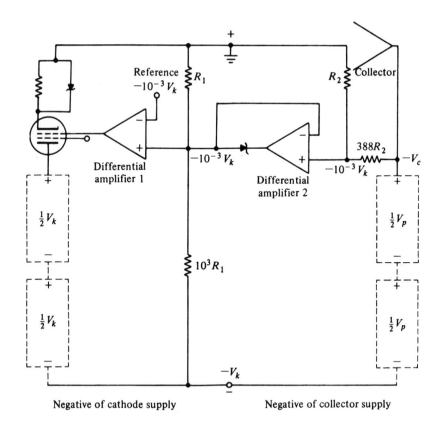

Figure 7-4-5   Regulator between cathode voltage and collector voltage.

7-5 STATE OF THE ART

The characteristics of a gridded traveling-wave tube are wide bandwidth, high gain, and high power. High power is accomplished by the high cathode voltage with shadow-grid control; high gain is achieved by the electron interaction in multicoupled cavities. Coupled-multicavity gridded traveling-wave tubes are operated with their cavities stagger-tuned in order to obtain greater bandwidth at some reduction in gain. This situation is analogous to the well-known design of wideband IF amplifiers in which each stage is tuned to a slightly different frequency in order to improve the overall gain-bandwidth product. Figure 7-5-1 shows the state of the art for U.S. high-power TWTs [8].

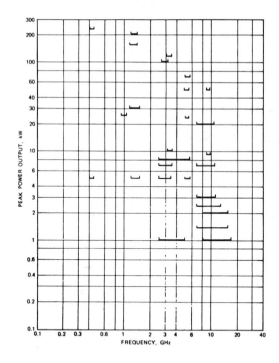

Figure 7-5-1   State of the art for U.S. high-power TWTs.

All traveling-wave tubes require some means of holding the electron beam together as it travels through the slow-wave circuit of the tube, for the beam tends to spread out as a result of the mutual repulsive forces between electrons. The magnetic circuit of a gridded traveling-wave tube consists of a solenoid structure and two soft iron pole pieces. The magnetic lines are parallel to the direction of propagation of the electron beam. The input and output pole pieces function as magnetic shields. In addition, a number of permalloy "field straightener" discs are mounted perpendicular to the axis of the tube. Since these discs act as equipotential planes with respect to the magnetic field, they force the magnetic field to be axially symmetric with respect to the axis of the tube. The magnetic circuit is surrounded by an external magnetic shield that reduces the leakage field outside the shield to a negligible amount. In certain tube structures in which the interaction structure is short enough, permanent magnet focusing is often used. In some high-power gridded tubes for airborne and space applications, periodic-permanent-magnet (PPM) focusing is utilized. Figure 7-5-2 illustrates four common methods of magnetic focusing.

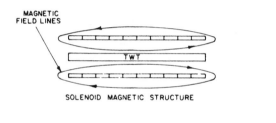

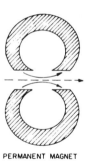

PERMANENT MAGNET

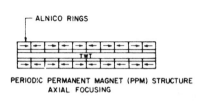

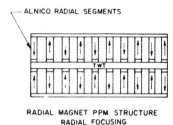

Figure 7-5-2   Methods of magnetic focusing (Courtesy of Hughes Aircraft Company, Electron Dynamics Division).

7-6 TWYSTRON HYBRID AMPLIFIERS

Several hybrid devices using combinations of klystron and traveling-wave tube components have been developed in order to achieve a better performance than each tube can obtain separately. The only widely used hybrid device is known under the tradename Twystron amplifier [9]. The Twystron amplifier consists of a multicavity klystron input section and a traveling-wave output section. Figure 7-6-1 illustrates a schematic circuit diagram of a Twystron amplifier in comparison with the diagrams of klystron, coupled-cavity traveling-wave tube, and extended interaction klystron, as a function of the interaction impedance. The chief feature of the Twystron amplifier is that it combines the advantages of klystrons and traveling-wave tubes. Figure 7-6-2 shows a cutaway of typical S-band Twystron amplifier.

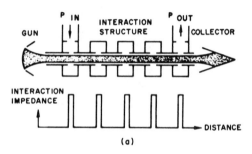

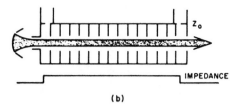

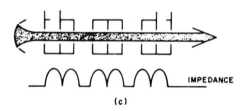

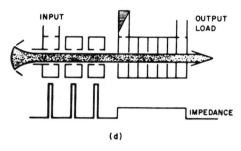

Figure 7-6-1 Schematic diagrams of high-power linear-beam tubes (Reprinted by permission of the IEEE, Inc.)

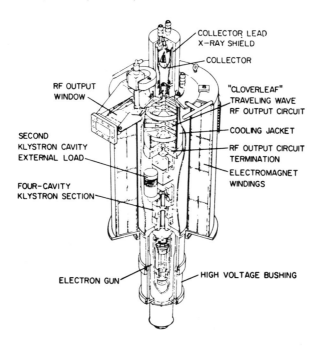

Figure 7-6-2   Cutaway of a twystron amplifier at S band (Reprinted by permission of the IEEE, Inc.)

The four-cavity stagger-tuned klystron driver section provides higher gain at the end bands than its midband because the driver section can be heavily loaded without excessive power loss. The traveling-wave output circuit (cloverleaf) is designed principally for high efficiency over the desired bandwidth. Its gain at the band edges is 6 to 10 dB below the values of the midband. The combination of klystron and traveling-wave tubes yields a relatively flat gain characteristic over the entire frequency range as shown in Fig. 7-6-3. Gain flatness and high efficiency are achieved in the Twystron amplifier because of the more efficient bunching in the klystron at the band edges where the output slow-wave circuit gain is low.

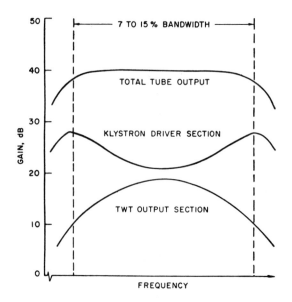

Figure 7-6-3 Relative gain of a twystron amplifier (Reprinted by permission of the IEEE, Inc.)

The principal applications of Twystron amplifiers are in land-based or shipboard high-power radar transmitters. Typical power output levels at S and C bands are from 1 to 10 MW for peak power and from 1 to 30 kW for average power [7]. Table 7-6-1 lists typical tube characteristics for several Twystron amplifiers. Figure 7-6-4 shows typical power output versus frequency at C-band Twystron amplifier.

TABLE 7-6-1. TYPICAL TWYSTRON AMPLIFIER CHARACTERISTICS

Tube type	VA-145	VA-915	VA-146	VA-145LV
Frequency bandwidth, GHz	2.7–2.9 2.9–3.1 3.0–3.2	3.1–3.6	5.4–5.9	3.1–3.5
Peak power output, MW	3.5	7.0	4.0	1.0

TABLE 7-6-1.   Continued

Tube type	VA-145	VA-915	VA-146	VA-145LV
Average power output, kW	7.0	28.0	10.0	1.0
Pulse width, μs	10.0	40.0	20.0	50.0
Efficiency, %	35.0	30.0	30.0	30.0
Beam voltage, kV	117.0	180.0	140.0	80.0
Beam current, A	80.0	150.0	95.0	45.0
Drive power, kW	0.3	3.0	2.0	1.0

From A. Staprans et al. [6]; reprinted with permission from The IEEE, Inc.

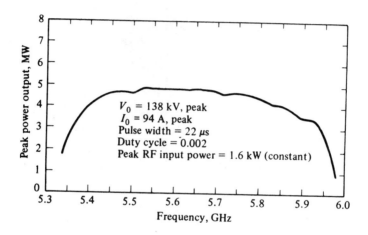

Figure 7-6-4   Power output versus frequency for C-band twystron amplifier VA-146 (Reprinted by permission of the IEEE, Inc.)

The Varian VA-145E as shown in Fig. 7-6-5 is a typical Twystron amplifier and has a peak output power of 2.5 MW at frequency range from 2.9 to 3.1 GHz. Its duty cycle is 0.002 and efficiency is 35%. The average power is 5 kW. The beam voltage is 117 kV dc and the beam current is 80 A dc.

Figure 7-6-5 Photograph of Varian VA-145E twystron amplifier (Courtesy of Varian Associates, Inc.).

7-7 BACKWARD-WAVE AMPLIFIER (BWA)

The O-type backward-wave amplifier (BWA) has an RF signal impressed upon the slow-wave structure near the collector; and its signal output at the gun end as shown in Fig. 7-7-1. Separate power supplies are provided for the anode and the helix. This arrangement makes the adjustment of the beam current independent of the helix voltage. The electron beam is assumed to be confined in the center of helix by an axial magnetic field. The electrons near the cathode are velocity-modulated as in the ordinary traveling-wave tube.

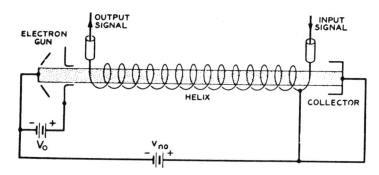

Figure 7-7-1 Schematic diagram of backward-wave amplifier.

   While the signal waves are traveling toward the cathode, the electrons are bunched somewhere near the collector. If the bunched electrons are decelerated by the microwave fields, the bunched electron beams transfer energy to the input microwave fields. The microwave propagates on the slow-wave structure toward the output, and the amplification of a microwave signal in the backward-wave tube is accomplished.

   The electronic equation as shown in Eq. (7-1-26) is derived under the assumption that the traveling waves move toward in the positive z direction. However, as described in Section 7-1-2, the direct current density J_0 is assumed to be a positive number in the negative z direction. Hence the convection-beam current in the electron induced by a spatial harmonic of the circuit field in the backward-wave amplifier is given by

$$i = -j \frac{\beta_e I_0}{2V_0 (j\beta_e - \gamma)^2} E_1 \qquad (7\text{-}7\text{-}1)$$

It should be noted that the electronic equation as shown in Eq. (7-7-1) for a backward-wave amplifier differs from the corresponding equation given by Eq. (7-1-26) for a forward-wave amplifier by a minus sign.

   The circuit equation shown in Eq. (7-1-35) is applicable to backward waves as well as forward waves. The equation is repeated here

$$E_1 = - \frac{\gamma^2 \gamma_0 Z_0}{\gamma^2 - \gamma_0^2} i \qquad (7\text{-}7\text{-}2)$$

   Combining the electronic Eq. (7-7-1) and the circuit Eq. (7-7-2) yields the wave-mode equation of the backward-wave amplifier as

$$(\gamma^2 - \gamma_0^2)(j\beta_e - \gamma)^2 = j\beta_e \gamma^2 \gamma_0 \frac{I_0 Z_0}{2V_0} = j\beta_e \gamma^2 \gamma_0 2C^3 \qquad (7\text{-}7\text{-}3)$$

A comparison of Eq. (7-7-3) with Eq. (7-1-36) indicates that the two wave-mode equations differ only by a negative sign. If the gain parameter C in the forward-wave Eq. (7-1-36) is changed to negative sign, the two wave-mode equations for the forward-wave amplifier and the backward-wave amplifier is identical. As a result, the four propagation constants for the backward-wave amplifier as shown in Eq. (7-7-3) can be written by just changing the sign of the C parameter of the four propagation constants for the forward-wave amplifier.

Then they are

$$\gamma_1 = -\beta_e C \frac{\sqrt{3}}{2} + j\beta_e \left(1 - \frac{C}{2}\right) \tag{7-7-4}$$

$$\gamma_2 = \beta_e C \frac{\sqrt{3}}{2} + j\beta_e \left(1 - \frac{C}{2}\right) \tag{7-7-5}$$

$$\gamma_3 = j\beta_e(1 + C) \tag{7-7-6}$$

$$\gamma_4 = -j\beta_e \left(1 + \frac{C^3}{4}\right) \tag{7-7-7}$$

where $\quad C = \left[\dfrac{I_0 Z_0}{4V_0}\right]^{1/3}$ is the traveling-wave tube gain parameter.

δ = integration factors are replaced and they are

$$\delta_1 = -\frac{\sqrt{3}}{2} - j\frac{1}{2} \tag{7-7-8}$$

$$\delta_2 = \frac{\sqrt{3}}{2} - j\frac{1}{2} \tag{7-7-9}$$

$$\delta_3 = j1 \tag{7-7-10}$$

$$\delta_4 = -j\frac{C^3}{4} \tag{7-7-11}$$

These four propagation constants represent four different modes of wave propagation in the O-type backward-wave tube. The predominant wave is the one that is represented by

$$V(z) = V_1 e^{-\gamma_1(\ell - z)}$$

$$= V_1 \exp\left[\frac{\sqrt{3}}{2}\beta_e C(\ell - z)\right] \exp\left[-j\beta_e \left(1 - \frac{C}{2}\right)(\ell - z)\right] \tag{7-7-12}$$

The input voltage is located at $z = \ell$, which means

$$V(\ell) = V_1 \tag{7-7-13}$$

The output voltage is located at $z = 0$—that is,

$$V(0) = V_1 \exp\left[\frac{\sqrt{3}}{2}\beta_e \ell C\right] \exp\left[-j\beta_e \ell \left(1 - \frac{C}{2}\right)\right] \tag{7-7-14}$$

The factor $\beta_e \ell$ is conventionally written as $2\pi N$, where N is the circuit length in electronic wavelength, that is,

$$N = \frac{\ell}{\lambda_e} \quad \text{and} \quad \beta_e \ell = \frac{2\pi \ell}{\lambda_e} = 2\pi N$$

Then the amplitude of the output voltage is then given by

$$V(0) = V_1 \exp\left[\frac{\sqrt{3}}{2} 2\pi NC\right] \exp\left[-j2\pi N\left(1 - \frac{C}{2}\right)\right] \quad (7\text{-}7\text{-}15)$$

The output power gain is then expressed as

$$\text{Power gain} = 20 \log \left|\frac{V(0)}{V(\ell)}\right|$$

$$= 20 \log \left|\exp\left[\frac{\sqrt{3}}{2} 2\pi NC\right] \exp\left[-j2\pi N\left(1 - \frac{C}{2}\right)\right]\right| \quad (7\text{-}7\text{-}16)$$

$$= -9.54 \text{ dB} + 47.3NC \quad \text{dB}$$

where 9.54 dB is the initial power loss at the input circuit due to the fact that the input voltage splits into three waves of equal magnitude and the growing-wave voltage is only one-third the total input voltage.

Example 7-7-1:   Wave Modes in a BWA

A O-type backward-wave amplifier (BWA) has the following parameters:

Beam voltage	$V_0 = 6 \text{ kV}$
Beam current	$I_0 = 1 \text{ A}$
Operating frequency	$f = 4 \text{ GHz}$
Length	$N = 15$ in wavelength
Characteristic impedance	$Z_0 = 20 \text{ }\Omega$

Determine:

(a) The dc electron velocity

(b) The phase constant of the electron beam

(c) The gain parameter

(d) The output power gain

(e) The propagation constants

(f) The backward-wave equations in exponential form

Solution:

(a) The dc electron velocity is

$$\mathcal{V}_0 = 0.593 \times 10^6 \times (6 \times 10^3)^{1/2} = 4.59 \times 10^7 \text{ m/s}$$

(b) The phase constant of the electron beam is

$$\beta_e = 2\pi \times 4 \times 10^9/(4.59 \times 10^7) = 547.56 \text{ rad/m}$$

(c) The gain parameter is

$$C = \left(\frac{1 \times 20}{4 \times 6 \times 10^3}\right)^{1/3} = 0.094$$

(d) The output power gain is

$$A_p = -9.54 + 47.3 \times 15 \times 0.094 = 57.15 \text{ dB}$$

(e) The propagation constants are

$$\gamma_1 = -547.56 \times 0.094 \times 0.866 + j547.56\left(1 - \frac{0.094}{2}\right)$$

$$= -44.57 + j521.82$$

$$\gamma_2 = +44.57 + j521.82$$

$$\gamma_3 = j547.56\,(1 + 0.094) = +j599.00$$

$$\gamma_4 = -j547.56\left(1 + \frac{0.094^3}{4}\right) = -j547.56$$

(f) The backward-wave equations for negative z direction are
  (1) Backward wave with decreasing amplitude is

$$E_1 \exp(j\omega t + \gamma_1 z) = E_1 \exp(-44.57z)\exp(j\omega t + j521.82z)$$

  (2) Backward wave with increasing amplitude is

$$E_1 \exp(j\omega t + \gamma_2 z) = E_1 \exp(44.57z)\exp(j\omega t + j521.82z)$$

  (3) Backward wave with constant amplitude is

$$\exp(j\omega t + \gamma_3 z) = \exp(j\omega t + j599.0z)$$

  (4) Forward wave with constant amplitude is

$$\exp(j\omega t + \gamma_4 z) = \exp(j\omega t - j547.56z)$$

7-8 BACKWARD-WAVE OSCILLATOR (BWO)

The most commonly used slow-wave structure for backward-wave oscillator (BWO) is the helix. Interaction takes place at $n = -1$ space harmonic as shown in the ω-β diagram (Brillouin diagram) of Fig. 4-4-5. When n is negative the phase velocity is negative. This means that the electron beam moves in the positive z direction, while the velocity coincides with the negative space harmonic's phase velocity. As a result, the microwave signal moves in the negative z direction. This type of tube is called a *backward-wave oscillator*. The frequency of oscillation is approximately given by the intersection of the voltage line with the -1 space harmonic. Figure 7-8-1 shows a schematic diagram of a SE 201 backward-wave oscillator.

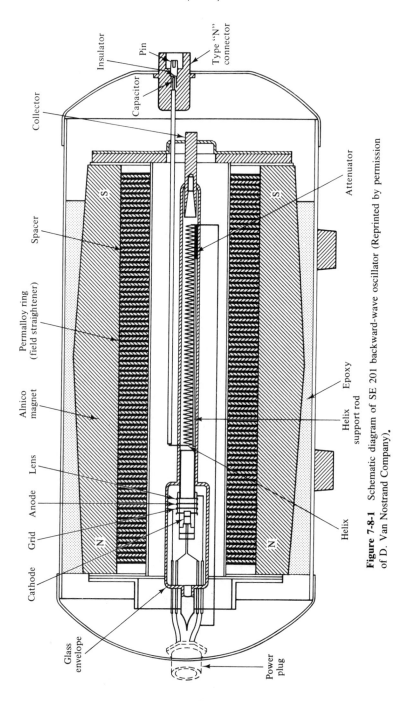

Figure 7-8-1 Schematic diagram of SE 201 backward-wave oscillator (Reprinted by permission of D. Van Nostrand Company).

At the highest frequency the tube has an efficiency of about 3%. The output power is about 10 mW over the frequency range from 7 to 12 GHz. However, the major characteristic of the tube is its relative strength of the desired output signal as compared with all other spurious frequencies. The desired signal in this tube is at least 60 dB larger than the total power in all spurious signals. The traveling-wave backward-wave oscillator in general produces extremely clean output signals.

The term *backward* actually means that the wave group velocity is directed opposite to the electron-beam velocity. If the positive z direction is assumed to be from the left to right, the electron beam flows from the left to right and acts as a generator at each point along the circuit. The power delivered by the beam at each such point is divided into two waves, which proceed respectively to the left and to the right. If the beam current I_0 and the circuit length N is so adjusted that the output voltage at the right end is zero, the output power gain becomes infinite, and the backward wave is starting oscillation. Thus the oscillation condition of a backward-wave oscillator (BWO) is given by

$$\frac{V(\ell)}{V(0)} = 0 \qquad\qquad (7\text{-}8\text{-}1)$$

Since the output voltage is an exponential function of the propagation constant γ_n and the circuit length N, the propagation constant γ_n is a function of the integration factor δ_n, and the gain parameter C is a function of the beam current I_0, the solutions of Eq. (7-8-1) are series of multiple values of NC. Therefore, Eq. (7-8-1) represents higher-order oscillations in a particular backward-wave tube, if the beam current is increased sufficiently far above the starting current.

The Varian VA-173Y as shown in Fig. 7-8-2 is a typical linear-beam backward-wave oscillator and has a minimum output power of 40 mW at frequency range from 7.0 to 12.4 GHz. The maximum anode voltage is 100 V dc, and the anode maximum anode current is 10 mA dc.

VA-173Y

Figure 7-8-2 Photograph of Varian BWO VA-173Y (Courtesy of Varian Associates, Inc.).

REFERENCES

[1] KOMPFNER, R., The Traveling-Wave Tube as Amplifier. *Proc. IRE,* 35, Feb. 1947, pp. 124–27.

[2] MENDEL, J. T., Helix and Coupled-Cavity Traveling-Wave Tubes. *Proc. IEEE,* 61, No. 3, Mar. 1973, pp. 280–98.

[3] CHODOROW, M., and CRAIG, R. A., Some New Circuits for High-Power Traveling-Wave Tubes. *Proc. IRE,* 45, Aug. 1957, pp. 1106–18.

[4] ROUMBANIS, T., et al., A Megawatt X-Band TWT Amplifier with 18% Bandwidth. *Proc. High-Power Microwave Tube Symposium,* vol. 1, The Hexagon, Fort Monmouth, N.J., September 25–26, 1962.

[5] RUETZ, A. J., and YOCOM, W. H., High-Power Traveling-Wave Tubes for Radar Systems. *IRE Trans. Mil. Electron. MIL-5,* Apr. 1961, pp. 39–45.

[6] STAPRANS, A., et al., High-Power Linear-Beam Tubes. *Proc. IEEE,* 61, No. 3, Mar. 1973, pp. 299–330.

[7] ROUMBANIS, T., Centipede Twystron Amplifier and Traveling-Wave Tubes for Broadband, High-Efficiency, Super-Power Amplification. *Proc. 7th Int. Conf. on Microwave and Optical Generation and Amplification,* Hamburg, September 16–20, 1968.

[8] LIAO, SAMUEL Y., The Effect of Collector Voltage Overdepression on Tube Performance of the Gridded Traveling-Wave Tubes. Report for Hughes Aircraft Company, El Segundo, Calif., August 1977.

[9] LaRUE, A. D., and RUBERT, R. R., Multi-Megawatt Hybrid TWTs at S-band and C-band. Presented to the IEEE Electron Devices Meeting, Washington, D.C., October 1964.

PROBLEMS

Helix Traveling-Wave-Tube Amplifiers (TWAs)

7-1-1. A CW helix traveling-wave-tube amplifier has the following parameters:

Beam voltage	$V_0 = 9$ kV
Operating frequency	$f = 8$ GHz
Helical-coil diameter	$d = 1.5$ cm

    (a) Compute the dc electron velocity.
    (b) Calculate the helical pitch for proper synchronism between the beam velocity and RF signal axial velocity.

7-1-2. A high-power CW helix traveling-wave amplifier has the following parameters:

Beam voltage	$V_0 = 10$ kV
Beam diameter	$d = 3$ mm
dc beam charge density	$\rho_0 = 2.5 \times 10^{-3}$ C/m³
Operating frequency	$f = 10$ GHz
Axial electric field	$E_1 = 35$ kV/m
Velocity perturbation	$\mathcal{V}_1 = 2 \times 10^6$ m/s
Charge density perturbation	$\rho_1 = 10^{-4}$ C/m³

Calculate:

(a) The dc electron velocity

(b) The phase constant of the dc electron beam

(c) The propagation constant for $\alpha_e = -50$

(d) The dc beam current density

(e) The beam current

(f) The current density perturbation

(g) The convection current

7-1-3. A pulsed-helix TWA has the following parameters:

Beam voltage	$V_0 = 13$ kV
Beam current	$I_0 = 1.8$ A
Operating frequency	$f = 9$ GHz
Characteristic impedance	$Z_0 = 10\ \Omega$

Determine:

(a) The dc electron velocity

(b) The phase constant of the electron beam

(c) The gain parameter

(d) The four propagation constants

(e) The four wave equations in exponential form in $+z$ direction

7-1-4. A pulsed-helix TWA has the following parameters:

Beam voltage	$V_0 = 15$ kV
Beam current	$I_0 = 3$ A
Length	$N = 10$ in wavelength
Characteristic impedance	$Z_0 = 50\ \Omega$

(a) Compute the gain parameter.

(b) Calculate the output power gain.

7-1-5. A multicoupled-cavity TWA has the following parameters:

Beam voltage	$V_0 = 33$ kV
Electron-gun perveance	$K = 1.42 \times 10^{-6}$ A/V$^{3/2}$

(a) Determine the beam current I_0.

(b) Find the beam power P_0.

7-1-6. A TWT operates under the following parameters:

Beam current	$I_0 = 50$ mA
Beam voltage	$V_0 = 2.5$ kV
Characteristic impedance of helix	$Z_0 = 6.75$ ohms
Circuit length	$N = 100$
Frequency	$f = 8$ GHz

Determine:

   (a) The gain parameter C

   (b) The output power gain A_p in dB

   (c) All four propagation constants

   (d) The wave equations for all four modes in exponential form in $+z$ direction

7-1-7. An O-type traveling-wave tube operates at 2 GHz. The slow-wave structure has a pitch angle of 5.7°. Determine the propagation constant of the traveling wave in the tube. It is assumed that the tube is lossless.

7-1-8. An O-type helix traveling-wave tube operates at 8 GHz. The slow-wave structure has a pitch angle of 4.4° and an attenuation constant of 2 nepers per meter. Determine the propagation constant γ of the traveling wave in the tube.

7-1-9. In an O-type traveling-wave tube, the acceleration voltage (beam voltage) is 3000 V. The characteristic impedance is 10 Ω. The operating frequency is 10 GHz and the beam current is 20 mA. Determine the propagation constants of the four modes of the traveling waves.

7-1-10. Describe the structure of an O-type traveling-wave tube (TWT) and its characteristics; then explain how it works.

7-1-11. In an O-type traveling-wave tube, the acceleration voltage is 4000 V and the magnitude of the axial electric field is 4 V/m. The phase velocity on the slow-wave structure is 1.10 times the average electron-beam velocity. The operating frequency is 2 GHz. Determine the magnitude of velocity fluctuation.

7-1-12. A traveling-wave tube (TWT) has the following characteristics:

Beam voltage	$V_0 = 2$ kV
Beam current	$I_0 = 4$ mA
Frequency	$f = 8$ GHz
Circuit length	$N = 50$ in wavelength
Characteristic impedance	$Z_0 = 20$ ohms

Determine:

   (a) The gain parameter C

   (b) The power gain in dB

Gridded TWAs

7-3-1. A GTWA has the following parameters:

Cathode voltage	$V_0 = -34$ kV
Cathode current	$I_0 = 8$ A
Output power	$P_{\text{out}} = 70$ kW
Collector voltage	$V_c = -17$ kV

Compute:

   (a) The electronic efficiency

   (b) The overall efficiency

7-3-2. A gridded TWA is operated under overdepression of collector voltage as follows:

Overdepression voltage	V_c	$= -10$ kV
Returned current	I_r	$= 0.82$ A
Mass of heated iron pole piece	Mass	$= 200$ milligrams
Specific heat H of iron at 20°C	H	$= 0.108$ calories/gram °C

Determine:

(a) The number of electrons returned per second

(b) The energy in eV associated with these returning electrons in 20 ms

(c) The power in watts for the returning electrons

(d) The heat in calories associated with the returning electrons (a factor for converting joules to calories is 0.238)

(e) The temperature T in degrees centigrade for the output iron pole piece [Hint: $T = 0.238VIt/(\text{mass} \times \text{specific heat.})$]

(f) Whether the output iron pole piece is melted (its melting temperature is 1535°C)

7-3-3. The current I_r caused by the returning electrons at an over depression voltage of -11.5 kV in a GTWT is about 0.973 A as shown by the spent-beam curve in Fig. 7-3-3.

(a) Calculate the number of electrons returned per second.

(b) Determine the energy in eV associated with these returning electrons in 1 ms for part (a).

(c) Find the power in watts for the returning electrons.

7-3-4. The output iron pole piece of a GTWT has the following parameters:

Specific heat H at 20°C is 0.108 calories/gram °C.
Factor for converting joules to calories is 0.238.
Mass of the heated iron pole piece is assumed to be 203.05 milligrams.
Duration time t of the collector depression transient voltage of -11.5 kV is 15 ms.
Melting point of iron is 1535°C.

(a) Calculate the heat in calories associated with the returning electrons at an overdepression voltage of -11.5 kV.

(b) Compute the temperature T in degrees centigrade for the output iron pole piece [Hint: $T = 0.238VIt/(\text{mass} \times \text{specific heat}).$]

(c) Determine whether the output iron pole piece is melted.

7-3-5. The efficiency of a gridded traveling-wave tube (GTWT) is expressed as

$$\eta = \frac{\text{RF } P_{ac}}{P_{dc}} = \frac{\text{RF } P_{ac}}{V_0 I_0}$$

If the cathode voltage is -18 kV and the collector voltage is depressed to -7.5 kV, determine the efficiency of the GTWT.

O-Type Backward-Wave Amplifiers (BWA)

7-7-1. A O-type backward-wave amplifier (BWA) has the following parameters:

Beam voltage	$V_0 = 8$ kV

Beam current	I_0	= 1.2 A
Operating frequency	f	= 8 GHz
Length	N	= 10 in wavelength
Characteristic impedance	Z_0	= 50 Ω

Compute:

(a) The dc electron velocity

(b) The phase constant of the electron beam

(c) The gain parameter

(d) The output power gain

(e) The propagation constants

(f) The backward-wave equations

CHAPTER 8

Microwave Crossed-Field Electron Tubes

8-0 INTRODUCTION

In previous chapters, several commonly used linear-beam tubes were described in detail. In these tubes, the dc magnetic field that is in parallel with the dc electric field is used merely to focus the electron beam. In crossed-field devices, however, the dc magnetic field and the dc electric field are perpendicular to each other. In all crossed-field tubes, the dc magnetic field plays a direct role in the RF interaction process.

Crossed-field tubes derive their name from the fact that the dc electric field and the dc magnetic field are perpendicular to each other. They are also called *M*-type tubes after the French TPOM (*tubes à propagation des ondes à champs magnétique:* tubes for propagation of waves in a magnetic field). In a crossed-field tube, the electrons emitted by the cathode are accelerated by the electric field and gain velocity, but the greater their velocity, the more their path is bent by the magnetic field. If an RF field is applied to the anode circuit, those electrons entering the circuit during the retarding field are decelerated and give up some of their energy to the RF field. Consequently, their velocity is decreased, and these slower electrons will then travel the dc electric field far enough to regain essentially the same velocity as before. Due to the crossed-field interactions, only those electrons that have given up sufficient energy to the RF field can travel all the way to the anode. This phenomenon would make the *M*-type devices relatively efficient. Those electrons entering the circuit during the accelerating field are accelerated by means of receiving enough energy from the RF field and returned back toward the cathode. This back bombardment of the cathode produces heat in the cathode and decreases the operational efficiency.

In this chapter, several commonly used crossed-field tubes such as magnetrons, forward-wave crossed-field amplifiers (FWCFAs), backward-wave crossed-field amplifiers (BWCFAs or amplitrons), and backward-wave crossed-field oscillators (BWCFOs or carcinotrons) are studied.

Cylindrical magnetron: The cylindrical magnetron was developed by Boot and Randall in early 1940.

Coaxial magnetron: The coaxial magnetron introduced the principle of integrating a stabilizing cavity into the magnetron geometry.

Voltage-tunable magnetron: The voltage-tunable magnetron has the cathode-anode geometry of the conventional magnetron, but its anode can be tuned easily.

Inverted magnetron: The inverted magnetron has the inverted geometry of the conventional magnetron with the cathode placed on the outside surrounding the anode and microwave circuit.

Forward-wave crossed-field amplifier (FWCFA): The forward-wave crossed-field amplifier is also called the *M*-type forward-wave amplifier.

Backward-wave crossed-field amplifier (BWCFA): The backward-wave crossed-field was developed by Raytheon Company in 1960 and its tradename is "amplitron." It is a broadband, high-power, high-gain and high-efficiency microwave tube, and it has many applications such as in airborne radar systems and spaceborne communications systems.

Backward-wave crossed-field oscillator (BWCFO): In 1960, the French Company developed a new type of crossed-field device in which an injection gun replaced the conventional cylindrical cathode of the magnetron and its tradename is "carcinotron." It is also called the *M-type backward-wave oscillator.*

All of these crossed-field electron tubes are tabulated in Table 8-0-1.

TABLE 8-0-1.   CROSSED-FIELD ELECTRON TUBES

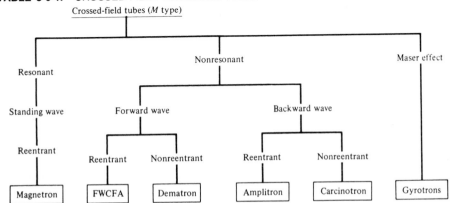

8-1 MAGNETRON OSCILLATORS

After Hull invented the magnetron in 1921 [1], it remained an interesting laboratory device until about 1940. During World War II, an urgent need for high-power microwave generators for radar transmitters led to a very rapid development of the magnetron to its present state.

All magnetrons consist of some form of anode and cathode operated in a dc magnetic field normal to a dc electric field between the cathode and anode. Due to the crossed-field between the cathode and anode, the electrons emitted from the cathode are influenced by the crossed-field to move in curved paths. If the dc magnetic field is strong enough, the electrons will not arrive in the anode, but return instead to the cathode. Consequently, the anode current is cut off. Magnetrons can be classified into three types:

1. *Split anode magnetron:* This type of magnetrons makes use of a static negative resistance between two anode segments.
2. *Cyclotron-frequency magnetrons:* This type of magnetrons operate under the influence of synchronism between an alternating component of electric field and a periodic oscillation of electrons in a direction parallel to the field.
3. *Traveling-wave magnetrons:* This type of magnetrons depends upon the interaction of electrons with a traveling electromagnetic field of linear velocity. They are customarily referred to as *magnetrons*.

Negative-resistance magnetrons ordinarily operate at frequencies below the microwave region. Although cyclotron-frequency magnetrons operate at frequencies in microwave range, their power output is very small (about 1 W at 3 GHz), and their efficiency is very low (about 10% in the split-anode type and 1% in the single-anode type). Thus, the first two types of magnetrons are not considered in this text. In this section, only the traveling-wave magnetrons such as cylindrical magnetron, linear (or planar) magnetron, coaxial magnetron, voltage-tunable magnetron, inverted coaxial magnetron, and frequency-agile magnetron will be discussed.

8-1-1 Cylindrical Magnetron

A schematic diagram of a cylindrical magnetron oscillator is shown in Fig. 8-1-1. This type of magnetron is also called a *conventional magnetron.*

In a cylindrical magnetron, several reentrant cavities are connected to the gaps. The dc voltage V_0 is applied between the cathode and the anode. The magnetic flux density B_0 is in the positive z direction. When the dc voltage and the magnetic flux are

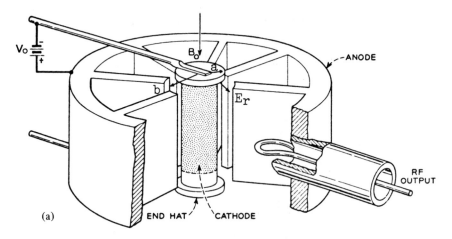

Figure 8-1-1   Schematic diagram of a cylindrical magnetron.

adjusted properly, the electrons will follow cycloidal paths in the cathode-anode space under the combined force of both electric and magnetic fields as shown in Fig. 8-1-2.

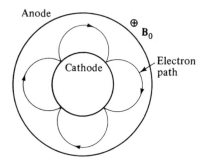

Figure 8-1-2   Electron path in a cylindrical magnetron.

Equations of electron motion.   The equations of motion for electrons in a cylindrical magnetron can be written with the aid of Eqs. (2-3-3a) and (2-3-3b) as

$$\frac{d^2r}{dt^2} - r\left(\frac{d\phi}{dt}\right)^2 = \frac{e}{m}E_r - \frac{e}{m}rB_z\frac{d\phi}{dt} \tag{8-1-1}$$

$$\frac{1}{r}\frac{d}{dt}\left(r^2\frac{d\phi}{dt}\right) = \frac{e}{m}B_z\frac{d\phi}{dt} \tag{8-1-2}$$

where $\frac{e}{m} = 1.759 \times 10^{11}$ C/kg is the charge-to-mass ratio of electron and $B_0 = B_z$ is assumed in the positive z direction.

Rearrangement of Eq. (8-1-2) results in the following form

$$\frac{d}{dt}\left(r^2\frac{d\phi}{dt}\right) = \frac{e}{m}B_z r\frac{dr}{dt} = \frac{1}{2}\omega_c\frac{d}{dt}(r^2) \tag{8-1-3}$$

where $\omega_c = \dfrac{e}{m}B_z$ is the cyclotron angular frequency. Integration of Eq. (8-1-3) yields

$$r^2\frac{d\phi}{dt} = \frac{1}{2}\omega_c r^2 + \text{constant} \tag{8-1-4}$$

at $r = a$, where a is the radius of the cathode cylinder, and $\dfrac{d\phi}{dt} = 0$, constant $= -\dfrac{1}{2}\omega_c a^2$. The angular velocity is expressed by

$$\frac{d\phi}{dt} = \frac{1}{2}\omega_c\left(1 - \frac{a^2}{r^2}\right) \tag{8-1-5}$$

Since the magnetic field does no work on the electrons, the kinetic energy of electron is given by

$$\frac{1}{2}m\mathcal{V}^2 = eV \tag{8-1-6}$$

However, the electron velocity has r and ϕ components such as

$$\mathcal{V}^2 = \frac{2e}{m}V = \mathcal{V}_r^2 + \mathcal{V}_\phi^2 = \left(\frac{dr}{dt}\right)^2 + \left(r\frac{d\phi}{dt}\right)^2 \tag{8-1-7}$$

at $r = b$, where b is the radius from the center of the cathode to the edge of the anode, $V = V_0$, and $dr/dt = 0$, when the electrons just graze the anode, Eqs. (8-1-5) and (8-1-7) become

$$\frac{d\phi}{dt} = \frac{1}{2}\omega_c\left(1 - \frac{a^2}{b^2}\right) \tag{8-1-8}$$

$$b^2\left(\frac{d\phi}{dt}\right)^2 = \frac{2e}{m}V_0 \tag{8-1-9}$$

Substitution of Eq. (8-1-8) into Eq. (8-1-9) results in

$$b^2\left[\frac{1}{2}\omega_c\left(1 - \frac{a^2}{b^2}\right)\right]^2 = \frac{2e}{m}V_0 \tag{8-1-10}$$

The electron will acquire a tangential as well as a radial velocity. Whether the electron will just graze at the anode and return back toward the cathode depends upon the relative magnitudes of V_0 and B_0. The *Hull cutoff magnetic equation* is obtained from Eq.

(8-1-10) as

$$B_{0c} = \frac{\left(8V_0 \frac{m}{e}\right)^{1/2}}{b\left(1 - \frac{a^2}{b^2}\right)}$$

(8-1-11)

This means that if $B_0 > B_{0c}$ for a given V_0, the electrons will not reach the anode. Conversely, the cutoff voltage is given by

$$V_{0c} = \frac{e}{8m} B_0^2 b^2 \left(1 - \frac{a^2}{b^2}\right)^2$$

(8-1-12)

This means that if $V_0 < V_{0c}$ for a given B_0, the electrons will not reach the anode. Equation (8-1-12) is often called the *Hull cutoff voltage equation*.

Example 8-1-1:   Conventional Magnetron

A X-band pulsed cylindrical magnetron has the following operating parameters:

Anode voltage	V_0	= 26 kV
Beam current	I_0	= 27 A
Magnetic flux density	B_0	= 0.336 Wb/m^2
Radius of cathode cylinder	a	= 5 cm
Radius of vane edge to center	b	= 10 cm

Compute:

(a) The cyclotron angular frequency

(b) The cutoff voltage for a fixed B_0

(c) The cutoff magnetic flux density for a fixed V_0

Solution:

(a) The cyclotron angular frequency is

$$\omega_c = \frac{e}{m} B_0 = 1.759 \times 10^{11} \times 0.336 = 5.91 \times 10^{10} \text{ rad}$$

(b) The cutoff voltage for a fixed B_0 is

$$V_{0c} = \frac{1}{8} \times 1.759 \times 10^{11} (0.336)^2 (10 \times 10^{-2})^2 \left(1 - \frac{5^2}{10^2}\right)^2$$

$$= 139.50 \text{ kV}$$

(c) The cutoff magnetic flux density for a fixed V_0 is

$$B_{0c} = \left(8 \times 26 \times 10^3 \times \frac{1}{1.759 \times 10^{11}}\right)^{1/2} \left[10 \times 10^{-2} \left(1 - \frac{5^2}{10^2}\right)\right]^{-1}$$

$$= 14.495 \text{ mWb/m}^2$$

Cyclotron angular frequency.   Since the magnetic field is normal to the motion of electrons that travel in a cycloidal path, the outward centrifugal force is equal to the pulling force. Hence

$$\frac{m\mathcal{V}^2}{R} = e\mathcal{V}B$$

(8-1-13)

where R = radius of the cycloidal path
      $\mathcal{V}$ = tangential velocity of the electron

The cyclotron angular frequency of the circular motion of the electron is then given by

$$\omega_c = \frac{\mathcal{V}}{R} = \frac{eB}{m}$$

(8-1-14)

The period for one complete revolution can be expressed as

$$T = \frac{2\pi}{\omega} = \frac{2\pi m}{eB}$$

(8-1-15)

Since the slow-wave structure is closed on itself, or "reentrant," oscillations are possible only if the total phase shift around the structure is an integral multiple of 2π radius. Thus, if there are N reentrant cavities in the anode structure, the phase shift between two adjacent cavities can be expressed as

$$\phi_n = \frac{2\pi n}{N}$$

(8-1-16)

where n is an integer indicating the nth mode of oscillation. In order for oscillations to be produced in the structure, the anode dc voltage must be adjusted so that the average rotational velocity of the electrons corresponds to the phase velocity of the field in the slow-wave structure. Magnetron oscillators are ordinarily operated in the π mode— that is,

$$\phi_n = \pi \qquad (\pi\text{-mode})$$

(8-1-17)

Figure 8-1-3 shows the lines of force in the π-mode of eight-cavity magnetron.

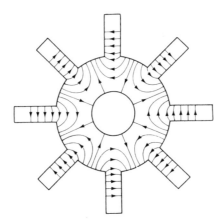

Figure 8-1-3 Lines of force in π mode of eight-cavity magnetron

It is evident that in the π-mode the excitation is largely in the cavities, having opposite phase in successive cavities. The successive rise and fall of adjacent anode-cavity fields may be regarded as a traveling wave along the surface of the slow-wave structure. For the energy to be transferred from the moving electrons to the traveling field, the electrons must be decelerated by a retarding field when they pass through each anode cavity. If L is the mean separation between cavities, the phase constant of the fundamental-mode field is given by

$$\beta_0 = \frac{2\pi n}{NL} \tag{8-1-18}$$

The traveling-wave field of the slow-wave structure may be obtained by solving Maxwell's equations subject to the boundary conditions. The solution for the fundamental ϕ component of the electric field is of the form [1]

$$E_{\phi 0} = jE_1 e^{j(\omega t - \beta_0 \phi)} \tag{8-1-19}$$

where E_1 is a constant and β_0 is given in Eq. (8-1-18). Thus, the traveling field of the fundamental mode will travel around the structure with angular velocity

$$\frac{d\phi}{dt} = \frac{\omega}{\beta_0} \tag{8-1-20}$$

where $\dfrac{d\phi}{dt}$ can be found from Eq. (8-1-19). When the cyclotron frequency of the electrons is equal to angular frequency of the field, the interaction between the field and electron occurs and the energy is transferred—that is,

$$\omega_c = \beta_0 \frac{d\phi}{dt} \tag{8-1-21}$$

Power output and efficiency.   The efficiency and power output of a magnetron depend upon the resonant structure and the dc power supply. Figure 8-1-3(a) shows an equivalent circuit for a resonator of a magnetron.

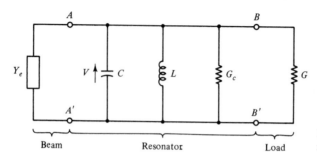

Figure 8-1-3A   Equivalent circuit for one resonator of a magnetron.

where Y_e = electronic admittance
V = RF voltage across the vane tips
C = capacitance at the vane tips
L = inductance of the resonator
G_r = conductance of the resonator
G = load conductance per resonator

Each resonator of the slow-wave structure is taken to comprise a separate resonant circuit as shown in Fig. 8-1-3(a). The unloaded quality factor of the resonator is given by

$$Q_{un} = \frac{\omega_0 C}{G_r} \tag{8-1-21a}$$

where $\omega_0 = 2\pi f_0$ is the angular resonant frequency. The external quality factor of the load circuit is

$$Q_{ex} = \frac{\omega_0 C}{G_\ell} \tag{8-1-21b}$$

Then the loaded Q_ℓ of the resonant circuit is expressed by

$$Q_\ell = \frac{\omega_0 C}{G_r + G_\ell} \tag{8-1-21c}$$

The circuit efficiency is defined as

$$\eta_c = \frac{G_\ell}{G_\ell + G_r}$$
$$= \frac{G_\ell}{G_{ex}} = \frac{1}{1 + Q_{ex}/Q_{un}} \tag{8-1-21d}$$

The maximum circuit efficiency is obtained when the magnetron is heavily loaded, that is, for $G_\ell \gg G_r$. However, heavy loading makes the tube operating quite sensitive to the load, which is undesirable in some cases. Therefore, the ratio of Q_ℓ/Q_{ex} chosen is often a compromise between the conflicting requirements for high circuit efficiency and frequency stability.

The electronic efficiency is defined as

$$\eta_e = \frac{P_{gen}}{P_{dc}} = \frac{V_0 I_0 - P_{lost}}{V_0 I_0} \tag{8-1-21e}$$

where P_{gen} = RF power induced into the anode circuit
$\quad P_{dc} = V_0 I_0$ power from the dc power supply
$\quad V_0$ = anode voltage
$\quad I_0$ = anode current
$\quad P_{lost}$ = power lost in the anode circuit

The RF power generated by the electrons can be written as

$$
\begin{aligned}
P_{gen} &= V_0 I_0 - P_{lost} \\
&= V_0 I_0 - I_0 \frac{m}{2e} \frac{\omega_0^2}{\beta^2} + \frac{E_{max}^2}{B_z^2} \\
&= \frac{1}{2} N |V|^2 \frac{\omega_0 C}{Q_\ell}
\end{aligned}
\tag{8-1-21f}
$$

where $\quad N$ = total number of resonators
$\quad\quad V$ = RF voltage across the resonator gap
$\quad E_{max} = M_1 |V|/L$ is the maximum electric field

$\quad M_1 = \sin\left(\beta_n \dfrac{\delta}{2}\right) \Big/ \left(\beta_n \dfrac{\delta}{2}\right) \simeq 1$ for small δ is the gap factor for the π-mode operation

$\quad\quad \beta$ = phase constant
$\quad\quad B_z$ = magnetic flux density
$\quad\quad L$ = center-to-center spacing of the vane tips

The power generated may be simplified to

$$P_{gen} = \frac{NL^2 \omega_0 C}{2M_1^2 Q_\ell} E_{max}^2 \tag{8-1-21g}$$

The electronic efficiency may be rewritten as

$$\eta_e = \frac{P_{gen}}{V_0 I_0} = \frac{1 - \dfrac{m\omega_0^2}{2eV_0\beta^2}}{1 + \dfrac{I_0 m M_1^2 Q_\ell}{B_z e\, NL^2 \omega_0 C}} \tag{8-1-21h}$$

Example 8-1-1A: Pulsed Magnetron

A X-band pulsed conventional magnetron has the following operating parameters:

Anode voltage	V_0	$= 5.5 \text{ kV}$
Beam current	I_0	$= 4.5 \text{ A}$
Operating frequency	f	$= 9 \times 10^9 \text{ Hz}$
Resonator conductance	G_r	$= 2 \times 10^{-4} \text{ mho}$
Loaded conductance	G_ℓ	$= 2.5 \times 10^{-5} \text{ mho}$
Vane capacitance	C	$= 2.5 \text{ pF}$
Duty cycle	DC	$= 0.002$
Power loss	P_{loss}	$= 18.50 \text{ kW}$

Compute:

(a) The angular resonant frequency

(b) The unloaded quality factor

(c) The loaded quality factor

(d) The external quality factor

(e) The circuit efficiency

(f) The electronic efficiency

Solution:

(a) The angular resonant frequency is

$$\omega_r = 2 \times 9 \times 10^9 = 56.55 \times 10^9 \text{ rad}$$

(b) The unloaded quality factor is

$$Q_{\text{un}} = \frac{56.55 \times 10^9 \times 2.5 \times 10^{-12}}{2 \times 10^{-4}} = 707$$

(c) The loaded quality factor is

$$Q_\ell = \frac{56.55 \times 10^9 \times 2.5 \times 10^{-12}}{2 \times 10^{-4} + 2.5 \times 10^{-5}} = 628$$

(d) The external quality factor is

$$Q_{\text{ex}} = \frac{56.55 \times 10^9 \times 2.5 \times 10^{-12}}{2.5 \times 10^{-5}} = 5655$$

(e) The circuit efficiency is

$$\eta_c = \frac{1}{1 + 5655/707} = 11.11\%$$

(f) The electronic efficiency is

$$\eta_e = \frac{5.5 \times 10^3 \times 4.5 - 18.5 \times 10^3}{5.5 \times 10^3 \times 4.5} = 25.25\%$$

State of the art.   For many years, magnetrons have been the high-power sources
in operating frequencies as high as 70 GHz. Military radar relies upon conventional
traveling-wave magnetrons to generate high-peak-power RF pulses. No other microwave
devices could perform the same function with the same size, weight, voltage, and effi-
ciency-range advantage as the conventional magnetrons do. At the present state of the
art, a magnetron can deliver a peak power output of up to 40 MW with the dc voltage
in the order of 50 kV at the frequency of 10 GHz. The average power outputs are up to
800 kW. Its efficiency is very high, ranging from 40 to 70%. Figure 8-1-4 shows the
state of the art for U.S. high-power magnetrons. The beacon magnetrons are miniature
conventional magnetrons which deliver peak outputs as high as 3.5 kW, yet weigh less
than 2 lb. These devices are ideal for use where a very-compact, low-voltage source of
pulsed power is required such as in airborne, missile, satellite, or Doppler systems. Most
of the beacon magnetrons exhibit negligible frequency shift and provide long-life performance under the most severe environmental and temperature conditions.

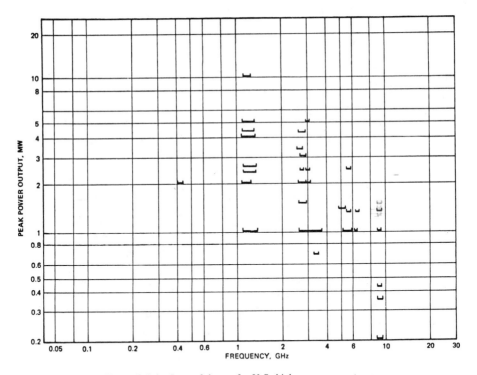

Figure 8-1-4   State of the art for U.S. high-power magnetrons.

The Litton L-5080 pulse magnetron as shown in Fig. 8-1-5 is a typical vane-strap
magnetron oscillator and has a maximum peak output power of 250 kW at frequency
range from 5.45 to 5.825 GHz. Its duty cycle is 0.0012.

Figure 8-1-5   Photograph of Litton L-5080 magnetron (Courtesy of Litton Electron Tube Division).

8-1-2 Linear Magnetron

The schematic diagram of a linear magnetron is shown in Fig. 8-1-5a.

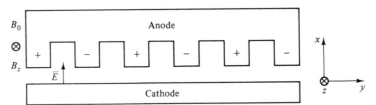

Figure 8-1-5a   Schematic diagram of a linear magnetron.

In the linear magnetron as shown in Fig. 8-1-5a, the electric field E_x is assumed in the $+x$ direction and the magnetic flux density B_z in the $+z$ direction. The differential equations of motion of electrons in the crossed electric and magnetic fields can be written from Eqs. (2-3-2) as

$$\frac{d^2x}{dt^2} = -\frac{e}{m}\left(E_x + B_z\frac{dy}{dt}\right) \qquad (8\text{-}1\text{-}22)$$

$$\frac{d^2y}{dt^2} = \frac{e}{m}B_z\frac{dx}{dt} \qquad (8\text{-}1\text{-}23)$$

$$\frac{d^2z}{dt^2} = 0 \qquad (8\text{-}1\text{-}24)$$

where $\dfrac{e}{m} = 1.759 \times 10^{11}$ C/kg is the charge-to-mass of an electron

B_z = magnetic flux density in $+z$ direction
E_x = electric field in $+x$ direction

In general, the presence of space charges causes the field to be a nonlinear function of the distance x, and the complete solution of Eqs. (8-1-22) through (8-1-24) is not simple. Equation (8-1-23), however, can be integrated directly. Under the assumption that the electrons emit from the cathode surface with zero initial velocity, and if origin is taken to be on the surface, Eq. (8-1-23) becomes

$$\frac{dy}{dt} = \frac{e}{m} B_z x \qquad (8\text{-}1\text{-}25)$$

Equation (8-1-25) shows that, regardless of space charges, the electron velocity parallel to the electrode surface is proportional to the distance of the electron from the cathode and to the magnetic flux density B_z. How far the electron moves from the cathode depends upon B_z and upon the manner in which the potential V varies with x, that in turn depends upon space-charge distribution, anode potential, and electrode spacing.

If the space-charge is assumed to be negligible, the cathode potential zero, and the anode potential V_0, the differential electric field becomes

$$\frac{dV}{dx} = \frac{V_0}{d} \qquad (8\text{-}1\text{-}26)$$

where V_0 = anode potential in volts
d = distance between cathode and anode in meters

Substitution of Eq. (8-1-26) into Eq. (8-1-22) yields

$$\frac{d^2x}{dt^2} = \frac{e}{m}\left(\frac{V_0}{d} - B_z\frac{dy}{dt}\right) \qquad (8\text{-}1\text{-}27)$$

Combination of Eqs. (8-1-25) and (8-1-27) results in

$$\frac{d^2x}{dt^2} + \left(\frac{e}{m}B_z\right)^2 x - \frac{e}{m}\frac{V_0}{d} = 0 \qquad (8\text{-}1\text{-}28)$$

Solution of Eq. (8-1-28) and substitution of the solution into Eq. (8-1-25) yield the following equations for the path of an electron with zero velocity at cathode (origin point) as

$$x = \frac{V_0}{B_z\omega_c d}[1 - \cos(\omega_c t)] \qquad (8\text{-}1\text{-}29)$$

$$y = \frac{V_0}{B_z\omega_c d}[\omega_c t - \sin(\omega_c t)] \qquad (8\text{-}1\text{-}30)$$

$$z = 0 \qquad (8\text{-}1\text{-}31)$$

where $\omega_c = \dfrac{e}{m} B_z$ is the cyclotron angular frequency

$f_c = 2.8 \times 10^6 \, B_z$ is the cyclotron frequency in Hz

Equations (8-1-29) through (8-1-31) are those of a cycloid generated by a point on a circle of radius $V_0/(B_z\omega_c d)$ rolling on the plane of the cathode with angular frequency ω_c. The maximum distance to which the electron moves in a direction normal to the cathode is $2V_0 m/(B_z^2 ed)$. When this distance is just equal to the anode-cathode distance d, the electrons just graze the anode surface and the anode current is just cut off. Then the cutoff condition is

$$\frac{2V_0 m}{B_z^2 ed} = d \qquad (8\text{-}1\text{-}32)$$

Let a constant K equal to

$$K = \frac{d^2 B_z^2}{V_0} = \frac{2m}{e} = 1.14 \times 10^{-11} \qquad (8\text{-}1\text{-}33)$$

When the value of K is less than 1.14×10^{-11}, electrons strike the anode; when the value is larger than 1.14×10^{-11}, they return to the cathode. Figure 8-1-6 shows the electron path.

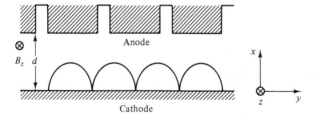

Anode

B_z d

Cathode

x, z, y

Figure 8-1-6 Electron path in a linear magnetron.

From Eq. (8-1-32), the Hull cutoff voltage for a linear magnetron is given by

$$V_{0c} = \frac{1}{2} \frac{e}{m} B_0^2 d^2 \qquad (8\text{-}1\text{-}34)$$

where $B_0 = B_z$ is the magnetic flux density in $+z$ direction. This means that if $V_0 < V_{0c}$ for a given B_0, the electrons will not reach the anode.

Similarly, the Hull cutoff magnetic flux density for a linear magnetron is expressed as

$$B_{0c} = \frac{1}{d} \sqrt{2 \frac{m}{e} V_0} \qquad (8\text{-}1\text{-}35)$$

This means that if $B_0 > B_{0c}$ for a given V_0, the electrons will not reach the anode.

Example 8-1-2:   Linear Magnetron

A linear magnetron has the following operating parameters:

Anode voltage	$V_0 = 10$ kV
Cathode current	$I_0 = 1$ A
Magnetic flux density	$B_0 = 0.01$ Wb/m^2
Distance between cathode and anode	$d = 5$ cm

Compute:

(a) The Hull cutoff voltage for a fixed B_0

(b) The Hull cutoff magnetic flux density for a fixed V_0

Solution:

(a) The Hull cutoff voltage is

$$V_{0c} = \tfrac{1}{2} \times 1.759 \times 10^{11} \times (0.01)^2 \times (5 \times 10^{-2})^2$$
$$= 22.00 \text{ kV}$$

(b) The Hull cutoff magnetic flux density is

$$B_{0c} = \frac{1}{5 \times 10^{-2}} \times \left(\frac{2 \times 10 \times 10^3}{1.759 \times 10^{11}} \right)^{1/2}$$
$$= 6.74 \text{ mWb/m}^2$$

Hartree condition.   The Hull cutoff condition determines the anode voltage or magnetic field necessary to obtain nonzero anode current as a function of the magnetic field or anode voltage in the absence of electromagnetic field. The Hartree condition can be derived as follows as shown in Fig. 8-1-6a.

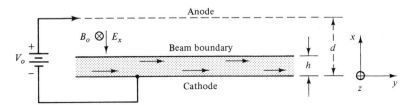

Figure 8-1-6a   Linear model of a magnetron.

The electron beam lies within a region extending a distance h from the cathode, where h is known as the hub thickness. The spacing between the cathode and anode is d. The electron motion is assumed to be in the positive y direction with a velocity

$$\mathcal{V}_y = -\frac{E_x}{B_0} = \frac{1}{B_0}\frac{dV}{dx} \qquad (8\text{-}1\text{-}35\text{a})$$

where $B_0 = B_z$ is the magnetic flux density in the positive z direction

V = potential

From the principle of energy conservation, we have

$$\tfrac{1}{2}m \, \mathcal{V}_y^2 = eV \tag{8-1-35b}$$

Combining Eqs. (8-1-35a) and (8-1-35b) yields

$$\left(\frac{dV}{dx}\right)^2 = \frac{2eV}{m} B_0^2 \tag{8-1-35c}$$

This differential equation may be rearranged as

$$\left(\frac{m}{2eB_0}\right)^{1/2} \frac{dV}{\sqrt{V}} = dx \tag{8-1-35d}$$

Integration of Eq. (8-1-35d) yields the potential within the electron beam as

$$V = \frac{eB_0^2}{2m} x^2 \tag{8-1-35e}$$

where the constant of integration has been eliminated for $V = 0$ at $x = 0$. The potential and electric field at the hub surface are given by

$$V(h) = \frac{e}{2m} B_0^2 h^2 \tag{8-1-35f}$$

and

$$E_x = -\frac{dV}{dx} = -\frac{e}{m} B_0^2 h \tag{8-1-35g}$$

The potential at the anode is thus obtained from Eq. (8-1-35g) as

$$
\begin{aligned}
V_0 &= -\int_0^d E_x \, dx \\
&= -\int_0^h E_x \, dx - \int_h^d E_x \, dx \\
&= V(h) + \frac{e}{m} B_0^2 h(d - h) \\
&= \frac{e}{m} B_0^2 h(d - h/2)
\end{aligned} \tag{8-1-35h}
$$

The electron velocity at the hub surface is obtained from Eqs. (8-1-35a) and (8-1-35g) as

$$\mathcal{V}_y(h) = \frac{e}{m} B_0 h \tag{8-1-35i}$$

For synchronism, this electron velocity is equal to the phase velocity of the slow-wave structure and that is

$$\frac{\omega}{\beta} = \frac{e}{m} B_0 h \tag{8-1-35j}$$

For the π-mode operation, the anode potential is finally given by

$$V_{0h} = \frac{\omega B_0 d}{\beta} - \frac{m}{2e} \frac{\omega^2}{\beta^2} \tag{8-1-35k}$$

This is the Hartree anode voltage equation that is a function of the magnetic flux density and the spacing between the cathode and anode.

Example 8-1-2a: Linear Magnetron

A linear magnetron has the following operating parameters:

Anode voltage	$V_0 = 15$ kV
Cathode current	$I_0 = 1.2$ A
Operating frequency	$f = 8$ GHz
Magnetric flux density	$B_0 = 0.015$ Wb/m^2
Hub thickness	$h = 2.77$ cm
Distance between anode and cathode	$d = 5$ cm

Calculate:

(a) The electron velocity at the hub surface
(b) The phase velocity for synchronism
(c) The Hartree anode voltage

Solution:

(a) The electron velocity is

$$\mathcal{V} = 1.759 \times 10^{11} \times 0.015 \times 2.77 \times 10^{-2}$$
$$= 0.73 \times 10^8 \text{ m/s}$$

(b) The phase velocity is

$$\mathcal{V}_{ph} = \frac{\omega}{\beta} = 0.73 \times 10^8 \text{ m/s}$$

(c) The Hartree anode voltage is

$$V_{0h} = 0.73 \times 10^8 \times 0.015 \times 5 \times 10^{-2}$$

$$- \frac{1}{2 \times 1.759 \times 10^{11}} \times (0.73 \times 10^8)^2$$

$$= 5.475 \times 10^4 - 1.515 \times 10^4$$

$$= 39.60 \text{ kV}$$

8-1-3 Coaxial Magnetron

The coaxial magnetron is composed of an anode resonator structure surrounded by an inner single high-Q cavity operating in the TE_{011} mode as shown in Fig. 8-1-7.

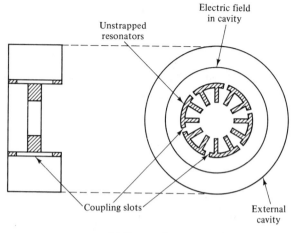

(a) Cross section

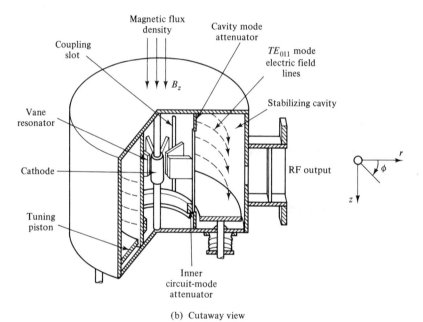

(b) Cutaway view

Figure 8-1-7 Schematic diagram of a coaxial magnetron (Courtesy of Varian Associates, Inc.).

The slots in the back walls of alternate cavities of the anode resonator structure tightly couple the electric fields in these resonators to the surrounding cavity. In the π-mode operation, the electric fields in every other cavity are in phase, and so they couple in the same direction into the surrounding cavity. As a result, the surrounding coaxial cavity stabilizes the magnetron in the desired π-mode operation.

In the desired TE_{011} mode, the electric fields follow a circular path within the cavity and reduce to zero at the walls of the cavity. Current flow in the TE_{011} mode is in the walls of the cavity in circular paths about the axis of the tube. The undesired modes are damped out by the attenuator within the inner slotted cylinder near the ends of the coupling slots. The tuning mechanism is simple and reliable. As the straps are not required, the anode resonator for the coaxial magnetron can be larger and less complex than for the conventional strapped magnetron. Thus cathode loading is lower and voltage gradients are reduced.

The Varian SFD-333TM magnetron as shown in Fig. 8-1-7a is typical X-band coaxial magnetron and has a minimum peak power of 400 kW at frequency range from 8.9 to 9.6 GHz. Its duty cycle is 0.0013. The nominal anode voltage is 32 kV and the peak anode current is 32 A.

Figure 8-1-7a   Photograph of Varian SFD-333TM coaxial magnetron (Courtesy of Varian Associates, Inc.).

8-1-4 Voltage-Tunable Magnetron

The voltage-tunable magnetron is a broadband oscillator with frequency changed by varying the applied voltage between the anode and sole. As shown in Fig. 8-1-8, the electric beam is emitted from a short cylindrical cathode at one end of the device. Electrons are formed into a hollow beam by the electric and magnetic forces near the cathode and then are accelerated radically outward from the cathode. The electron beam is then injected into the region between the sole and the anode; and rotates about the sole at the rate controlled by the axial magnetic field and the dc voltage applied between the anode and sole.

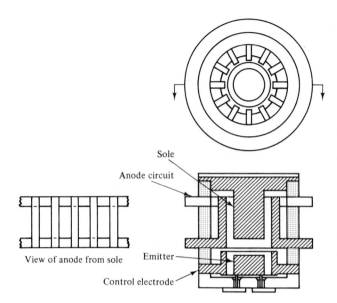

View of anode from sole

Sole

Anode circuit

Emitter

Control electrode

Figure 8-1-8 Cross-section view of a voltage-tunable magnetron (Courtesy of Varian Associates, Inc.)

The voltage-tunable magnetron uses a low-Q resonator and its bandwidth may exceed 50% at low power levels. In the π-mode operation, the bunch process of the hollow beam occurs in the resonator and the frequency of oscillation is determined by the rotational velocity of the electron beam. In other words, the oscillation frequency can be controlled by varying the applied dc voltage between the anode and sole. Power output can be adjusted to some extent through the use of the control electrode in the electron gun. At high power levels and high frequencies, the bandwidth is limited to a few percent. However, at low power levels and low frequencies, the bandwidth may approach 70%.

8-1-5 Inverted Coaxial Magnetron

Magnetron could be built with the anode and cathode inverted—that is, with the cathode surrounding the anode. The basic problem of mode suppression had prevented its use. In an inverted coaxial magnetron, the cavity was located inside a slotted cylinder and a resonator vane array was arranged on the outside. The cathode was built as a ring around the anode. Figure 8-1-9 shows the schematic diagram of an inverted coaxial magnetron.

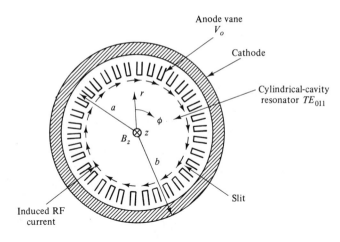

Figure 8-1-9 Schematic diagram of an inverted coaxial magnetron (Courtesy of Varian Associates, Inc.)

Mathematically, the motion equations of the electrons in an inverted coaxial magnetron can be written from Eqs. (2-2-5) as

$$\frac{d^2 r}{dt^2} - r\left(\frac{d\phi}{dt}\right)^2 = \frac{e}{m} E_r - \frac{e}{m} r B_z \frac{d\phi}{dt} \qquad (8\text{-}1\text{-}36)$$

$$\frac{1}{r}\frac{d}{dt}\left(r^2 \frac{d\phi}{dt}\right) = \frac{e}{m} B_z \frac{d\phi}{dt} \qquad (8\text{-}1\text{-}37)$$

where $\dfrac{e}{m} = 1.759 \times 10^{11}$ C/kg is the charge-to-mass ratio of electron

$B_0 = B_z$ is assumed in the positive z direction

Rearrangement of Eq. (8-1-37) results in the following form

$$\frac{d}{dt}\left(r^2 \frac{d\phi}{dt}\right) = \frac{e}{m} B_z r \frac{dr}{dt} = \frac{1}{2}\omega_c \frac{d}{dt}(r^2) \qquad (8\text{-}1\text{-}38)$$

where $\omega_c = \dfrac{e}{m} B_z$ is the cyclotron angular frequency. Integration of Eq. (8-1-38) yields

$$r^2 \frac{d\phi}{dt} = \frac{1}{2}\omega_c r^2 + \text{constant} \qquad (8\text{-}1\text{-}39)$$

at $r = b$, where b is the radius of the cathode cylinder, and $\dfrac{d\phi}{dt} = 0$, constant $= -\dfrac{1}{2}\omega_c b^2$.

The angular velocity is expressed by

$$\frac{d\phi}{dt} = \frac{1}{2}\omega_c\left(1 - \frac{b^2}{r^2}\right) \qquad (8\text{-}1\text{-}40)$$

Since the magnetic field does no work on the electrons, the kinetic energy of electron is given by

$$\tfrac{1}{2}m\mathcal{V}^2 = eV \tag{8-1-41}$$

However, the electron velocity has r and ϕ components such as

$$\mathcal{V}^2 = \frac{2e}{m} V = \mathcal{V}_r^2 + \mathcal{V}_\phi^2 = \left(\frac{dr}{dt}\right)^2 + \left(r\frac{d\phi}{dt}\right)^2 \tag{8-1-42}$$

at $r = a$, where a is the radius from the center of the cylinder to the edge of the anode, $V = V_0$, and $dr/dt = 0$, when the electrons just graze the anode, Eqs. (8-1-40) and (8-1-42) become

$$\frac{d\phi}{dt} = \frac{1}{2}\omega_c\left(1 - \frac{b^2}{a^2}\right) \tag{8-1-43}$$

$$a^2\left(\frac{d\phi}{dt}\right)^2 = \frac{2e}{m} V_0 \tag{8-1-44}$$

Substitution of Eq. (8-1-43) into Eq. (8-1-44) results in

$$a^2\left[\frac{1}{2}\omega_c\left(1 - \frac{b^2}{a^2}\right)\right]^2 = \frac{2e}{m} V_0 \tag{8-1-45}$$

The electron will acquire a tangential as well as a radial velocity. Whether the electron will just graze at the anode and return back toward the cathode depends upon the relative magnitudes of the anode voltage V_0 and the magnetic flux density B_0. The cutoff condition can be obtained from Eq. (8-1-45) as

$$V_{0c} = \frac{e}{8m} B_0^2 a^2\left(1 - \frac{b^2}{a^2}\right)^2 \tag{8-1-46}$$

This means that if $V_0 < V_{0c}$ for a given B_0, the electrons will not reach the anode. Equation (8-1-46) is often called the *Hull cutoff voltage equation*. Similarly, the magnetic cutoff condition is expressed by

$$B_{0c} = \frac{-\left(8V_0\dfrac{m}{e}\right)^{1/2}}{a\left(1 - \dfrac{b^2}{a^2}\right)} \tag{8-1-47}$$

This means that if $B_0 > B_{0c}$ for a given V_0, the electrons will not reach the anode. Equation (8-1-47) is called the *Hull cutoff magnetic equation*.

The advantage of an inverted coaxial magnetron design is that the cathode current density can be reduced to one-tenth of that used in cathode-centered magnetrons. Thus the millimeter magnetron is a practical and long-life device. The output waveguide can be in the circular electric mode that has extremely low transmission loss. Figure 8-1-10 shows a comparison between the inverted coaxial magnetron and a conventional magnetron designed for the same frequency. It should be noted that the cathode sizes are quite different.

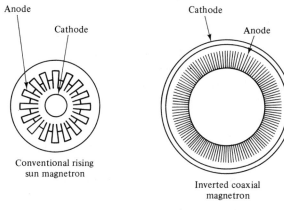

Conventional rising
sun magnetron

Inverted coaxial
magnetron

Figure 8-1-10 Comparison between an inverted coaxial magnetron and conventional magnetron (Courtesy of Varian Associates, Inc.).

Example 8-1-5:   Inverted Coaxial Magnetron

An inverted coaxial magnetron has the following parameters:

Anode voltage	$V_0 = 10$ kV
Cathode current	$I_0 = 2$ A
Anode radius	$a = 3$ cm
Cathode radius	$b = 4$ cm
Magnetic flux density	$B_0 = 0.01$ Wb/m^2

Determine:

  (a) The cutoff voltage for a fixed B_0

  (b) The cutoff magnetic flux density for a fixed V_0

Solution:

  (a) The cutoff voltage is

$$V_{0c} = \frac{1}{8} \times 1.759 \times 10^{11} \times (0.01)^2 \times (3 \times 10^{-2})^2$$

$$\times \left(1 - \frac{4^2}{3^2}\right)^2$$

$$= 1.20 \text{ kV}$$

  (b) The cutoff magnetic flux density is

$$B_{0c} = -\left(\frac{8 \times 10 \times 10^3}{1.759 \times 10^{11}}\right)^{1/2} \left[3 \times 10^{-2}\left(1 - \frac{4^2}{3^2}\right)\right]^{-1}$$

$$= 0.0289 \text{ Wb/m}^2$$

8-1-6 Frequency-Agile Coaxial Magnetron

A frequency-agile coaxial magnetron differs from a standard tunable magnetron. The frequency agility (FA) of a coaxial magnetron is defined as the capability to tune the output frequency of the radar with sufficiently high speed to produce a pulse-to-pulse frequency change greater than the amount required effectively to obtain decorrelation of adjacent radar echoes. The frequency-agile magnetron, together with appropriate receiver integration circuits, can reduce target scintillation, increase the detectability of target in a clutter environment, and improve resistance to electronic countermeasures (ECM). The increase of the pulse-to-pulse frequency separation will improve the radar system performance. Furthermore, the greater the pulse-to-pulse frequency separation, the more difficult it will be to center a jamming transmitter on the radar frequency for effective interference with system operation.

The frequency-agile coaxial magnetrons are classified into three types:

1. *Dither magnetrons:* The output RF frequency varies periodically with a constant excursion, constant rate and a fixed center frequency.

2. *Tunable/dither magnetrons:* The output RF frequency varies periodically with a constant excursion and constant rate, but the center frequency can be manually tuned by hand or mechanically tuned by a servomotor.

3. *Accutune magnetrons:* The output RF frequency variations are determined by the waveforms of an externally generated, low-level voltage signal. With proper selection of a tuning waveform, the accutune magnetron combines the features of dither and tunable/dither magnetrons, together with a capability for varying the excursion, rate, and tuning waveform. Figure 8-1-11 shows a picture of accutune magnetron.

Figure 8-1-11   X-band accutune magnetron VMX-1430 (Courtesy of Varian Associates, Inc.).

The X-band frequency-agile coaxial magnetron VMX-1430 is a typical agile magnetron. Its pulse voltage is 15 kV and pulse current is 15 A. Its maximum duty cycle is

0.0011 and accutune range is 1 GHz. Its center frequency is 9.10 GHz and peak output power is 90 kW. The agile rate and agile excursion are shown in Fig. 8-1-12.

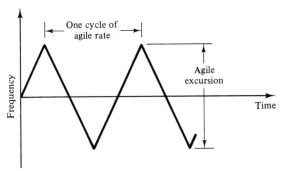

Figure 8-1-12   Agile rate and agile excursion.

From Fig. 8-1-12, it can be seen that the agile rate is the number of times per second that the transmitter frequency traverses the agile excursion and returns to its starting frequency. Similarly, the agile excursion is defined as the total frequency variation of the transmitter during agile operation.

The number of pulses that can be effectively integrated cannot be greater than the number of pulses placed on the target during one scan of the antenna, and, therefore, the antenna beamwidth and scan rate become factors that must also be considered in determining the integration period of the radar. Consequently, a design value for agile excursion can now be expressed in terms of radar operating parameters as

$$\text{Agile excursion} = \frac{N}{\tau} \qquad (8\text{-}1\text{-}48)$$

where N = number of pulses placed on the target during one radar scan, say, 20, whichever is smaller

τ = shortest pulse duration used in the system

The frequency, or *pulse repetition rate (PRR)*, is given by

$$f = \frac{\text{DC}}{\tau} \qquad (8\text{-}1\text{-}49)$$

where DC = duty cycle is the ratio of the pulse duration over repetition period for a pulse. The duty cycle is defined as

$$\text{Duty cycle} = \frac{\text{Pulse duration}}{\text{Pulse repetition period}} = \frac{\tau}{T} = \tau f \qquad (8\text{-}1\text{-}50)$$

Hence, the agile rate can be written as

$$\text{Agile rate} = \frac{1}{2T} \qquad (8\text{-}1\text{-}51)$$

where the 2 in the denominator is counted for the fact that two excursions through the agile frequency range occur during each cycle of agile rate.

Example 8-1-6:   Frequency-agile Magnetron

A frequency-agile coaxial magnetron has the following operating parameters:

$$\begin{array}{lll}
\text{Pulse duration} & \tau & = 0.20, \ 0.40, \ 0.80 \ \mu s \\
\text{Duty cycle} & DC & = 0.001 \\
\text{Pulse rate on target} & N & = 14 \text{ per scan}
\end{array}$$

Determine:

(a) The agile excursion

(b) The pulse-to-pulse frequency separation

(c) The signal frequency

(d) The time for N pulses

(e) The agile rate

Solution:

(a) The agile excursion is

$$\text{Agile excursion} = \frac{14}{0.2 \times 10^{-6}} = 70 \text{ MHz}$$

(b) The pulse-to-pulse frequency separation is

$$f_p = \frac{1}{\tau} = \frac{1}{0.20 \times 10^{-6}} = 5 \text{ MHz}$$

(c) The signal frequency is

$$f = \frac{DC}{\tau} = \frac{0.001}{0.20 \times 10^{-6}} = 5 \text{ kHz}$$

(d) The time for 14 pulses per second is

$$\text{Time} = \frac{N}{f} = \frac{14}{5000} = 2.8 \text{ ms}$$

(e) The agile rate is

$$\text{Agile rate} = \frac{1}{2 \times 0.0028} = 178.57 \text{ Hz}$$

8-2 FORWARD-WAVE CROSSED-FIELD AMPLIFIER (FWCFA OR CFA)

The crossed-field amplifier (CFA) is an outgrowth of the magnetron. CFAs can be grouped by their mode of operation as forward-wave or backward-wave types and by their electron stream source as emitting-sole or injected-beam types. The first group concerns the direction of the phase and group velocity of the energy on the microwave circuit. This can be seen from the ω-β diagrams of Fig. 4-4-5 as discussed in Sec. 4-4-2. Since the electron stream reacts to the RF electric field forces, the behavior of the phase velocity with frequency is of prime concern. The second group emphasizes the method by which electrons reach the interaction region and how they are controlled. This can be seen in the schematic diagrams of Fig. 8-2-1.

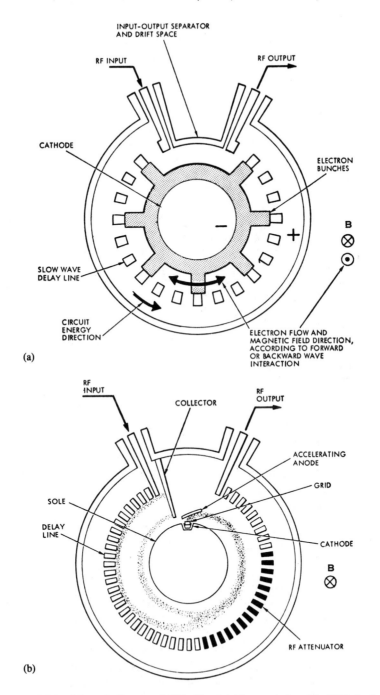

Figure 8-2-1   Schematic diagrams of CFAs (Reprinted by permission of the IEEE, Inc.).

In the forward-wave mode the helix-type slow-wave structure is often selected as the microwave circuit for the crossed-field amplifier, whereas in the backward-wave mode the strapped bar line represents a satisfactory choice. A structure of strapped crossed-field amplifier is shown in Fig. 8-2-2.

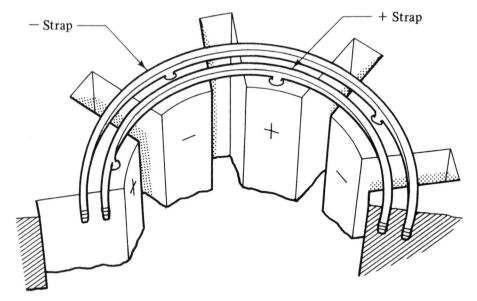

Figure 8-2-2   Diagram of a strapped CFA (Reprinted by permission of the IEEE, Inc.),

8-2-1 Principles of Operation

In the emitting-sole tube, the current emanated from the cathode is in response to the electric field forces in the space between cathode and anode. The amount of current is a function of the dimension, the applied voltage, and the emission properties of the cathode. The perveance of the interaction geometry tends to be quite high, about 5 to 10×10^{-10}, which results in a high-current and high-power capability at relative low voltage. In the injected-beam tube the electron beam is produced in a separate gun assembly and is injected into the interaction region.

The beam-circuit interaction features are similar in both the emitting-sole and the injected-beam tubes. Favorably phased electrons continue toward the positively polarized anode and are ultimately collected, whereas unfavorably phased electrons are directed toward the negative polarized electrode.

In linear-beam interaction, as discussed for traveling-wave tubes in Sec. 7-1, the electron stream is first accelerated by an electric gun to the full dc velocity; the dc velocity is approximately equal to the axial phase velocity of the RF field on the slow-wave structure. After interaction occurs, the spent electron beam leaves the interaction region with a low-average velocity. The difference in velocity is accounted for by the RF energy created on the microwave circuit. In the crossed-field amplifier, the electron is exposed to the dc electric field force, magnetic field force, and the electric field force of the RF field, and even the space-charge force from other electrons. The last force is normally not considered in analytic approaches because of its complexity. Under the influence of the three forces the electrons travel in spiral trajectories in a direction tending along equipotentials. The exact motion has been subject to much analysis by means of a computer. Figure 8-2-3 shows the pattern of the electron flow in the crossed-field amplifier by computerized techniques [2]. It can be seen that when the spoke is positively polarized or the RF field is in the positive half-cycle, the electron speeds up toward the anode; while the spoke is negatively polarized or the RF field is in the negative half-cycle, the electrons are returned toward the cathode. Consequently, the electron beam moves in a spiral path in the interaction region.

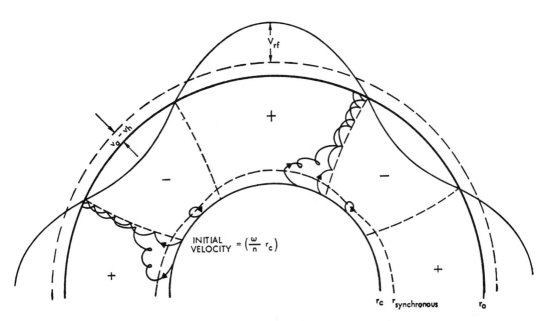

Figure 8-2-3 Motion of electrons in CFA (Reprinted by permission of the IEEE, Inc.).

The total power generated in a given crossed-field amplifier (CFA) is independent of the RF input power, as long as the input power exceeds the threshold value for spoke stability at the input. The power generated can be increased only by increasing the anode voltage and current. Neglecting circuit attenuation, the output power of a CFA is equal to the sum of the input power and the power generated in the interaction region. That is, the power gain of a CFA is given by

$$g = \frac{P_{out}}{P_{in}} = \frac{P_{in} + P_{gen}}{P_{in}} = 1 + \frac{P_{gen}}{P_{in}} \tag{8-2-1}$$

where $P_{out} = P_{in} + P_{gen}$
P_{in} = RF input power
P_{gen} = RF power induced into the anode circuit by electrons

Therefore, the CFA is not a linear amplifier, but rather is termed as a saturated amplifier.

The efficiency of a CFA is defined as the product of the electronic efficiency η_e and the circuit efficiency η_c. The electronic efficiency η_e is defined as in Eq. (8-1-21h). The overall efficiency is then expressed as

$$\eta = \eta_c \eta_e = \frac{P_{out} - P_{in}}{V_{a0} I_{a0}} \qquad (8\text{-}2\text{-}2)$$

where P_{out} = RF output power
P_{in} = RF input power
$\eta_e = P_{gen}/P_{dc}$
$P_{dc} = V_{a0}/I_{a0}$ is dc power
V_{a0} = anode dc voltage
I_{a0} = anode dc current

The circuit efficiency is defined as

$$\eta_c = \frac{\eta}{\eta_e} = \frac{P_{out} - P_{in}}{P_{gen}} \qquad (8\text{-}2\text{-}3)$$

where $P_{gen} = \eta_e V_{a0} I_{a0}$

Since the power generated per unit length is constant, the output power is given by

$$P_{out} = P_{in} e^{-2\alpha \ell} + \int_0^\ell \frac{P_{gen}}{\ell} e^{-2\alpha(\ell - \phi)} \, d\phi$$
$$= P_{in} e^{-2\alpha \ell} + \frac{P_{gen}}{2\alpha \ell} (1 - e^{-2\alpha \ell}) \qquad (8\text{-}2\text{-}4)$$

where α = circuit attenuation constant
ℓ = circuit length in ϕ direction

Substitution of Eq. (8-2-4) in Eq. (8-2-3) results in the circuit-efficiency equation as

$$\eta_c = \left(\frac{1}{2\alpha \ell} - \frac{P_{in}}{P_{gen}} \right) (1 - e^{-2\alpha \ell}) \qquad (8\text{-}2\text{-}5)$$

The term P_{in}/P_{gen} becomes negligible for high-gain CFA.

It is assumed that the input signal is sufficiently strong for spoke stability, that the RF power grows linearly with distance along the circuit, that the dc current for spoke is constant, and that the back-bombardment loss is not considered, the electronic-efficiency equation can be derived as follows. The average drift electron velocity at any position is given by

$$\mathcal{V} = \frac{E}{B} \tag{8-2-6}$$

where E = total electric field at the position under consideration
B = magnetic flux density at the position

The power flow at any position is related to the RF field and the beam-coupling impedance of the circuit by

$$P = \frac{E_{max}^2}{2\beta^2 Z_c} \tag{8-2-7}$$

where E_{max} = peak electric field
$\beta = \omega/\mathcal{V}$ is phase constant
Z_c = beam-coupling impedance

The power loss per spoke due to the electron motion toward the anode at any position is derived from Eq. (8-2-6) as

$$P_s = V_{s0} I_{s0} = I_{s0} \frac{m}{e} \frac{\beta^2 Z_c P}{B^2} \tag{8-2-8}$$

where V_{s0} = dc voltage per spoke
I_{s0} = dc current per spoke

Since the power varies linearly with position, the average power loss over the entire circuit length is

$$P_{s,avg} = I_{s0} \frac{m}{2e} \frac{\beta^2 Z_c}{B^2} (P_{in} + P_{out}) \tag{8-2-9}$$

By using Eqs. (8-2-1), (8-2-9), and $\mathcal{V}\beta = \omega$, the total power loss for all the spokes is given by

$$P_{lost} = I_{a0} \frac{m}{2e} \left(\frac{\omega}{\beta}\right)^2 + I_{a0} \frac{m}{2e} \frac{\beta^2 Z_c}{B^2} \left(\frac{g + 1}{g - 1}\right) P_{gen} \tag{8-2-10}$$

where g = power gain. Finally, the electronic efficiency is given by

$$\eta_e = \frac{P_{gen}}{V_{a0} I_{a0}} = \frac{1 - \dfrac{m\omega^2}{2eV_{a0}\beta^2}}{1 + \dfrac{I_{a0}}{B^2} \dfrac{m\beta^2 Z_c}{2e} \left(\dfrac{g + 1}{g - 1}\right)} \tag{8-2-11}$$

8-2-2 Microwave Characteristics

The crossed-field amplifier (CFA) is characterized by its low or moderate power gain, moderate bandwidth, high efficiency, saturated amplification, small size, low weight, and high perveance. These features have allowed the crossed-field amplifier to be used in a variety of electronic systems ranging from low-power and high-reliability space communications to multimegawatt, high-average-power, coherent pulsed radar. Figure 8-2-4 shows the present state of the art for the U.S. high-power forward-wave crossed-field amplifiers.

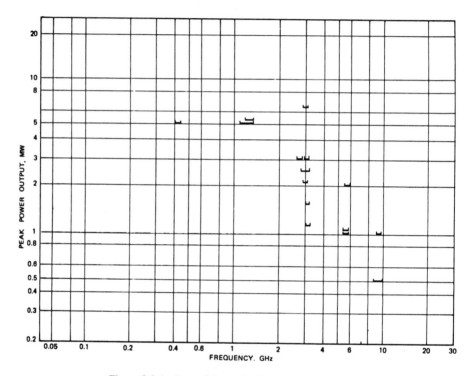

Figure 8-2-4 State of the art for U.S. high-power CFAs.

The Raytheon QKS-1541 amplifier as shown in Fig. 8-2-5 is a typical forward-wave crossed-field amplifier (CFA) and has an average power of 14 kW at frequency range from 2.9 to 3.1 GHz. Its pulse width is 28 s. Its typical peak anode voltage is 44 to 56 kV and peak anode current is 58 A.

Figure 8-2-5 Photograph of Raytheon QKS-1541 CFA (Courtesy of Raytheon Company, Microwave Tube Operation).

Example 8-2-1:   Crossed-Field Amplifier

A CFA operates under the following parameters:

Anode dc voltage	V_{a0} = 2 kV
Anode dc current	I_{a0} = 1.5 A
Electronic efficiency	η_e = 20%
RF input power	P_{in} = 80 W

Calculate:

(a) The induced RF power

(b) The total RF output power

(c) The power gain in decibels

Solution:

(a) The induced RF power is

$$P_{gen} = 0.20 \times 2 \times 10^3 \times 1.5 = 600 \text{ W}$$

(b) The RF output power is

$$P_{out} = P_{in} + P_{gen} = 80 + 600 = 680 \text{ W}$$

(c) The power gain is

$$g = \frac{P_{out}}{P_{in}} = \frac{680}{80} = 8.50 = 9.3 \text{ dB}$$

8-3 BACKWARD-WAVE CROSSED-FIELD AMPLIFIER (AMPLITRON)

The tradename of the backward-wave crossed-field amplifier is "amplitron" and its schematic diagram is shown in Fig. 8-3-1.

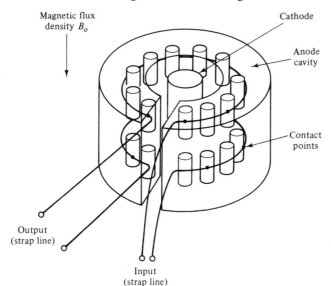

Magnetic flux density B_o

Cathode

Anode cavity

Contact points

Output (strap line)

Input (strap line)

Figure 8-3-1   Schematic diagram of Amplitron.

The anode cavity and pins comprise the resonator circuits. A pair of pins and the cavity are excited in opposite phase by the strap line. The electron beam and the electromagnetic waves interact in the resonant circuits. The amplitron can deliver 3-MW pulse with 10-μs duration at S-band and the tube gain reaches 8 dB.

The highly successful QK434 amplitron produced by Raytheon exhibited stable gain as high as 16 dB, power output levels ranging from a few hundred kilowatts to 3 MW, and efficiencies ranging from 60% for normal power levels to 76% for high-power, low-gain operation. Tube is commonly used in air surveillance radar and military pulsed radar. Figure 8-3-2 shows the photo of QK434 amplitron.

Figure 8-3-2   Photograph of QK-434 amplitron (Courtesy of IEEE or Brown [3]).

The two-stage superpower amplitron also manufactured by Raytheon generated 425 kW of CW power at an efficiency of 76%. The gain was 9 dB and the bandwidth was 5% at a mean frequency of 3 GHz. The tube was used for all high-data-rate transmission in the Apollo program. Figure 8-3-3 shows the picture of a two-stage superpower amplitron.

Figure 8-3-3   Photograph of a two-stage superpower amplitron (Courtesy of IEEE or Brown [3]).

The circuit and electronic equations for the M-type amplifier and oscillator were developed by several authors [5,6]. The basic secular equation including the space-charge effect is given by

$$(\gamma^2 - \gamma_0^2)(j\beta_e - \gamma)\,[(j\beta_e - \gamma)^2 + \beta_m^2]$$
$$= -j\beta_e\gamma_0\gamma^2\left[(j\beta_e - \gamma) + j\,\frac{2\alpha}{1+\alpha^2}\,\beta_m\right]H^2 \qquad (8\text{-}3\text{-}1)$$

where $\gamma_0 = $ circuit propagation constant

$\gamma = $ harmonic wave propagation constant

$\beta_e = \dfrac{\omega}{\mathcal{V}_0}$ is the electron-beam phase constant

$\mathcal{V}_0 = \sqrt{\dfrac{2e}{m}}\,V_0$ is the dc electron-beam velocity

$\beta_m = \dfrac{\omega_c}{\mathcal{V}_0} = \dfrac{e}{\mathcal{V}_0 m}\,B_0$ is the cyclotron phase constant

$\omega_c = \dfrac{e}{m}\,B_0$ is the cyclotron angular frequency

$B_0 = $ crossed magnetic flux density

$H^2 = 2(1 + \alpha^2)\phi^2 C^3$

$C = \left(\dfrac{I_0 Z_0}{4V_0}\right)^{1/3}$ is the gain parameter

$\phi = A\exp(-j\gamma y) + B\exp(j\gamma y)$ is the wave equation

$\alpha = \dfrac{A\exp(-j\gamma y) - B\exp(j\gamma y)}{A\exp(-j\gamma y) + B\exp(j\gamma y)} = j\,\dfrac{1}{\gamma\phi}\,\dfrac{d\phi}{dy}$ is a factor

In general, there are five solutions of γ from Eq. (8-3-1). Let

$$\gamma_0 = j\beta \qquad (8\text{-}3\text{-}2)$$

$$\gamma = j\beta(1 + p) \qquad (8\text{-}3\text{-}3)$$

where p is a very small constant ($p \ll 1$). Substituting Eqs. (8-3-2) and (8-3-3) in Eq. (8-3-1) we have

$$p\left(\frac{\beta_e}{\beta} - 1 - p\right)\left[\left(\frac{\beta_e}{\beta} - 1 - p\right)^2 - \left(\frac{\beta_m}{\beta}\right)^2\right]$$
$$= \frac{-\beta_e}{2\beta}\left[\left(\frac{\beta_e}{\beta} - 1 - P\right) + \frac{2\alpha\beta_m}{(1+\alpha^2)\beta}\right]H^2 \qquad (8\text{-}3\text{-}4)$$

The right-hand term is small, as it is not much different from C, so that if either one of the left-hand terms is also small, the Eq. (8-3-4) will be satisfied.

If $\beta_e \doteq \beta$ is assumed, the first factor is small and the solution for p becomes

$$p = \pm j \left(\frac{\alpha}{1 + \alpha^2} \right)^{1/2} \left(\frac{\beta}{\beta_m} \right)^{1/2} H \tag{8-3-5}$$

If $\beta_e \pm \beta_m \doteq \beta$ is assumed, the second factor is small and the solution for p is given by

$$p^2 = \pm \frac{1}{4} \frac{(1 \pm \alpha)^2}{1 + \alpha^2} \left(\frac{\beta}{\beta_m} \pm 1 \right) H^2$$

From the definition of p, gain is only obtained when p is imaginary. From Eq. (8-3-5), this only happens when α is positive. In Eq. (8-3-6), p is imaginary when $\beta_e + \beta_m = \beta$ and α is less than unity. This last condition comes from rewriting Eq. (8-3-6) as

$$p = \pm \frac{1}{2} \frac{1 - \alpha}{(1 + \alpha)^{1/2}} \left(\frac{\beta_e}{\beta_m} \right)^{1/2} H \tag{8-3-7}$$

The solution for a M-type backward-wave amplifier can be obtained by setting

$$\gamma = -j\beta(1 + p) \tag{8-3-8}$$

The only possible solution for $\beta + \beta_e = \beta_m$ is

$$p = \pm j \frac{1}{2} \frac{1 + \alpha}{(1 + \alpha^2)^{1/2}} \left(\frac{\beta_e}{\beta_m} \right)^{1/2} H \tag{8-3-9}$$

This gives rise to increasing backward waves for all values of α except $\alpha = -1$. Because $H \propto C^{3/2}$, Eqs. (8-3-5), (8-3-7), and (8-3-9) show that the gain per unit length in M-type devices is lower than in O-type devices using similar circuits due to $C^{3/2} < C$.

Example 8-3-1:   Amplitron Characteristics

An amplitron has the following operating parameters:

Anode voltage	$V_0 = 15$ kV
Anode current	$I_0 = 3$ A
Magnetic flux density	$B_0 = 0.2$ Wb/m^2
Operating frequency	$f = 8$ GHz
Characteristic impedance	$Z_0 = 50$ Ω

Determine:

(a) The dc electron-beam velocity
(b) The electron-beam phase constant
(c) The cyclotron angular frequency
(d) The cyclotron phase constant
(e) The gain parameter

Solution:

(a) The dc electron-beam velocity is

$$\mathcal{V}_0 = 0.593 \times 10^6 \times (15 \times 10^3)^{1/2} = 0.762 \times 10^8 \text{ m/s}$$

(b) The electron-beam phase constant is

$$\beta_e = \frac{\omega}{\mathcal{V}_0} = \frac{2\pi \times 8 \times 10^9}{0.762 \times 10^8} = 692.36 \text{ rad/m}$$

(c) The cyclotron angular frequency is

$$\omega_c = \frac{e}{m} B_0 = 1.759 \times 10^{11} \times 0.2 = 35.18 \times 10^9 \text{ rad/s}$$

(d) The cyclotron phase constant is

$$\beta_m = \frac{\omega_c}{\mathcal{V}_0} = \frac{35.18 \times 10^9}{0.726 \times 10^8} = 484.57 \qquad \text{rad/m}$$

(e) The gain parameter is

$$C = \left(\frac{I_0 Z_0}{4V_0}\right)^{1/3} = \left(\frac{3 \times 50}{4 \times 15 \times 10^3}\right)^{1/3} = 0.136$$

8-4 BACKWARD-WAVE CROSSED-FIELD OSCILLATOR (CARCINOTRON)

The backward-wave crossed-field oscillator or *M*-carcinotron has two configurations: linear *M*-carcinotron and circular *M*-carcinotron.

Linear *M*-carcinotron.   The *M*-carcinotron oscillator is an *M*-type backward-wave oscillator. The interaction between the electrons and the slow-wave structure takes place in a space of crossed field. A linear model of the *M*-carcinotron oscillator is shown in Fig. 8-4-1.

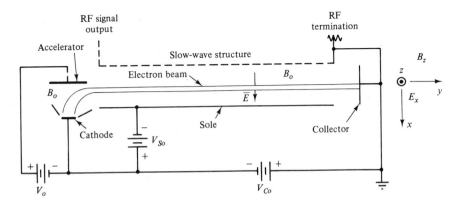

Figure 8-4-1   Linear model of an M-carcinotron oscillator (Reprinted by permission of Van Nostrand Company).

The slow-wave structure is in parallel with an electrode known as the sole. A dc electric field is maintained between the grounded slow-wave structure and the negative sole. A dc magnetic field is directed into the page. The electrons emitted from the cathode are bent through a 90° angle by the magnetic field. The electrons interact with a backward-wave space harmonic of the circuit, and the energy in the circuit flows opposite to the direction of the electron motion. The slow-wave structure is terminated at the collector end, and the RF signal output is removed at the electron-gun end. Since the *M*-carcinotron is a crossed-field device, its efficiency is very high, ranging from 30 to 60%.

The perturbed electrons moving in synchronism with the wave in a linear *M*-carcinotron is shown in Fig. 8-4-2.

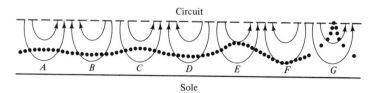

Figure 8-4-2   Beam electrons and electric field lines in an M-carcinotron.

Electrons at position *A* near the beginning of the circuit are moving toward the circuit, whereas electrons at position *B* are moving toward the sole. Farther down the circuit, electrons at position *C* are closer to the circuit, and electrons at position *D* are closer to the sole. However, electrons at position *C* have departed a greater distance from the unperturbed path than have electrons at position *D*. Thus, the electrons have lost a net amount of potential energy, this energy having been transferred to the RF field. The reason for the greater displacement of the electrons moving toward the circuit is that these electrons are in stronger RF fields, since they are closer to the circuit. Electrons at position *E* and position *F* further demonstrated this behavior. Electrons at position *G* have moved so far from the unperturbed position that some of them are being intercepted on the circuit. The length from position *A* through position *G* is a half-cycle of the electron motion.

Circular *M*-carcinotron.   The *M*-carcinotrons are generally constructed in the circular reentrant form as shown in Fig. 8-4-3. The slow-wave structure and sole are circular and nearly reentrant to conserve magnet weight. The sole has the appearance of the cathode in a magnetron.

The Litton L-3721 M-BWO as shown in Fig. 8-4-4 is a typical *M*-type backward-wave oscillator (M-BWO or carcinotron) and has a minimum power of 200 W at frequency range from 1.0 to 1.4 GHz.

The carcinotron can exist in both linear and circular configurations. The delay line is terminated at the collector end by spraying attenuating material on to the surfaces of the conductors. The output is taken from the gun end of the delay line that is an interdigital line. Clearly, in this case, the electron drift velocity has to be in synchronism with a

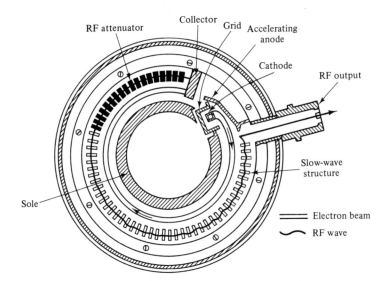

Figure 8-4-3   Schematic diagram of a circular M-carcinotron (Courtesy of Raytheon Company, Microwave Tube Operation).

L-3721 7⅝" Wide

Figure 8-4-4   Photograph of Litton L-3721 BWO (Courtesy of Litton Company, Electron Tube Division).

backward-space harmonic. As in the case of O-type devices, the only modification in the secular equation is a change of sign in the circuit equation. If this change is made in Eq. (8-3-1), we write

$$\gamma_0 = j\beta \tag{8-4-1}$$

$$\gamma = jk + \epsilon \tag{8-4-2}$$

We obtain for the carcinotron on eliminating negligible terms,

$$(\beta^2 - k^2 + j2k\epsilon)\,[\,j(\beta_e - k) - \epsilon][\beta_m^2 - (\beta_e - k)^2 - j2(\beta_e - k)\epsilon]$$
$$= j\beta\beta_e k^2 \left(\beta_e - k + \frac{2\alpha}{1 + \alpha^2}\beta_m\right)H^2 \tag{8-4-3}$$

A solution of Eq. (8-4-3) for synchronism can be obtained by setting $\beta = \beta_e$ and $\beta_e - k = \beta_e b'$, where b' is a small number so that terms like b'^2 and $b'\epsilon$ may be neglected. This yields

$$2\epsilon(j\beta_e b' - \epsilon) = \beta\beta_e k \left(\frac{b'}{\beta_m^2} + \frac{2\alpha}{1 + \alpha^2}\frac{1}{\beta_m}\right)H^2$$

$$\doteq \beta_e k \frac{2\alpha}{1 + \alpha^2}\frac{\beta}{\beta_m}H^2 \tag{8-4-4}$$

$$= 2\beta_e kD^2$$

where

$$D^2 = \frac{\alpha}{1 + \alpha^2}\frac{\beta}{\beta_m}H^2 \tag{8-4-5}$$

$$\epsilon = \beta_e D\delta \tag{8-4-6}$$

$$b' = bD \tag{8-4-7}$$

$$\delta(\delta - jb) = -1 \text{ or } \delta^2 - jb\delta + 1 = 0 \tag{8-4-8}$$

As a result, we have reduced the number of waves to two, with propagation constants given by

$$\gamma_1 = j(\beta_e + b) + \beta_e D\delta_1 \tag{8-4-9}$$

$$\gamma_2 = j(\beta_e + b) + \beta_e D\delta_2 \tag{8-4-10}$$

where the δ's are the roots of Eq. (8-4-8) and they are

$$\delta_1 = j\,\frac{b - \sqrt{b^2 + 4}}{2}$$

$$\delta_2 = j\,\frac{b + \sqrt{b^2 + 4}}{2}$$

To determine the amplification of the growing waves, the input reference point is set at $y = 0$ and the output reference point is taken at $y = \ell$. It follows that at $y = 0$, the voltage at the input point can be computed as follows:

$$V_1(0) + V_2(0) = V(0) \tag{8-4-11}$$

$$\frac{V_1(0)}{\delta_1} + \frac{V_2(0)}{\delta_2} = 0 \tag{8-4-12}$$

Solving Eqs. (8-4-11) and (8-4-12) simultaneously we have

$$V_1(0) = \frac{V(0)}{1 - \delta_2/\delta_1} = \frac{\delta_1 V(0)}{\delta_2} \tag{8-4-13}$$

$$V_2(0) = \frac{-V(0)}{1 - \delta_1/\delta_2} = \frac{-\delta_2 V(0)}{\delta_1} \tag{8-4-14}$$

Then the voltage at the output point $y = \ell$ is given by

$$V(0) = V_1(0) \exp(-\gamma_1\ell) + V_2(0) \exp(-\gamma_2\ell) \tag{8-4-15}$$
$$= V(0)[\delta_1 \exp(-\gamma_1\ell) - \delta_2 \exp(-\gamma_2\ell)]/(\delta_1 - \delta_2)$$

The term in the square bracket is the inverse of the voltage gain of the device. Oscillation takes place when this is zero. That is,

$$\delta_1 \exp(-\gamma_1\ell) = \delta_2 \exp(-\gamma_2\ell) \tag{8-4-16}$$

or

$$\delta_1/\delta_2 = \exp(\gamma_1\ell - \gamma_2\ell) \tag{8-4-17}$$
$$= \exp[-\beta_e D\ell(\delta_2 - \delta_1)]$$

From Eq. (8-4-8) we have

$$\delta_1/\delta_2 = \frac{b - \sqrt{b^2 + 4}}{b + \sqrt{b^2 + 4}} \tag{8-4-18}$$

and

$$\delta_2 - \delta_1 = j\sqrt{b^2 + 4} \tag{8-4-19}$$

Then

$$\delta_1/\delta_2 = \exp\left(-j\beta_e D\ell\sqrt{b^2 + 4}\right) \tag{8-4-20}$$

Equations (8-4-18) and (8-4-20) can only be satisfied simultaneously if $b = 0$ and $\delta_1 = -\delta_2$. Then

$$2\beta_e D\ell = (2n + 1)\pi \tag{8-4-21}$$

where n = any integer numbers. If we introduce N as usual defined through $\beta_e \ell = 2\pi N$, the oscillation condition becomes

$$DN = \frac{2n + 1}{4} \tag{8-4-22}$$

Example 8-4-1: Carcinotron Characteristics

A circular carcinotron has the operating parameters:

Anode voltage	$V_0 = 20$ kV
Anode current	$I_0 = 3.5$ A
Magnetic flux density	$B_0 = 0.3$ Wb/m^2
Operating frequency	$f = 4$ GHz
Characteristic impedance	$Z_0 = 50$ Ω
D factor	$D = 0.8$
b factor	$b = 0.5$

Compute:

(a) The dc electron velocity

(b) The electron-beam phase constant

(c) The delta differentials

(d) The propagation constants

(e) The oscillation condition

Solution:

(a) The dc electron velocity is

$$\mathcal{V}_0 = 0.593 \times 10^6 \times (20 \times 10^3)^{1/2} = 0.8386 \times 10^8 \text{ m/s}$$

(b) The electron-beam phase constant is

$$\beta_e = \frac{\omega}{\mathcal{V}_0} = \frac{2\pi \times 4 \times 10^9}{0.8386 \times 10^8} = 300 \text{ rad/m}$$

(c) The delta differentials are

$$\delta_1 = j\frac{0.5 - \sqrt{(0.5)^2 + 4}}{2} = -j0.78$$

$$\delta_2 = j\frac{0.5 + \sqrt{(0.5)^2 + 4}}{2} = j1.28$$

(d) The propagation constants are

$$\gamma_1 = j(\beta_e + b) + \beta_e D\delta_1$$
$$= j(300 + 0.5) + 0.5 \times 0.8 \times (-j0.78) = j300.20$$
$$\gamma_2 = j(300 + 0.5) + 0.5 \times 0.8 \times (+j1.28) = j301.00$$

(e) The oscillation occurs at

$$DN = 1.25 \quad \text{for } n = 1$$

then

$$N = 1.5625$$

and

$$\ell = \frac{2\pi N}{\beta_e} = \frac{2\pi \times 1.5625}{300} = 3.27 \text{ cm}$$

REFERENCES

[1] HUTTER, R. G. E., *Beam and Wave Electrons in Microwave Tubes*. D. Van Nostrand Company, Princeton, N.J., 1960.

[2] SKOWRON, JOHN F., The Continuous-Cathode (Emitting-Sole) Crossed-Field Amplifier. *Proc. IEEE,* 61, No. 3, Mar. 1973, pp. 330–56.

[3] BROWN, WILLIAM C., The Microwave Magnetron and Its Derivatives. *IEEE Trans. on Electron Devices.* Vol. ED-31, No. 11, Nov. 1984, pp. 1595–1605.

[4] GEWARTOWSKI, J. V., and WATSON, H. A., *Principles of Electron Tubes*. D. Van Nostrand Company, Princeton, N.J., 1965, p. 391.

[5] PIERCE, J. R., *Traveling-Wave Tubes*. D. Van Nostrand Company, Princeton, N.J., 1950. p. 210.

[6] BECK, A. H. W., *Space-Charge Waves and Slow Electromagnetic Waves*. Pergamon Press, New York, 1958, p. 250.

PROBLEMS

Magnetrons

8-1-1. Describe the principle of operation for a normal cylindrical magnetron and its characteristics.

8-1-2. A normal cylindrical magnetron has the following parameters:

Inner radius	$R_a = 0.15$ meter
Outer radius	$R_b = 0.45$ meter
Magnetic flux density	$B_0 = 1.2$ milliwebers/m^2

(a) Determine the Hull cutoff voltage.
(b) Determine the cutoff magnetic flux density if the beam voltage V_0 is 6000 V.

8-1-3. It is assumed that in a normal cylindrical magnetron the inner cylinder of radius a carries a current of I_0 in the z direction (i.e., $I = I_0 u_z$) and the anode voltage is V_0. The outer radius is b. Determine the differential equation in terms of the anode voltage V_0 and the current I_0.

8-1-4. Compare the cutoff conditions for an inverted cylindrical magnetron (i.e., the inner cathode voltage is V_0 and the outer anode is grounded) with a normal cylindrical magnetron. It is assumed that the magnetic field does no work on the electrons.

8-1-5. It is assumed that electrons in an inverted cylindrical magnetron leave the interior of the coaxial cathode with initial velocity due to thermal voltage V_t in volts. Find the initial velocity required for the electrons to just hit the anode at the center conductor.

8-1-6. It is assumed that electrons in an inverted cylindrical magnetron leave the interior of the coaxial cathode with zero initial velocity. Find the minimum velocity for an electron to just graze the anode at the center conductor.

8-1-7. In a linear magnetron the electric and magnetic field intensities as shown in Fig. P8-1-7 are given by

$$\mathbf{E} = E_2\mathbf{u}_z = -\frac{V_0}{d}\mathbf{u}_z$$

$$\mathbf{B} = B_y\mathbf{u}_y$$

Figure p8-1-7   Problem 8-1-7

Determine the trajectory of an electron with an initial velocity $\mathcal{V}_0$ in the z direction.

8-1-8. A X-band pulsed cylindrical magnetron has the following parameters:

Anode voltage	$V_0 = 32$ kV
Anode current	$I_0 = 84$ A
Magnetic flux density	$B_0 = 0.01$ Wb/m^2
Radius of cathode cylinder	$a = 6$ cm
Radius of vane edge to center	$b = 12$ cm

Compute:

(a) The cyclotron angular frequency
(b) The cutoff voltage for a fixed B_0
(c) The cutoff magnetic flux density for a fixed V_0

8-1-9. A X-band pulsed conventional magnetron has the following parameters:

Anode voltage	$V_0 = 22$ kV
Anode current	$I_0 = 28$ A
Operating frequency	$f = 10$ GHz
Resonator conductance	$G_r = 3 \times 10^{-4}$ $\mho$
Loaded conductance	$G_\ell = 3 \times 10^{-5}$ $\mho$
Vane capacitance	$C = 3$ pF
Duty cycle	$DC = 0.001$
Power loss	$P_{\text{loss}} = 200$ kW

Compute:

(a) The angular resonant frequency

(b) The unloaded quality factor Q_{un}

(c) The loaded quality factor Q_ℓ

(d) The external quality factor Q_{ex}

(e) The circuit efficiency

(f) The electronic efficiency

8-1-10. A linear magnetron has the following parameters:

Anode voltage	V_0	= 20 kV
Cathode current	I_0	= 17 A
Magnetic flux density	B_0	= 0.01 Wb/m^2
Distance between cathode and anode	d	= 6 cm

Calculate:

(a) The Hull cutoff voltage for a fixed B_0

(b) The Hull cutoff magnetic flux density for a fixed V_0

8-1-11. A linear magnetron has the following parameters:

Anode voltage	V_0	= 32 kV
Cathode current	I_0	= 60 A
Operating frequency	f	= 10 GHz
Magnetic flux density	B_0	= 0.01 Wb/m^2
Hub thickness	h	= 3 cm
Distance between anode and cathode	d	= 6 cm

Compute:

(a) The electron velocity at the hub surface

(b) The phase velocity for synchronism

(c) The Hartree anode voltage

8-1-12. An inverted coaxial magnetron has the following parameters:

Anode voltage	V_0	= 30 kV
Cathode current	I_0	= 25 A
Anode radius	a	= 2.5 cm
Cathode radius	b	= 5 cm
Magnetic flux density	B_0	= 0.01 Wb/m^2

Determine:

(a) The cutoff voltage for a fixed B_0

(b) The cutoff magnetic flux density for a fixed V_0

8-1-13. A frequency-agile coaxial magnetron has the following parameters:

Pulse duration	τ	= 0.30, 0.60, 0.90 μs
Duty cycle	DC	= 0.0011
Pulse rate on target	N	= 15 per scan

Determine:

(a) The agile excursion

(b) The pulse-to-pulse frequency separation

(c) The signal frequency

(d) The time for N pulse

(e) The agile rate

Crossed-Field Amplifiers (CFAs)

8-2-1. Derive Eq. (8-2-4).

8-2-2. Derive the circuit-efficiency Eq. (8-2-5).

8-2-3. Derive Eqs. (8-2-8), (8-2-10), and (8-2-11).

8-2-4. A CFA has the following operating parameters:

Anode dc voltage	$V_{a0} = 3$ kV
Anode dc current	$I_{a0} = 3$ A
Electronic efficiency	$\eta_e = 25\%$
RF input power	$P_{in} = 100$ W

Compute:

(a) The induced RF power

(b) The total RF output power

(c) The power gain in dB

8-2-5. A CFA operates under the following parameters:

Anode dc voltage	$V_{a0} = 1.80$ kV
Anode dc current	$I_{a0} = 1.30$ A
Electronic efficiency	$\eta_e = 22\%$
RF input power	$P_{in} = 70$ W

Calculate:

(a) The induced RF power

(b) The total RF output power

(c) The power gain in dB

Amplitrons

8-3-1. An amplitron operates under the following parameters:

Operating frequency	$f = 9$ GHz
Anode voltage	$V_0 = 20$ kV

Anode current	$I_0 = 3.5$ A
Magnetic flux density	$B_0 = 0.3$ Wb/m^2
Characteristic impedance	$Z_0 = 50$ Ω

Compute:

(a) The dc electron-beam velocity

(b) The electron-beam phase constant

(c) The cyclotron angular frequency

(d) The cyclotron phase constant

(e) The gain parameter

8-3-2. An amplitron has the following operating parameters:

Operating frequency	$f = 10$ GHz
Anode voltage	$V_0 = 25$ kV
Anode current	$I_0 = 4$ A
Magnetic flux density	$B_0 = 0.35$ Wb/m^2
Characteristic impedance	$Z_0 = 50$ Ω

Calculate:

(a) The dc electron-beam velocity

(b) The electron-beam phase constant

(c) The cyclotron angular frequency

(d) The cyclotron phase constant

(e) The gain parameter

Carcinotrons

8-4-1. A circular carcinotron operates under the following parameters:

Operating frequency	$f = 8$ GHz
Anode voltage	$V_0 = 25$ kV
Anode current	$I_0 = 4$ A
Magnetic flux density	$B_0 = 0.35$ Wb/m^2
Characteristic impedance	$Z_0 = 50$ ohms
D factor	$D = 0.75$
b factor	$b = 0.50$

Calculate:

(a) The dc electron velocity

(b) The electron-beam phase constant

(c) The delta differentials

(d) The propagation constants

(e) The oscillation condition

8-4-2. A circular carcinotron has the following operating parameters:

Operating frequency	$f = 6$ GHz
Anode voltage	$V_0 = 30$ kV
Anode current	$I_0 = 3.5$ A
Magnetic flux density	$B_0 = 0.28$ Wb/m^2
Characteristic impedance	$Z_0 = 50$ ohms
D factor	$D = 0.78$
b factor	$b = 0.55$

Compute:

(a) The dc electron velocity

(b) The electron-beam phase constant

(c) The delta differentials

(d) The propagation constants

(e) The oscillation condition

CHAPTER 9

Microwave Fast-Wave Electron Tubes

9-0 INTRODUCTION

Conventional microwave electron tubes are commonly referred to as slow-wave electron tubes. The RF-signal circuit is designed in such way so that the phase velocity of the RF signal wave is slowed down to the electron-beam velocity for electron interactions. For example, the helix traveling-wave tube uses a helical structure to slow the wave velocity to the beam velocity for signal amplification.

In fast-wave electron tubes, however, the RF circuit is usually a smooth waveguide or a large resonator in which no attempt is made to reduce the signal-wave velocity. Instead, the electron beam is injected into the RF signal field in a manner such that electron interactions can take place. The fast-wave electron tubes are also referred to as free-electron devices.

In recent years, many fast-wave electron tubes such as gyrotrons, Ubitrons, and Peniotrons are designed, developed, and tested. A gyrotron can generate a peak power of a megawatt (MW) or more at 100 GHz. Ubitron, like the conventional traveling-wave tube, is an O-type traveling-wave amplifier and can produce a peak power of 1.6 and 0.15 MW at 16 and 54 GHz, respectively. Peniotron is a potential high-power tube in the gyrotube family, but it operates on nonrelativistic mass change process. All those fast-wave electron tubes are analyzed in this chapter.

9-1 PHASE BUNCHING PROCESS IN GYROTRON

The gyrotron is a new type of microwave device utilizing the electron cyclotron maser (CM) interaction between an electromagnetic wave and an electron beam moving along helical trajectories in an applied magnetic field. Its radiation is generated by the relativistically stimulated electrons. The device has become a very promising high-power source at microwave frequencies. To date with gyrotron oscillators (gyromonotrons) a CW power of 212 kW at 28 GHz and a pulsed power of 1 MW at 100 GHz have been achieved [1]. The efficiency is up to 50%. Potential applications of the gyrotrons include radar, communications, plasma heating of controlled nuclear fusion devices, and others.

The linear mechanism of the cyclotron maser was first proposed by Twiss [2] classically in 1958 and by Schneider [3] quantum mechanically in 1959. In 1959, Gaponov published a paper on the classical theory of the cyclotron maser [4]; in the same year, Pantell published what was perhaps the first experimental work concerning the electron cyclotron maser mechanism [5].

The gyrating electrons in a magnetron are not in phase while the smooth anodes extract the absorbing (or unfavorable) electrons from the interaction space. But the electron bombardment of the walls limits the high-power generation.

The physical mechanism responsible for the microwave radiation in a gyrotron has its origin in a relativistic effect. Initially, the phases of the electrons in their cyclotron orbits are random, but the phase bunching of electrons necessary for transfer of energy from the electron beam to the field occurs because the dependence of the effective relativistic-electron mass on energy. Those electrons that emit energy to the radiation wave become lighter, rotate faster, and, hence, advance ahead in phase (leading phase); those electrons that absorb energy from the wave become heavier, rotate more slowly, and recede in phase (lagging phase). As a result, phase bunching is developed in the interaction space, and the electrons radiate coherently and amplify the radiation wave [6]. Figure 9-1-1 shows a diagram for the instantaneous vector relationship of two electrons in the phase space of the Doppler-shifted cyclotron oscillation.

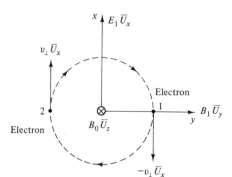

Figure 9-1-1 Instantaneous vector diagram of perturbed electrons in a gyrotron.

The perturbed electric field is indicated by $E_1 \mathbf{U}_x$ and the magnetic flux density by $B_1 \mathbf{U}_y$. The uniform magnetic flux density B_0 is in the positive z axis (into the page). Points 1 and 2 indicate the positions of two perturbed electrons with perpendicular velocities $\mathbf{v}$ at the $x = 0$ at $t = 0$. If the two electrons are not perturbed, they will follow the projection of the dashed circle on the x-y plane. When they are perturbed, they will be phase-bunched in the x direction.

If the radiation wave is assumed to propagate in the positive z direction, the Doppler-shifted cyclotron angular frequency (or gyrofrequency) of the propagating wave is defined as [7].

$$\Omega_D = k_z \mathcal{V}_z + \Omega_e / \gamma \tag{9-1-1}$$

where $\Omega_e / \gamma = \Omega_c$ is the relativistic gyrofrequency of the electron

$k_z \mathcal{V}_z$ = Doppler-shifted frequency

k_z = axial wave number in radians per meter

$\mathcal{V}_z$ = axial drift velocity of the electron

$\Omega_e = \dfrac{e}{m} B_0$ is the nonrelativistic gyrofrequency of the electron

B_0 = external magnetic flux density

$\gamma = \left(1 - \dfrac{\mathcal{V}_\perp^2 + \mathcal{V}_z^2}{c^2} \right)^{-1/2}$ is the relativistic factor

$\mathcal{V}_\perp$ = azimuthal component of the electron drift velocity

$\mathcal{V}_z$ = axial component of the electron drift velocity

The instantaneous values of Ω_D can be expressed as

$$\Omega_D(0) = k_z \mathcal{V}_z(0) + \frac{\Omega_e}{\gamma(0)} \qquad \text{at } t = 0 \tag{9-1-2}$$

and

$$\Omega_D(\Delta t) = k_z \mathcal{V}_z(\Delta t) + \frac{\Omega_e}{\gamma(\Delta t)} \qquad \text{at } t = \Delta t \tag{9-1-3}$$

Subtracting Eq. (9-1-2) from Eq. (9-1-3) yields the differential Doppler-shifted wave gyrofrequency as

$$\Delta \Omega_D = k_z \Delta \mathcal{V}_z + \Omega_e \left(\frac{1}{\gamma(0) + \Delta \gamma} - \frac{1}{\gamma(0)} \right) \tag{9-1-4}$$

or

$$\Delta \Omega_D \doteq k_z \Delta \mathcal{V}_z - \frac{\Omega_e \Delta \gamma}{\gamma^2(0)} \tag{9-1-5}$$

where $\Delta \Omega_D = \Omega_D(\Delta t) - \Omega_D(0)$

$\Delta \mathcal{V}_z = \mathcal{V}_z(\Delta t) - \mathcal{V}_z(0)$

$\Delta \gamma = \gamma(\Delta t) - \gamma(0)$

From the relativistic equation of electron motion

$$\frac{d}{dt} m \gamma \mathbf{v} = -e[E + \mathbf{v} \times (B_0 \mathbf{U}_z + B_1 \mathbf{U}_y)] \tag{9-1-6}$$

the differential components of the drift velocity and the relativistic factor can be derived, respectively, as

$$\Delta \mathbf{v}_z \simeq \frac{-e}{\gamma(0)m} \, (\mathbf{v}_\perp \times B_1 \mathbf{U}_y) \, \Delta t \tag{9-1-7}$$

where B_1 is the instantaneous magnetic flux density in the positive y direction and

$$\Delta \gamma \simeq \frac{-e}{mc} \, (\mathbf{v}_\perp \cdot E_1 \mathbf{U}_x) \, \Delta t \tag{9-1-8}$$

where E_1 is the instantaneous electric field intensity in the positive x direction. By applying Faraday's law

$$B_1 \mathbf{U}_y = \frac{c}{\omega} \, \mathbf{K} \times E_1 \mathbf{U}_x \tag{9-1-9}$$

Equation (9-1-7) becomes

$$\Delta \mathcal{V}_z = \frac{-ek_z c}{\gamma_0 m \omega} \, (\mathbf{v}_\perp \cdot E_1 \mathbf{U}_x) \, \Delta t \tag{9-1-10}$$

where $\quad \gamma_0 = \gamma(0) = \left(1 - \dfrac{\mathcal{V}_{\perp 0}^2}{c^2}\right)^{-1/2}$ is the relativistic factor at $t = 0$

$\mathcal{V}_{\perp 0}$ = perpendicular (or azimuthal) drift velocity of the electron

Substitution of Eqs. (9-1-8) and (9-1-10) into Eq. (9-1-5) yields the differential Doppler-shifted wave gyrofrequency [7] as

$$\Delta \Omega_D = \frac{-ek_z^2 c}{\gamma_0 m \omega} \left(1 - \frac{\omega \Omega_e}{\gamma_0 k_z^2 c^2}\right) (\mathbf{v}_\perp \cdot E_1 \mathbf{U}_x) \, \Delta t \tag{9-1-11}$$

Equation (9-1-11) indicates the phase-bunching (or phase modulation) mechanism of the gyrotron. Specifically, it can be explained as follows:

1. If electron 1 recedes in phase, electron 2 will advance ahead in phase, and vice versa. As a result, the two electrons will bunch together in phase at some point in the x direction.

2. The first term on the right-hand side of the equation results in axial bunching due to the effect of the perturbed axial velocity $\Delta \mathcal{V}_z$, and, similarly, the second term results in azimuthal bunching due to the effect of the instantaneous relativistic factor $\Delta \gamma$.

3. The axial bunching process is dominant if the value of the bracketed factor is greater than zero—that is, if

$$\left(1 - \frac{\omega \Omega_e}{\gamma_0 k_z^2 c^2}\right) > 0 \tag{9-1-12}$$

4. The azimuthal bunching process is dominant if the value of the bracketed factor is smaller than zero—that is, if

$$\left(1 - \frac{\omega \Omega_e}{\gamma_0 k_z^2 c^2}\right) < 0 \qquad (9\text{-}1\text{-}13)$$

5. For a given k_z, the relativistic azimuthal bunching process may be dominant at nonrelativistic electron energies, while the nonrelativistic axial bunching process may be dominant at relativistic electron energies.

In a gyrotron, the normally used electron gun is a magnetron injection gun as shown in Fig. 9-1-2 [10].

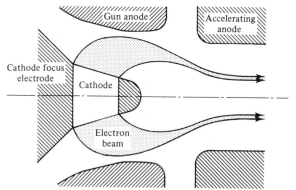

Figure 9-1-2 Magnetron injection gun for gyrotron (Courtesy of Varian Associates, Inc.).

The gun is so-called because of the magnetron-like shape of the cathode structure. The electron beam is formed into a hollow beam by the electric and magnetic fields. Due to the interactions between the electrons and the electromagnetic field the electrons follow a cyclotron path in a waveguide cavity. The tangential velocity of the electron is normally two times the axial velocity so that the electron path is rotational as shown in Fig. 9-1-3 [10].

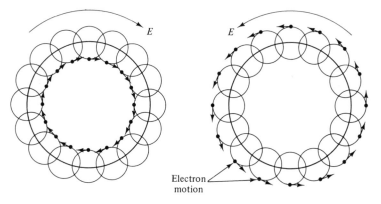

Figure 9-1-3 Electron motion in a gyrotron (Courtesy of Varian Associates, Inc.).

Since the magnetic field is very strong, the orbit diameter for the electrons is small. As a result, the thickness of the hollow electron beam may be several times the diameter of the electron orbit. Consequently, the hollow beam contains a large number of small beams, referred to as beamlets by Dohler [9] as shown in Fig. 9-1-4. The bunched electrons are all moving in the direction of the electric field, and so they all give up energy to the field on each half-cycle of rotation.

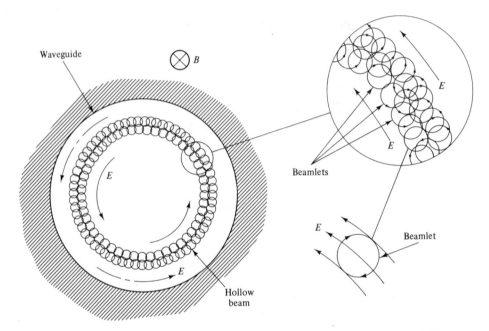

Figure 9-1-4 Electron trajectories in a gyrotron (Reprinted by permission of the IEEE, Inc.).

Example 9-1-1:   Gyro-Traveling-Wave Amplifier

A gyro-traveling-wave amplifier is operating under the following parameters:

Magnetic flux density	$B_0 = 14$ kilogausses
Axial wave number	$k_z = 2.00$ rad/cm
Axial component of the electron drift velocity	$\mathcal{V}_z = 0.25c$
Azimuthal component of the electron drift velocity	$\mathcal{V}_\perp = 0.50c$

Determine:

  (a) The nonrelativistic gyrofrequency of the electron

  (b) The Doppler-shifted frequency

  (c) The relativistic factor

  (d) The Doppler-shifted gyrofrequency

Solution:

(a) From Eq. (9-1-1), the nonrelativistic gyrofrequency is

$$\Omega_e = \frac{e}{m} B_0 = 1.759 \times 10^{11} \times \frac{14 \times 10^3}{10^4} = 2.46 \times 10^{11} \text{ rad/s}$$

(b) The Doppler-shifted frequency is

$$k_z \mathcal{V}_z = 2.00 \times 10^2 \times 0.25 \times 3 \times 10^8 = 1.5 \times 10^{10} \text{ rad/s}$$

(c) The relativistic factor is

$$\gamma = \left[1 - \frac{\mathcal{V}_\perp^2 + \mathcal{V}_z^2}{c^2} \right]^{-1/2} = \left[1 - \frac{(0.50c)^2 + (0.25c)^2}{c^2} \right]^{-1/2} = 1.21$$

(d) The Doppler-shifted gyrofrequency is

$$\Omega_D = k_z \mathcal{V}_z + \Omega_e / \gamma = 1.5 \times 10^{10} + \frac{2.46 \times 10^{11}}{1.21}$$

$$= 2.18 \times 10^{11} \text{ rad/s} = 34.7 \text{ GHz}$$

9-2 RESONANCE CONDITIONS FOR GYROTRON

In the unstable cyclotron maser, electromagnetic radiation is emitted by the relativistically stimulated electrons gyrating about an external magnetic field. The bunching process in the maser instability is a relativistic effect due to the energy-dependent electron cyclotron frequency. If the electrons are uniformly distributed in the phase space, there is no net energy extracted from the electrons through their interactions with the wave electric field. If the electrons are bunched in the phase space, a net beam energy can be extracted from the electrons. The gyrotron emits radiation energy near the frequency at [7]

$$\omega \simeq \Omega_D \simeq \Omega_e / \gamma_0 \qquad (9\text{-}2\text{-}1)$$

In the azimuthal-bunching dominant region, which is due to the relativistic effect, the requirement is therefore

$$\frac{\omega}{k_z} > c \qquad (9\text{-}2\text{-}2)$$

This means that the axial drift velocity of the cyclotron electron is greater than the velocity of light in a vacuum. Similarly, in the axial-bunching dominant region, which is due to the nonrelativistic effect, the requirement is

$$\frac{\omega}{k_z} < c \qquad (9\text{-}2\text{-}3)$$

In other words, the axial drift velocity of the cyclotron electrons is smaller than the velocity of light in a vacuum. The wavelength is determined primarily by the applied magnetic field and is not restricted necessarily by the dimensions of a resonant structure.

Example 9-2-1:   Gyrotron Characteristics

A gyrotron has the following operating parameters:

Magnetic flux density	B_0	$= 1.5$ Wb/m^2
Axial wave number	k_z	$= 210$ rad/m
Azimuthal drift velocity of electron	$\mathcal{V}_{\perp 0}$	$= 0.26c$

Determine:

(a) The nonrelativistic gyrofrequency

(b) The Doppler shifted frequency

(c) The relativistic factor at $t = 0$

(d) The gyrofrequency

(e) The dominant bunching region

Solution:

(a) The nonrelativistic gyrofrequency is

$$\Omega_e = 1.759 \times 10^{11} \times 1.5 = 2.639 \times 10^{11} \text{ rad/s}$$

(b) The Doppler-shifted frequency is

$$k_z \mathcal{V}_z = 210 \times 0.26 \times 3 \times 10^8 = 1.638 \times 10^{10} \text{ rad/s}$$

(c) The relativistic factor at $t = 0$ is

$$\gamma_0 = \left[1 - \frac{(0.26)^2 \, c^2}{c^2} \right]^{-1/2} = 1.0356$$

(d) The gyrofrequency is

$$\omega = \frac{2.639 \times 10^{11}}{1.0356} = 2.548 \times 10^{11} \text{ rad/s}$$

(e) The dominant bunching region is in

$$\frac{\omega}{k_z} = \frac{2.548 \times 10^{11}}{210} = 12.135 \times 10^8 \text{ m/s} > c$$

in the azimuthal region.

9-3 GYROMONOTRON OSCILLATOR

Most of the work on gyrotrons, both theoretical and experimental, has been carried out for two basic structures: gyromonotron (or gyrotron oscillator) and gyrotron traveling-wave amplifiers (gyro-TWA amplifiers).

Operational principles.   The gyromonotron is a single-cavity gyrotron oscillator. The nature of the cyclotron maser interaction in a cavity is much more complicated than that in a waveguide mainly because the electromagnetic wave in a cavity consists of both a forward and a backward component as compared with a single forward component in the waveguide. As a result, nonlinear analysis of the gyrotron oscillator is required. Figure 9-3-1 shows the side view and end view of a single-cavity gyrotron oscillator [8].

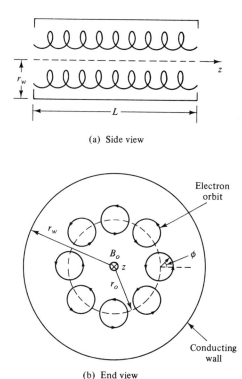

(a) Side view

(b) End view

Figure 9-3-1 Electron motion in a single-cavity gyrotron (Reprinted by permission of the IEEE, Inc.).

The single cavity is constructed by a circular tube of length L with a radius r_w. An annular electron beam is injected into an open-end cavity from the left-hand side and propagates to the right under the guidance of an applied magnetic field B_0. The electrons move along helical trajectories and have a substantial part of their kinetic energy in the form of transverse motion because their azimuthal velocity $\mathcal{V}_\perp$ is twice their axial velocity $\mathcal{V}_z$. Inside the cavity, the electron beam gives up a portion of its energy through interaction with the RF fields. If the average power diffracted out of the cavity, a steady state is then established.

Since the cyclotron maser interaction takes place between the electron beam and the RF-mode field, the $TE_{0n\ell}$ wave is assumed to be in the cavity, where n and ℓ are, respectively, the radial and axial eigenmode numbers. Under these assumptions, the relativistic equation of electron motion in cylindrical coordinates is given by

$$\frac{d}{dt}\,\gamma m\mathbf{v} = -e(\mathbf{E} + \mathbf{v} \times \mathbf{B}) \qquad (9\text{-}3\text{-}1)$$

where m = electron mass

$$\gamma = \left(1 - \frac{\mathcal{V}_\perp^2 + \mathcal{V}_z^2}{c^2}\right)^{-1/2} \text{ is the relativistic factor}$$

$E = E_\phi = E_{\phi 0}J_1(k_n r) \sin(k_z z) \cos(\omega t)$

$\mathbf{B} = \mathbf{B}_0 + B_r \mathbf{r} + B_z \mathbf{z}$

B_0 = applied magnetic flux density predominantly in the positive z direction

$B_r = (k_z/\omega)E_{\phi 0} J_1(k_n r) \cos(k_z z) \sin(\omega t)$

$B_z = -(k_n/\omega)E_{\phi 0} J_0(k_n r) \sin(k_z z) \sin(\omega t)$

$k_z = \ell\pi/L$ is the wave number in z direction

$k_n = X_n/r_w$ is the wave number in r_w direction

X_n = nth nonvanishing root of the Bessel functions

r_w = inner radius of the cavity

$\omega = (k_z^2 + k_n^2)^{1/2} c$ is the wave frequency

Output power and efficiency.   The total wave energy stored in the cavity can be written as

$$W = 0.25\pi\epsilon_0 \, E_{\phi 0}^2 \, J_0^2 (X_n) \, r_w^2 \, L \tag{9-3-2}$$

where $\epsilon_0 = 8.854 \times 10^{-12}$ F/m is the free-space permittivity

$J_0^2(X_n) = 0.16, 0.09, 0.0625$ for $n = 1, 2, 3$

If the quality factor Q of the cavity is assumed to be entirely due to diffraction loss (i.e., neglecting wall loss), the wave power generated by the cavity is given by

$$P_w = \omega W/Q \tag{9-3-3}$$
$$= 0.25\pi\epsilon_0 \, \omega E_{\phi 0}^2 J_0^2(X_n) \, r_w^2 \, L/Q$$

Thus, the efficiency for a beam power P_b required to sustain a steady-state oscillation in the cavity is

$$\eta = \frac{P_w}{P_b} \tag{9-3-4}$$

where $P_b = V_b I_b$

V_b = beam voltage

I_b = beam current

Table 9-3-1 lists data for a 35-GHz gyrotron oscillator.

TABLE 9-3-1. DATA OF 35-GHz GYROTRON
OSCILLATOR [8]

Mode		TE_{011}
Frequency	f	$= 35$ GHz
Beam voltage	V_b	$= 70$ kV
Beam current	I_b	$= 2.31$ A
Velocity ratio	$\mathcal{V}_\perp/\mathcal{V}_z$	$= 2$
Cavity radius	r_w	$= 0.526$ cm
Cavity length	L	$= 4.208$ cm
Beam quiding center position	r_0	$= 0.252$ cm
Magnetic flux density	B_0	$= 12.84$ kG
Efficiency	η	$= 67\%$
Output wave power	P_w	$= 108.7$ kW

After K. R. Chu; reprinted by permission of the IEEE, Inc.

Characteristics.   The gyrotron has emerged in recent years as a new power source of coherent millimeter and submillimeter radiation. Potential applications of the gyrotron include radar, communications, and plasma heating of controlled nuclear fusion devices. To reach the fusion ignition temperature, a great amount of energy (many mega joules) has to be injected for plasma heating. Furthermore, this should be done with the maximum efficiency to alleviate the energy break-even condition. A highly efficient gyrotron has been recognized as one of the most promising sources to meet those requirements. Figure 9-3-2 shows a schematic diagram for a 240-GHz gyromonotron with an output power of 3.5 kW. Table 9-3-2 lists the device parameters of the 240-GHz gyromonotron.

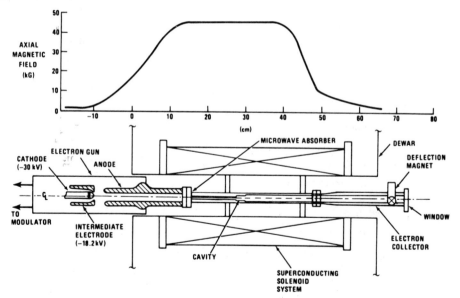

Figure 9-3-2   Schematic diagram of 240-GHz gyromonotron and plot of axial magnetic field (Reprinted by permission of the IEEE, Inc.).

**TABLE 9-3-2.   PARAMETERS OF 240-GHz
GYROMONOTRON [12]**

Frequency	240 GHz
Cavity mode	TE_{051}
Output power	3.5 kW
Cavity diameter × length	6.6 × 26.2 mm
Beam voltage	30 kV
Beam current	0.9 A
Beam geometry	annular
Mean beam diameter	1.2 mm
Initial ratio of electron perpendicular to parallel momentum in cavity	1.5
Beam thickness	0.2 mm
Magnetic field	45 kG
Magnetic field homogeneity in cavity	0.2%
Beam temperature:	
Axial component	10%
Transverse component	5%

After J. D. Silverstein; reprinted by permission of the IEEE, Inc.

Example 9-3-1:   Gyromonotron Oscillator

A gyromonotron oscillator has the following operating parameters:

Mode	TE_{011}	
Operating frequency	f	= 40 GHz
Beam voltage	V_b	= 80 kV
Beam current	I_b	= 3 A
Cavity radius	r_w	= 0.535 cm
Cavity length	L	= 4.35 cm
Electric field in ϕ direction	$E_{\phi 0}$	= 5 MV/m
Cavity quality factor	Q	= 60
Integer	n	= 1

Compute:

(a) The total wave energy stored in the cavity

(b) The wave power generated in the cavity

(c) The overall efficiency

Solution:

(a) The total stored energy is

$$W = 0.25\pi\epsilon_0 \, E_{\phi 0}^2 J_0^2(X_n) \, r_w^2 \, L$$
$$= 0.25 \times 3.1416 \times 8.854 \times 10^{-12} \, (5 \times 10^6)^2 \, (0.16)$$
$$\times (0.535 \times 10^{-2})^2 \times 4.35 \times 10^{-2}$$
$$= 3.46 \times 10^{-5} \, J$$

(b) The generated power is

$$P_w = \omega W/Q = \frac{2\pi \times 40 \times 10^9 \times 3.46 \times 10^{-5}}{60}$$

$$= 145 \text{ kW}$$

(c) The overall efficiency is

$$\eta = \frac{P_w}{P_{dc}} = \frac{145 \times 10^3}{80 \times 10^3 \times 3} = 60 \%$$

9-4 GYROTRON TRAVELING-WAVE AMPLIFIER (GYRO-TWA)

A new type of high-power and high-efficiency generator at microwave frequency range is the gyrotron traveling-wave amplifier (gyro-TWA). The wave amplification results from the interaction of a TE_{01}-mode wave in a circular waveguide with the fundamental cyclotron harmonic of an electron beam. The typical configuration of a gyro-TWA consists of an annular electron beam propagating inside a circular waveguide as shown in Fig. 9-4-1 [9].

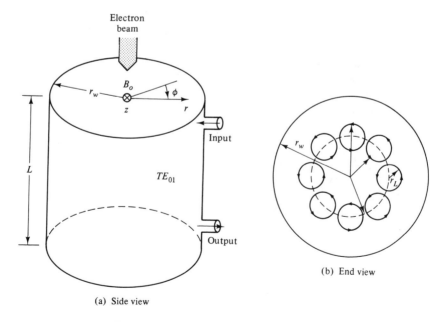

(a) Side view

(b) End view

Figure 9-4-1   Schematic diagram of a gyro-TWA.

In Fig. 9-4-1, the symbols are defined as follows:

$$r_\perp = \mathcal{V}_\perp/\Omega_c = \mathcal{V}_\perp\gamma\bigg/\left(\frac{e}{m}B_0\right) \text{ is Larmor radius}$$

$\dfrac{e}{m} = 1.759 \times 10^{11}$ C/kg is the charge-to-mass ratio of electron

$B_0 = B_z$ is the magnetic flux density in positive z direction

$$\gamma = \left(1 - \frac{\mathcal{V}_\perp^2 + \mathcal{V}_z^2}{c^2}\right)^{-1/2} \text{ is the relativistic factor}$$

r_w = radius of circular waveguide
r_0 = beam guiding center position
$r_1 = r_0 + r_\perp$ is the outer distance of the electron orbit
$r_2 = r_0 - r_\perp$ is the inner distance of the electron orbit
$\mathcal{V}_\perp$ = azimuthal velocity of electron
$\mathcal{V}_z$ = axial velocity of electron

$\Omega_c = \dfrac{e}{m}B_0/\gamma$ is the relativistic cyclotron angular frequency of electron

Operational mechanism. The physical mechanism responsible for the amplification in a gyro-TWA has its origin in a relativistic effect. When the azimuthal velocity of the electrons is twice larger than their axial velocity, the electron beam will be in phase bunching and they will transfer their energy to the field wave for amplification. The wave amplification takes place in a TE$_{011}$-mode circular cavity at the angular frequency

$$\omega^2 = k_z^2 c^2 + \omega_{cu}^2 \tag{9-4-1}$$

where　ω = wave angular frequency
$k_z = \ell\pi/L$ is the axial wave number
ℓ = mode number such as TE$_{0n\ell}$
L = length of the cavity
c = velocity of light in vacuum
ω_{cu} = cutoff frequency of the TE$_{01}$ mode

The fundamental cyclotron harmonic frequency of the electron beam is expressed as

$$\omega = k_z \mathcal{V}_z + \Omega_c \qquad (9\text{-}4\text{-}2)$$

where $\Omega_c = \Omega_e / \gamma$ is the relativistic cyclotron frequency

$\Omega_e = \dfrac{e}{m} B_0$ is the nonrelativistic cyclotron frequency

$B_0 = B_z$ is the magnetic flux density in positive z direction

$\gamma = \left(1 - \dfrac{\mathcal{V}_\perp^2 + \mathcal{V}_z^2}{c^2}\right)^{-1/2}$ is the relativistic factor

In order to achieve high amplification, it is desirable to select the magnetic field such that the group velocity of the waveguide mode nearly equals to the beam velocity—that is,

$$k_z c^2 / \omega - \mathcal{V}_z = 0 \qquad (9\text{-}4\text{-}3)$$

Output power and power gain.   The output power of an amplified wave for a linear-growth gyro-TWA is given by

$$P_{\text{out}} = P_{\text{in}} \exp(2\omega_i \tau) \qquad (9\text{-}4\text{-}4)$$

where $P_{\text{in}} = $ input power

$\omega_i = \overline{\omega}_i c / r_w$ is the linear growth rate

$\overline{\omega}_i = 0.0342$ is the normalized growth rate

$\tau = L / \mathcal{V}_z$ is the transit time of electron

The total power gain is then expressed by

$$\begin{aligned} G &= 8.686 \, \omega_i \tau \quad \text{dB} \\ &= 8.686 \, \omega_i / \mathcal{V}_z \quad \text{dB/unit length} \end{aligned} \qquad (9\text{-}4\text{-}5)$$

Characteristics.   The recent work on gyro-TWA has demonstrated that the device could produce an output power of 0.34 MW at 35 GHz with a power gain of 2 dB per centimeter. Figure 9-4-2 shows the schematic diagram of 35-GHz gyro-TWA and Table 9-4-1 lists its parameters.

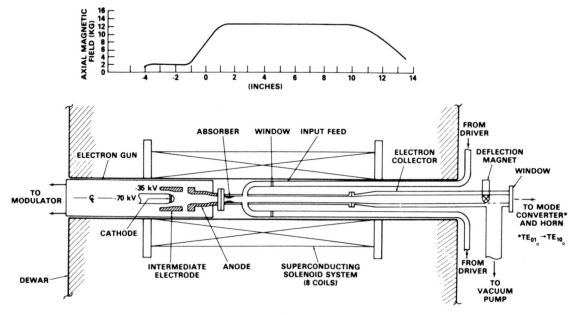

Figure 9-4-2 Schematic diagram of 35-GHz gyro-TWA (Reprinted by permission of the IEEE, Inc.).

TABLE 9-4-1.   CHARACTERISTICS OF 35-GHz GYRO-TWA [6]

n (density parameter)	2.076×10^{-3}
eV (electron energy)	70.82 keV
I_b (beam current)	9.48 A
η (efficiency)	51%
P_b (beam power)	671.5 kW
P_w (wave power)	342.5 kW
B_0 (magnetic flux density)	12.87 kG
k_z	1.96 rad/cm
r_w	5.37 mm
r_0	2.52 mm
r_ℓ	0.61 mm
r_1	1.91 mm
r_2	3.13 mm
$\mathcal{V}_{\perp 0}/c$	0.401
$\mathcal{V}_{z0}/c$	0.268
G (power gain)	2 dB/cm
Bandwidth (for gain = 20 dB)	2.6%

After K. R. Chu; reprinted by permission of the IEEE, Inc.

Varian Associates have demonstrated that the gyro-TWA can achieve a stable power gain of 17 dB at 4.9 GHz. Figure 9-4-3 shows a schematic diagram and photograph for their experimental gyro-TWT amplifier [11].

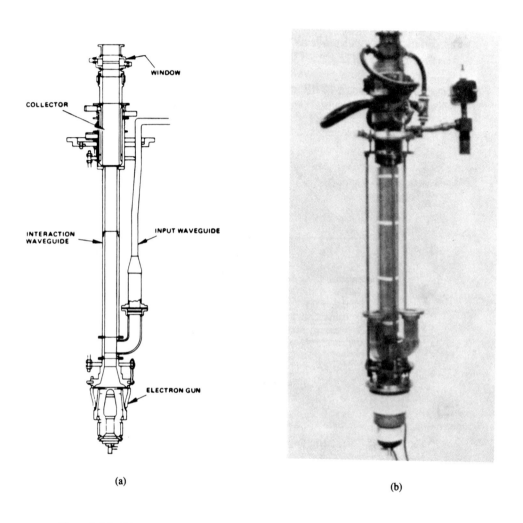

(a)

(b)

Figure 9-4-3   Diagram and photograph of Varian gyro-TWA (Reprinted by permission of the IEEE, Inc.).

Example 9-4-1:   Gyro-TWA

A gyro-TWA has the following operating parameters:

Operating frequency	f	$= 40$ GHz
Beam voltage	V_b	$= 70$ kV
Beam current	I_b	$= 2$ A
Cavity radius	r_w	$= 4.00$ mm
Cavity length	L	$= 5$ cm
Axial velocity	$\mathcal{V}_z$	$= 0.20c$
Input power	P_{in}	$= 2$ W

Determine:

(a) The linear growth rate

(b) The electron transit time

(c) The output power

(d) The power gain in dB

Solution:

(a) The linear growth rate is

$$\omega_i = \frac{\overline{\omega}_i c}{r_w} = \frac{0.0342 \times 3 \times 10^8}{4.00 \times 10^{-3}} = 2.565 \times 10^9 \text{ rad/s}$$

(b) The electron transit time is

$$\tau = \frac{L}{\mathcal{V}_z} = \frac{5 \times 10^{-2}}{0.2 \times 3 \times 10^8} = 0.833 \text{ ns}$$

(c) The output power is

$$
\begin{aligned}
P_{out} &= P_{in} \exp(2\omega_i \tau) \\
&= 2 \exp(2 \times 2.565 \times 10^9 \times 0.833 \times 10^{-9}) \\
&= 2 \times \exp(4.2732) \\
&= 2 \times 71.76 \\
&= 143.51 \text{ W}
\end{aligned}
$$

(d) The power gain is

$$
\begin{aligned}
g &= 8.686 \times 2.565 \times 10^9 \times 0.833 \times 10^{-9} \\
&= 18.558 \text{ dB}
\end{aligned}
$$

9-5 GYROKLYSTRON AMPLIFIER

In addition to the two types of gyrotron structures (gyromonotron and gyro-TWA) described previously, another structure is the gyroklystron amplifier. The two-cavity gyroklystron amplifier induces a transverse phase bunching in the input cavity, allows the bunching to continue ballistically between the input and output cavities, and then converts the transverse electron energy to the wave energy in the output cavity. Figure 9-5-1 shows the schematic diagrams for all three types of gyrotron structures for comparison [6].

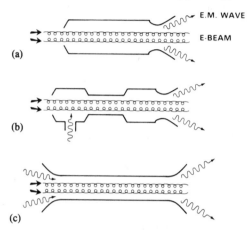

(a)

(b)

(c)

Figure 9-5-1 Types of gyrotrons: (a) Gyromonotron oscillator, (b) Gyroklystron amplifier, (c) Gyro-TWT amplifier (Reprinted by permission of the IEEE, Inc.).

The operation of a gyroklystron amplifier is similar to a conventional klystron except that electron bunching occurs in the azimuthal direction rather than in the axial direction. In a gyroklystron amplifier a signal is fed into the input cavity where the phase bunching is started. Then, the electron beam is permitted to drift to the output cavity. As the beam passes through the second cavity, the signal is amplified and removed. Figure 9-5-2 shows the cross-sectional view of a two-cavity gyroklystron amplifier.

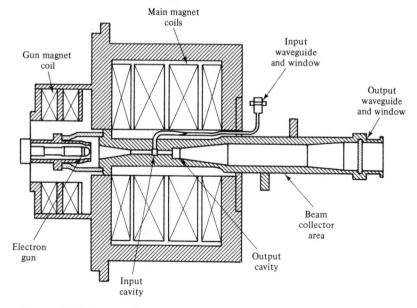

Figure 9-5-2 Cross-sectional view of a gyroklystron amplifier (Courtesy of Varian Associates, Inc.).

9-6 UBITRON

The Ubitron is an *undulated-beam-interaction electron* device. Like the conventional traveling-wave tube, it is an *O*-type traveling-wave amplifier and can produce peak power of 1.60 MW and 0.15 MW at 16 and 54 GHz, respectively.

9-6-1 Operational Mechanism

The Ubitron amplifier operates by interacting an undulating electron beam with the transverse electric field of a circular or rectangular waveguide. The fast electric wave of the waveguide passes through the electron beam, which changes azimuthal direction when the wave changes direction, to give constructive accumulative interaction. The force on the electron is then periodic and the motion of the electron is then undulated. The frequency of operation depends upon the phase velocity of the electric field, the velocity of the electron beam, and the pitch of the magnetic field. The RF TE_{01}-mode wave has no axial electric field component and the wave velocity in the waveguide is greater than the velocity of light ($\mathcal{V}_p > c$). The electron beam is undulated by the periodic magnetic field as shown in Fig. 9-6-1 [13].

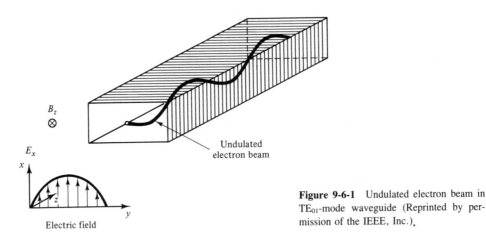

Figure 9-6-1 Undulated electron beam in TE_{01}-mode waveguide (Reprinted by permission of the IEEE, Inc.).

Bunching process.   If the velocity of the electron beam is properly adjusted the electron beam is synchronous with the RF field wave and the beam can give up its kinetic energy to the wave for amplification. The electron bunching process can be explained as follows. The electric field E_x is assumed in the $+x$ direction. The periodic magnetic field B_z is in the positive z direction and has a stationary sinusoidal component plus an RF magnetic field as

$$B_z = B_0 \sin (\beta_0 z) + \mu H_z \qquad (9\text{-}6\text{-}1)$$

The Lorentz force acting on an electron due to the presence of both the electric field $\mathbf{E}$

and the magnetic field **B** is shown in Eq. (2-3-1) as

$$\mathbf{F} = -e(\mathbf{E} + \mathbf{v} \times \mathbf{B}) = m\frac{d\mathbf{v}}{dt} \tag{9-6-2}$$

The periodic magnetic field causes the electron beam to be undulated. When the electron beam is undulated the electrons are classified statistically into three groups: most favorable phase, equal favorable phase, and least favorable phase with respect to the electric field. When the electron has a most favorable phase with respect to the electric field wave, its positions are indicated at three times, t_1, t_2, and t_3 as shown in Fig. 9-6-2 [13].

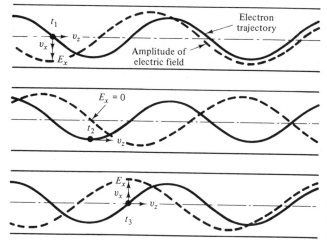

Figure 9-6-2 Electron positions with respect to traveling electric field (Reprinted by permission of the IEEE, Inc.).

At time t_1, the electron has a velocity component $\mathbf{v}_x$ in the negative x direction and the electric field E_x is also in the negative x direction. From Newton's second law of motion, Eq. (2-1-14) is shown as

$$\mathbf{F} = -e\mathbf{E}_x = m\mathbf{a} = m\frac{d\mathbf{v}_x}{dt} \tag{9-6-3}$$

This means that the force is toward the $-x$ direction. As a result, the electron is being decelerated by the electric field.

At time t_2, the electron is moving in the positive z direction with zero electric field, the electron velocity has no change. At time t_3, both the electron and the electric field are in the positive x direction, the electron is again decelerated. On the contrary, those electrons in the least favorable phase with respect to the electric field will be accelerated by the electric field, and those electrons in equal phase with respect to the field will have no effect. Consequently, the transverse velocity modulation in conjunction with the periodic magnetic field results in bunching of the electron beam. This bunching process is similar to the one described in Chapter 5, Sec. 5-4 for klystron amplifier. When the bunched electron beam moves into the decelerating field regions, the beam energy is converted to RF energy, and so amplification is achieved.

9-6-2 Characteristics

The most efficient model of the Ubitron uses a TE_{01}-mode circular waveguide as shown in Fig. 9-6-3 [14].

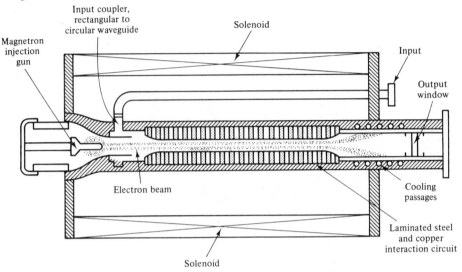

Figure 9-6-3   Schematic diagram of an Ubitron (Reprinted by permission of the IEEE, Inc.).

The circular waveguide is better than rectangular or coaxial waveguide from the standpoint of beam formation, beam control, and interaction efficiency. The circular waveguide is formed from alternate steel and copper discs. When placed in the magnetic field of a solenoid, the waveguide produces both a periodic magnetic field and a dc field that consequently undulates and focuses the electron beam. The waveguide is longitudinally slotted to bring the maximum electric field close to the wall of the waveguide, which is the region of maximum undulating magnetic field. A hollow electron beam from a magnetron injection gun is placed close to the wall of the waveguide to interact with the maximum electric and magnetic fields. The tube uses the rectangular waveguide input, which is transformed to a coaxial and then to a circular waveguide. The output of the tube is in circular waveguide, through a circular ceramic window.

For V-band operation, the beam voltage is 70 kV, the beam current is 35 A, and the magnetic flux density is 10 kG (kilogausses). The average power is about 250 W and the peak power is up to 150 kW at 54 GHz.

9-7 PENIOTRON

The Peniotron, which derives from *penio-*, meaning "spool" in Greek, is a spooling electron tube capable of efficiently generating high power at microwave frequency range.

The device is a member of the gyrodevice family, but it is different from the gyrodevice by operating on nonrelativistic mass change process. The peniotron can operate on two modes: peniotron oscillator and peniotron amplifier.

9-7-1 Peniotron Oscillator

The Peniotron oscillator is a traveling-wave gyrotron and has the following features:

1. The required dc magnetic flux density is lower than that in a gyrotron.
2. High order of spatial harmonic operation is feasible.
3. The dimensions of the structure can be chosen to sustain TE_{10}-mode operation.
4. High efficiency of 95 percent is achievable.

The RF resonant cavity consists of a section of TE_{10}-mode rectangular waveguide with a double pair of ridges and closed at both ends as shown in Fig. 9-7-1 [15].

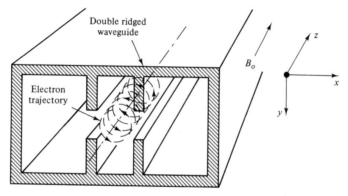

Figure 9-7-1   Waveguide with double pair of ridges for a Peniotron (Reprinted by permission of the IEEE, Inc.).

At each end there is a hole for beam entrance or beam exit that is below waveguide cutoff at operating frequencies. A hollow electron beam is immersed in a magnetic field in the waveguide. The waveguide is ridged to help concentrate the electric field near the electron orbits.

Spooling process.   The axial velocity of the electrons is such that the phasing of the electrons with respect to the RF field is similar to that in the Ubitron (Fig. 9-6-2). As a result, an electron with the most favorable phase follows a helical path, which, from the side, looks the same as the electron trajectory for the Ubitron in Fig. 9-6-2. When the cyclotron frequency is nearly the frequency of the RF field, those favorably phased electrons move up and down in the direction of the electric field as shown in Fig. 9-7-2 [15].

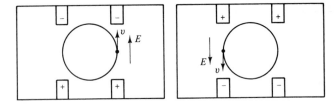

Figure 9-7-2 Electron motion with respect to electric field (Reprinted by permission of the IEEE, Inc.).

It is seen that the electric field moves one complete cycle and the electron advances only one half-cycle. Mathematically, the operating angular frequency can be expressed in terms of the cyclotron angular frequency as

$$\omega = 2p\omega_c \tag{9-7-1}$$

where $p = 1, 2, 3, 4, \ldots$ is the order of operation

$\omega_c = \dfrac{e}{m} B_0$ is the cyclotron angular frequency

$B_0 =$ axial magnetic flux density

The orbit diagrams of electrons in the waveguide are shown in Fig. 9-7-3 [15].

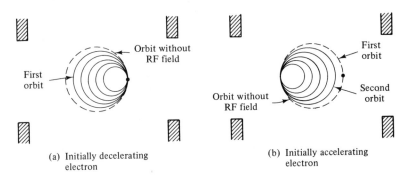

Figure 9-7-3 Orbit diagrams of electrons (Reprinted by permission of the IEEE, Inc.).

When an electron is initially decelerated by the RF field as indicated in Fig. 9-7-3(a), the electron speed is decelerated and its orbit becomes smaller. When the electron has moved one half-cycle and the RF field has advanced one complete cycle, the electron will be decelerated further. However, the electron will be closer to the ridges than before, and its continuing deceleration will result in an overall deceleration for that orbit. The electron will continue to spiral into even smaller orbits with its center moving to the left. By the same analysis, when an electron is initially accelerated its first orbit becomes larger. Then the electron speed is decelerated by the RF field. As a result, the electrons will spiral into smaller orbits and move to the right as shown in Fig. 9-7-3(b).

Operating frequency.   The Doppler-shifted interaction angular operating frequency is related to the RF field angular frequency by [15–17]

$$\omega = \omega_0 \left(1 \pm \frac{\mathcal{V}_\parallel}{\mathcal{V}_{ph}} \right)$$

$$= \omega_0 \pm \beta_0 \mathcal{V}_\parallel$$

(9-7-2)

where $\omega_0 = 2\pi f_0$ is the RF field angular frequency
  $\mathcal{V}_\parallel$ = axial velocity of the electron
  $\mathcal{V}_{ph}$ = phase velocity of the RF field wave

  $\beta_0 = \dfrac{\omega_0}{\mathcal{V}_{ph}}$ is the phase constant of the RF field wave

The phase constant of the Doppler shift is given by

$$\beta_e = \frac{\omega}{\mathcal{V}_\parallel} = \frac{\omega_0 - 2p\omega_c}{\mathcal{V}_\parallel}$$

(9-7-3)

9-7-2 Peniotron Amplifier

For a small-signal operation, the gain and bandwidth of the Peniotron amplifier obey the same relationships as in CFA. However, nonlinear effects will limit the final output power. Therefore, large-signal theory must be employed to find the efficiency and bandwidth in terms of final output condition. The gain parameter is expressed as

$$G = \left[\frac{I_0}{2V_0} \left(\frac{\mathcal{V}_\perp^2 + \mathcal{V}_\parallel^2}{\mathcal{V}_\parallel^2} \right) Z_{2p} \right]^{1/2}$$

(9-7-4)

where I_0 = beam current
  V_0 = beam voltage
  $\mathcal{V}_\perp$ = azimuthal velocity of the electron
  $\mathcal{V}_\parallel$ = axial velocity of the electron

  $Z_{2p} = \dfrac{\omega\mu}{\beta_g} = \eta / \sqrt{1 - (f_c/f_0)^2}$

  $\eta = 377\Omega$ in air waveguide
  f_c = cutoff frequency of the waveguide
  f_0 = RF field wave frequency

Example 9-7-1:   Peniotron Oscillator

A Peniotron oscillator operates in an air-filled waveguide under the following parameters:

Operating frequency	f	= 45 GHz
Cutoff frequency	f_c	= 40 GHz
Phase constant	β_0	= 2 rad/cm
Axial velocity	$\mathcal{V}_\parallel$	= 0.2c

Calculate:

(a) The phase velocity of the RF field wave

(b) The Doppler angular phase shift

(c) The wave impedance in the waveguide

Solution:

(a) The phase velocity is

$$\mathcal{V}_{ph} = \frac{\omega_0}{\beta_0} = \frac{2\pi \times 45 \times 10^9}{200} = 1.413 \times 10^9 \text{ m/s}$$

(b) The Doppler angular phase shift is

$$\omega = \omega_0 \pm \beta_0 \mathcal{V}_{\parallel} = 2\pi \times 45 \times 10^9 \pm 200 \times 0.2 \times 3 \times 10^8$$
$$= 2.827 \times 10^{11} \pm 0.120 \times 10^{11}$$
$$= 2.947 \times 10^{11} \text{ to } 2.707 \times 10^{11} \text{ rad/s}$$

(c) The wave impedance is

$$Z_{2p} = \frac{377}{[1 - (40/45)^2]^{1/2}} = 823 \ \Omega$$

Example 9-7-2:   Peniotron Amplifiers

A Peniotron amplifier operates in an air-filled waveguide under the following parameters:

Operating frequency	f	= 56 GHz
Cutoff frequency	f_c	= 59 GHz
Beam voltage	V_0	= 10 KV
Beam current	I_0	= 3 A
Azimuthal velocity	$\mathcal{V}_\perp$	= 0.4c
Axial velocity	$\mathcal{V}_\parallel$	= 0.1c
Input power	P_{in}	= 100 W

Calculate:

(a) The wave impedance in the waveguide

(b) The gain parameter

(c) The output power

Solution:

(a) The wave impedance in the waveguide is

$$Z_{2p} = \frac{377}{\sqrt{1 - (56/59)^2}} = 1193 \ \Omega$$

(b) The gain parameter is

$$G = \left[\frac{3}{2 \times 10^4} \left(\frac{(0.4c)^2 + (0.1c)^2}{(0.1c)^2} \right) \times 1193 \right]^{1/2}$$
$$= 30.42$$

(c) The output power is

$$P_{out} = 100 \times (30.42)^2 = 92.5 \text{ kW}$$

REFERENCES

[1] READ, M. E., et al., Experimental Examination of the Enhancement of Gyrotron Efficiencies by Use of Profiled Magnetic Fields. *IEEE,* Vol. MTT-30, No. 1, Jan. 1982, pp. 42–46.

[2] TWISS, R. Q., Radiation Transfer and the Possibility of Negative Absorption in Radio Astronomy. *Australian J. Phys.,* Vol. 11, 1958, pp. 564–79.

[3] SCHNEIDER, J., Stimulated Emission of Radiation by Relativistic Electrons in a Magnetic Field. *Phys. Rev. Lett.,* Vol. 2, 1959, pp. 504–5.

[4] GAPONOV, A. V., Interaction Between Electron Fluxes and Electromagnetic Waves in Waveguides. *Izv. VUZ.,* Radiofizika, Vol. 2, 1959, pp. 450–62.

[5] PANTELL, R. H., Backward-Wave Oscillations in an Unloaded Guide. *Proc. IRE,* Vol. 47, 1959, p. 1146.

[6] CHU, KWO RAY, et al., Characteristics and Optimum Operating Parameters of a Gyrotron Traveling-Wave Amplifier. *IEEE Trans. on Microwave Theory and Techniques,* Vol. MTT-27, No. 2, Feb. 1979, pp. 178–89.

[7] CHU, KWO RAY, and HIRSHFIELD, J. L., Comparative Study of the Axial and Azimuthal Bunching Mechanisms in Electromagnetic Cyclotron Instabilities. *Phys. Fluids,* Vol. 21, No. 3, Mar. 1978, pp. 461–66.

[8] CHU, KWO RAY, et al., Methods of Efficiency Enhancement and Scaling for the Gyrotron Oscillator. *IEEE Trans. on Microwave Theory and Techniques,* Vol. MTT-28, No. 4, April 1980, pp. 318–25.

[9] DOEHLER, GUNTER, Gyro-TWA, *IEEE Int. J. Electronics,* p. 50, April 1984.

[10] VARIAN ASSOCIATES, INC., Microwave Bulletin, p. 10, 1985. Palo Alto, CA.

[11] SYMONS, ROBERT S., et al., An Experimental Gyro-TWT. *IEEE Trans. on Microwave Theory and Techniques,* Vol. MTT-29, No. 3, Mar. 1981, pp. 181–84.

[12] SILVERSTEIN, J. D., et al., Practical Considerations in the Design of a High-Power 1-mm Gyromonotron. *IEEE Trans. on Microwave Theory and Techniques,* Vol. MTT-28, No. 9, Sept. 1980, pp. 962–66.

[13] PHILLIPS, R. M., The Ubitron, a High-Power Traveling-Wave Tube Based on a Periodic Beam Interaction in Unloaded Waveguide. *IRE Trans. on Electron Devices,* Vol. ED-7, No. 7, Oct. 1960, pp. 231–41.

[14] ENDERBY, C. E., and PHILLIPS, R. M., The Ubitron Amplifier—A High-Power Millimeter-Wave TWT. *Proc. IEEE,* Vol. 53, No. 10, Oct. 1965, pp. 1648–49.

[15] DOHLER, GUNTER, et al., The Peniotron: A Fast-Wave Device for Efficient High-Power Mm-Wave Generation. *IEDM Technical Digest,* 1978, pp. 400–403.

[16] YAMANOUCHI, K., et al., Cyclotron Fast-Wave tube. The Double Ridge Traveling-Wave Peniotron. *Int. Conference on Microwave Tubes,* Vol. 5, 1964, pp. 96–102.

[17] DOHLER, GUNTER, et al., Peniotron Oscillator Operating Performance. *1981 Int. Electron Devices Meeting,* 1981, pp. 328–31.

PROBLEMS

9-1. Explain the phase-bunching mechanism of the gyrotron and compare it with the velocity-modulation process in the klystron.

9-2. Derive Eqs. (9-1-4) and (9-1-5) from Eq. (9-1-1).

9-3. Derive Eqs. (9-1-7) from Eq. (9-1-6).

9-4. Derive Eq. (9-1-10) from Eqs. (9-1-7) and (9-1-9).

9-5. Derive Eq. (9-1-11) from Eqs. (9-1-5), (9-1-8), and (9-1-10).

9-6. Derive Eq. (9-2-2) from Eq. (9-1-12).

9-7. For a gyrotron-TWT amplifier, $\mathcal{V}_{\perp 0} = 0.401c$ and $\mathcal{V}_{z0} = 0.268c$, calculate the relativistic factor γ.

9-8. If the magnetic flux density B_0 is 12.87 kG (kilogausses) and the azimuthal component of the electron drift velocity $\mathcal{V}_{\perp 0}$ is 0.401c, compute the effective electron mass and the gyrofrequency in GHz.

9-9. If $k_z = 1.96$ per cm and $\mathcal{V}_{z0} = 0.268c$ determine the Doppler-shifted frequency.

9-10. A gyrotron-TWT amplifier has a beam power of 671.5 kW and a wave power of 342.5 kW. Find the efficiency of the device.

9-11. A gyro-TWA has the following parameters:

Magnetic flux density	$B_0 = 12$ kilogausses
Axial wave number	$k_z = 2.5$ rad/cm
Axial component of the electron drift velocity	$\mathcal{V}_z = 0.30c$
Azimuthal component of the electron drift velocity	$\mathcal{V}_\perp = 0.60c$

Compute:

  (a) The nonrelativistic gyrofrequency of the electron

  (b) The Doppler-shifted frequency

  (c) The relativistic factor

  (d) The Doppler-shifted gyrofrequency

9-12. A gyrotron has the following operating parameters:

Magnetic flux density	$B_0 = 1$ Wb/m^2
Axial wave number	$k_z = 200$ rad/m
Azimuthal drift velocity of electron	$\mathcal{V}_{\perp 0} = 0.25c$

Determine:

  (a) The nonrelativistic gyrofrequency

  (b) The Doppler-shifted frequency

  (c) The relativistic factor at $t = 0$

  (d) The gyrofrequency

  (e) The dominant bunching region

9-13. Describe the operational principles of the Ubitron and its potential applications.

9-14. Describe the operational principles of the Peniotron and its potential applications.

9-15. A gyromonotron oscillator operates under the following parameters:

Mode	TE_{011}	
Operating frequency	f	= 35 GHz
Beam voltage	V_b	= 75 kV
Beam current	I_b	= 3.2 A
Cavity radius	r_w	= 0.51 cm
Cavity length	L	= 4.47 cm
Electric field in ϕ direction	$E_{\phi 0}$	= 4.8 MV/m
Cavity quality factor	Q	= 200
Integer	n	= 1

Compute:

(a) The total wave energy stored in the cavity

(b) The wave power generated in the cavity

(c) The overall efficiency

9-16. A gyromonotron oscillator has the following operating parameters:

Mode	TE_{011}	
Operating frequency	f	= 45 GHz
Beam voltage	V_b	= 85 kV
Beam current	I_b	= 4 A
Cavity radius	r_w	= 0.55 cm
Cavity length	L	= 4.5 cm
Electric field in ϕ direction	$E_{\phi 0}$	= 6 MV/m
Cavity quality factor	Q	= 700
Integer	n	= 1

Calculate:

(a) The total wave energy stored in the cavity

(b) The wave power generated in the cavity

(c) The overall efficiency

9-17. A gyro-TWA has the following operating parameters:

Operating frequency	f	= 45 GHz
Beam voltage	V_b	= 80 kV
Beam current	I_b	= 2.5 A
Cavity radius	r_w	= 4.5 cm
Cavity length	L	= 10 cm
Axial velocity	$\mathcal{V}_z$	= 0.20c
Input power	P_{in}	= 3 W

Compute:

(a) The linear growth rate

(b) The electron transit time

(c) The output power

(d) The power gain in decibels

9-18. A gyro-TWA operates under the following parameters:

Operating frequency	f	$= 35$ GHz
Beam voltage	V_b	$= 85$ kV
Beam current	I_b	$= 3$ A
Cavity radius	r_w	$= 5$ cm
Cavity length	L	$= 6$ cm
Axial velocity	$\mathcal{V}_z$	$= 0.28c$
Input power	P_{in}	$= 4$ W

Calculate:

(a) The linear growth rate

(b) The electron transit time

(c) The output power

(d) The power gain in decibels

9-19. A Peniotron oscillator operates in an air-filled waveguide under the following parameters:

Operating frequency	f	$= 43$ GHz
Cutoff frequency	f_c	$= 40$ GHz
Phase constant	β_0	$= 3$ rad/cm
Axial velocity	$\mathcal{V}_\parallel$	$= 0.25c$

Determine:

(a) The phase velocity of the RF field wave

(b) The Doppler angular phase shift

(c) The wave impedance in the waveguide

9-20. A Peniotron amplifier operates in an air-filled waveguide under the following parameters:

Operating frequency	f	$= 57$ GHz
Cutoff frequency	f_c	$= 59$ GHz
Beam voltage	V_0	$= 9$ kV
Beam current	I_0	$= 2.8$ A
Azimuthal velocity	$\mathcal{V}_\perp$	$= 0.38c$
Axial velocity	$\mathcal{V}_\parallel$	$= 0.12c$
Input power	P_{in}	$= 150$ kW

Compute:

(a) The wave impedance in the waveguide

(b) The gain parameter

(c) The output power

CHAPTER 10

Light Image Tubes

10-0 INTRODUCTION

In previous chapters, conventional linear-beam tubes, traveling-wave tubes, crossed-field electron tubes, and fast-wave electron tubes were discussed in detail. Some microwave image tubes, however, are also very useful in commercial and military applications. In Chapters 10 and 11, image tubes such as FLIR tube, Vidicon, Thermicon, Orthicon, Isocon, CCI tube, SEC tube, and SIT tube are described.

In a radar system, the antenna acts as a transmitter and receiver for the electromagnetic radiation. But in an image system, the optical image tubes collect and transmit the image radiant flux. Therefore, the optical image tubes are simply analogous to the antennas in a radar system. This is why the image tubes are equally important as the conventional microwave tubes in commercial and military applications.

Image camera tubes are electrooptical devices that are used to detect the optical image of a scene and convert the detected image into electrical signals suitable to be televised. The most commonly used image tubes operate on the charge image principle and they consist of three main sections, as shown in Fig. 10-0-1 [1].

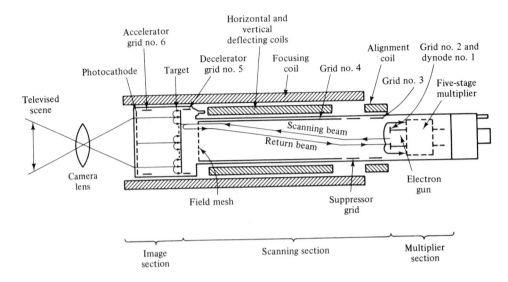

Figure 10-0-1 Schematic diagram of a light image tube (Reprinted by permission of Howard W. Sams & Co., Inc.)

1. *Image Section:* The image storage section stores incident radiant energy of the imaged scene and converts it into an electrical charge image pattern.

2. *Scanning Section:* The electron-beam scanning section reads out the charge pattern as a time-based video signal.

3. *Multiplier Section:* The electron-multiplier section raises the energy level of the small signal above the noise level for adequate display.

Image tubes operate on the infrared and visible-light frequency range by using detectors and converters for display of the scene image. High-image resolution and quality are the most desired requirements for the tube designers. They are commonly used for commercial, military, and space communications systems.

All image tubes use light or infrared radiation as a transmitting medium for signal or image transmission. The infrared and light frequencies are listed in Table 10-0-1.

TABLE 10-0-1. LIGHT AND INFRARED FREQUENCIES

Light frequency

Colors	Frequency	Wavelength	
Infrared	3×10^{12}–4.29×10^{14} Hz	100–0.7	μm
Red	4.29×10^{14}–4.92×10^{14} Hz	0.7–0.61	μm
Orange	4.92×10^{14}–5.08×10^{14} Hz	0.61–0.59	μm
Yellow	5.08×10^{14}–5.26×10^{14} Hz	0.59–0.57	μm
Green	5.26×10^{14}–6.00×10^{14} Hz	0.57–0.50	μm
Blue	6.00×10^{14}–6.67×10^{14} Hz	0.50–0.45	μm
Violet	6.67×10^{14}–6.98×10^{14} Hz	0.45–0.43	μm
Ultraviolet	6.98×10^{14}– 10^{17} Hz	0.43–0.003	μm

Infrared frequency

Near infrared (NIR)	4.29×10^{14}–1×10^{14} Hz	0.7–3	μm
Middle infrared (MIR)	1×10^{14}–5×10^{13} Hz	3–6	μm
Far infrared (FIR)	5×10^{13}–2×10^{12} Hz	6–15	μm
Extreme infrared (XIR)	2×10^{12}–3×10^{11} Hz	15–1000	μm

10-1 IMAGE SECTION

The image section of an image tube consists of two major parts: the tube target and the photocathode. The image detector is the heart of the tube target.

10-1-1 Image Detectors

Image-detecting devices are those that detect quantized radiation and convert it into electrical energy. There are two basic types of radiation-detecting devices: photon detectors and thermal detectors. Incident radiation changes the electrical properties in each of the detectors. Both photon and thermal detectors are quantum detectors, since radiation is quantized. The third type of radiation detector is a combination of the first two, and it may be called the *photothermic detector* or the *imaging detector*.

Operational mechanisms.   From modern solid-state theory, all solids are operated on two thermodynamic mechanisms: electronic and lattice. The electronic mechanism is characterized by the concept of three-energy bands: the valence band, the forbidden band, and the conduction band, as described previously. Radiation is absorbed directly by the electronic system of the solid and produces changes in the electrical properties. The unit of radiation for the electronic mechanism in a solid is a photon. The lattice of a solid is composed of atoms or molecules, and the lattice mechanism is characterized by lattice vibrations. The radiation is absorbed by the lattice, thereby producing heat changes in the lattice. The change in temperature of the lattice causes a change in the electronic system. The unit of radiation for the lattice mechanism is a phonon.

From quantum theory the unit of a photon is $h\nu$ and the unit of a phonon is kT. Because both units are quantized, the quantum condition is determined by

$$h\nu = kT \qquad \text{eV} \tag{10-1-1}$$

where $h = 6.6256 \times 10^{-34}$ watt-sec^2 (or joule-sec) is Planck's constant
  ν = frequency in hertz
  $k = 1.38 \times 10^{-23}$ W-s/°K is Boltzmann's constant
  T = absolute temperature in degrees Kelvin

The quantum-condition temperature is given by

$$T_D = \frac{h\nu}{k} \simeq \frac{\nu}{20} \qquad \text{°K} \tag{10-1-2}$$

where ν = frequency in GHz
  T_D = Debye temperature in degrees Kelvin

For example, if $\nu = 1000$ GHz, the Debye temperature T_D is 50 °K.

At very-low-temperature kT, energy is so small compared to the $h\nu$ of most radiation energies that there are few phonons. At somewhat higher temperatures, there are many low-frequency phonons but few high-frequency ones. At high temperatures, kT becomes greater than $h\nu$ for even the highest-frequency radiations. Therefore, the quantum region is at

$$T \le \frac{h\nu}{k} \qquad \text{°K} \tag{10-1-3}$$

and the classic region is at

$$T > \frac{h\nu}{k} \qquad \text{°K} \tag{10-1-4}$$

Figures of merit.   There are six figures of merit to describe the quality and performance of radiation detectors: (1) responsivity R; (2) time constant τ; (3) noise equivalent power (NEP); (4) detectivity D, specific detectivity D^*, and background-limited detectivity D^{**}; (5) quantum efficiency η_d; and (6) photocurrent gain G.

Responsivity R.   The *responsivity* is a measure of the dependence of the signal output of a detector upon the input radiant power and it is commonly expressed as voltage responsivity in volts per watt or current responsivity in amperes per watt. The voltage responsivity is defined as

$$R(T,f) = \frac{V_s}{P} = \frac{V_s}{HA_d} \qquad \text{volts/watt} \tag{10-1-5}$$

where V_s = rms signal voltage at the output of a detector of area A_d measured at frequency f in response to incident radiation power P (rms) at temperature T
  P = incident radiation rms power

H = irradiance in watts/cm^2
A_d = sensitive area of the detector in cm^2
f = modulation frequency in hertz
T = absolute temperature of the blackbody

Sometimes the responsivity can be expressed as $R(\lambda, f)$, where the output signal is measured at frequency f in response to monochromatic radiation of wavelength λ modulated at frequency f.

The current responsivity is defined as

$$R(T,f) = \frac{\eta_d q}{h\nu} = \frac{\eta_d \lambda_0}{1.24} \qquad \text{amperes/watt} \qquad (10\text{-}1\text{-}6)$$

where η_d = quantum efficiency of the detector
    q = charge of the carrier in coulombs
    $h\nu$ = photon energy in electron volts
    λ_0 = free-space wavelength in micrometers

The voltage or current sensitivity of a microwave solid-state detector is identical to the voltage or current responsivity defined in Eqs. (10-1-5) and (10-1-6).

Alternatively, the responsivity of a radiation detector can be expressed in terms of modulation frequency f and time constant τ—that is,

$$R(f) = \frac{R(0)}{[1 + (\omega\tau)^2]^{1/2}} \qquad (10\text{-}1\text{-}7)$$

where $R(0)$ = responsivity at zero frequency
    τ = time constant of the detector in seconds
    ω = $2\pi f$
    f = modulation frequency in hertz

The cutoff frequency f_c is defined as the frequency at which the responsivity falls to a value $R(0)/\sqrt{2}$ and is thus given by $f_c = 1/(2\pi\tau)$.

Time Constant.   The time constant τ, also known as response time, is a measure of the radiation detector's speed of response and it is defined as

$$\tau = \frac{1}{2\pi f_{3\text{dB}}} \qquad \text{seconds} \qquad (10\text{-}1\text{-}8)$$

where $f_{3\text{dB}}$ is the frequency at which the signal power is 3 dB below the value at zero frequency—that is, the voltage is 0.707 that of the final value or the power is 0.5 that of the final power. In practice, the time constant τ of a detector is the time required for the detector output to reach to 63% of its final value or to drop to 37% from its peak value.

Noise Equivalent Power (NEP).   The *noise equivalent power* is defined as the rms incident radiant power falling on the detector that is required to produce an rms signal voltage or current equal to the rms noise voltage or current at the detector output.

It is expressed as

$$\text{NEP} = \frac{HA_d}{V_s/V_n} = \frac{HA_d V_n}{V_s} = \frac{V_n}{R(T,f)} \qquad \text{watts} \qquad (10\text{-}1\text{-}9)$$

where V_n is the rms noise voltage at the detector output.

The postdetector electrical bandwidth, also called *noise bandwidth,* should be specified. The NEP varies in proportion to the square root of the electrical bandwidth. The smaller the bandwidth, the lower the NEP, and the better the detector sensitivity.

Detectivity D.   The sensitivity of a detector is described by its detectivity. *Detectivity D* is defined as the signal-to-noise ratio per unit incident radiation power and it is then the reciprocal of the noise equivalent power (NEP)—that is,

$$D = \frac{V_s/V_n}{HA_d} = \frac{1}{\text{NEP}} \qquad \text{watt}^{-1} \qquad (10\text{-}1\text{-}10)$$

Specific Detectivity D*.   The *specific detectivity D*,* also known as normalized detectivity, is a reciprocal of the NEP, normalized to a detector area of 1 cm^2 and an electrical bandwidth of 1 Hz. Then, assuming the detector noise varies with $A^{1/2}$ and $B^{1/2}$, the specific detectivity is expressed as

$$D^* = \frac{V_s/V_n B^{1/2}}{HA_d^{1/2}} = \frac{(A_d B)^{1/2}}{\text{NEP}} = D(A_d B)^{1/2} \qquad \text{cm-(Hz)}^{1/2}/\text{watt} \qquad (10\text{-}1\text{-}11)$$

where A_d = detector area in square centimeters (note that for a square detector, $A^{1/2}$ is its linear dimension)

   B = postdetection electrical bandwidth in Hz

Specific detectivity D^* is usually specified by two parameters, such as the source temperature and the modulation frequency. For instance, $D^*(400°K, 900)$ means a value of D^* measured with a 400°K blackbody at a spatial frequency of 900 Hz.

The significant feature of D^* is that the value of D^* is independent of the size of the detector and the NEP for a truly background-limited detector. Therefore, the detectivity of a detector can be determined if the background irradiance and the detector area are specified.

The value of D^* for a specific detector will depend on the wavelength of the signal radiation and the frequency at which it is modulated. Figure 10-1-1 shows the curves of D^* as a function of wavelength for several photodetectors, where PC stands for photoconductive and PV for photovoltaic.

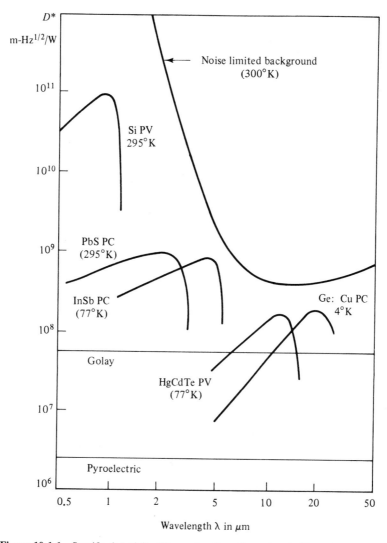

Figure 10-1-1 Specific detectivity D* as a function of wavelength (From J. Wilson, reprinted by permission of Prentice-Hall, Inc.).

Background-Limited Detectivity D**. The detectivity D^{**} (dee-double-star) is a figure of merit that includes the reference conditions of D^* and a reference to a field of view having a solid angle of π steradians (which is equal to a hemisphere). Thus, D^{**} is defined as

$$D^{**} = \left(\frac{\Omega}{\pi}\right)^{1/2} D^* = D^* \sin \theta \qquad \text{cm-(Hz-ster)}^{1/2}/\text{watt} \qquad (10\text{-}1\text{-}12)$$

where $\Omega = \pi \sin^2 \theta$ is the effective solid angle

θ = a half-angle subtended by the detector aperture

For a detector without shield, $\theta = 90°$ and $\Omega = \pi$; then $D^{**} = D^*$. Figure 10-1-2 shows a limited background for a detector.

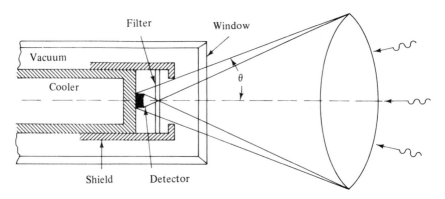

Figure 10-1-2   Detector with limited-background view.

Quantum Efficiency.   When a photon is absorbed by a detector, its quantum energy is transferred to a single electron within the surface. Therefore, the quantum efficiency of a radiation detector is defined as

$$\eta_d = \frac{\text{number of electrons collected}}{\text{number of incident electrons}} \qquad (10\text{-}1\text{-}13)$$

The quantum efficiency for an ideal detector is unity. This means that an ideal detector absorbs the incident radiation and converts it into a voltage or current that is proportional to the total radiation power incident on its surface.

Photocurrent Gain G.   The photocurrent gain G of a detector is defined as the number of charge carriers flowing between the two contact electrodes of a detector per second for each photon absorbed per second. That is,

$$G = \frac{I_p}{I_{\text{ph}}} = \frac{\tau}{\tau_t} \qquad (10\text{-}1\text{-}14)$$

where I_p = photocurrent flowing between the electrodes

I_{ph} = photocurrent

τ = carrier lifetime or recombination time

τ_t = carrier transit time

Table 10-1-2 lists some typical values of figures of merit for several commonly used radiation detectors.

TABLE 10-1-2. FIGURES OF MERIT OF COMMONLY USED
RADIATION DETECTORS.

Detectors	Time constant τ (seconds)	D^{**} [cm-(Hz-ster)$^{1/2}$/W]	Gain G	Operating temperature (°K)
Avalanche photodiodes	10^{-10}		10^2–10^4	300
p-i-n photodiodes	10^{-8}–10^{-12}		10^0	300
Photovoltaic detectors	10^{-6}–10^{-7}	10^8–10^{11}		12–100
Photoconductive detectors	10^{-3}–10^{-8}	10^9–10^{10}	10^6	4.2–300
Bolometers	10^{-3}	10^8–10^{10}		300
Thermocouples	10^{-3}–10^{-5}	10^9		300
Pyroelectric detectors	10^{-1}	10^9		300

Example 10-1-1:   Merit Figures of a Photon Detector

A certain photon detector has the following parameters:

Detecting area	$A_d = 2$ cm^2
Wavelength	$\lambda_0 = 1$ μm
Quantum efficiency	$\eta_d = 20$ %
Bandwidth	$B = 1$ Hz
Noise current	$I_n = 10$ pA

(a) Determine the responsivity.

(b) Find the noise equivalent power.

(c) Calculate the detectivity.

(d) Compute the specific detectivity.

Solution:

(a) From Eq. (10-1-6), the responsivity is

$$R(f) = \frac{\eta_d \lambda_0}{1.24} = \frac{0.2 \times 1}{1.24} = 0.1613 \text{ A/W}$$

(b) Then from Eq. (10-1-9) the noise equivalent power is

$$\text{NEP} = \frac{I_n}{R(f)} = \frac{10 \times 10^{-12}}{161.2 \times 10^{-3}} = 62 \text{ pW}$$

(c) From Eq. (10-1-10) the detectivity is

$$D = \frac{1}{\text{NEP}} = \frac{1}{62 \times 10^{-12}} = 1.61 \times 10^{10} \text{ W}^{-1}$$

(d) Finally, from Eq. (10-1-11) the specific detectivity is

$$D^* = D(A_d B)^{1/2} = 1.61 \times 10^{10}(2 \times 1)^{1/2}$$
$$= 2.28 \times 10^{10} \text{ cm-Hz}^{1/2}/\text{W}$$

Modes of operation.   All radiation detectors can be classified into three modes according to their radiation effects.

Photon-Effect Mode.   The photon effects are characterized by changes in certain properties of a material due to the interactions of the incident photon with the electrons. The photon effects can be subdivided into two types: internal and external.

Internal Photon Effects: The internal photon effects are those in which the photoexcited carriers (electrons or holes) remain within the sample such as photodiodes, as well as photovoltaic and photoconductive detectors.

External Photon Effects: The external photon effect, also known as the photoemission effect, is one in which an incident photon causes the emission of an electron from the surface of the absorbing material. The photoemissive detector is an example.

Thermal-Effect Mode.   The thermal effects are characterized by changes in certain properties of a material resulting from temperature changes caused by the heating effect of the incident radiation. Bolometers and thermocouples are included in this group. Generally they are wavelength independent.

Photothermic-Effect Mode.   The photothermic effect is characterized by a combination of photon and thermal effects. The Vidicon, for example, uses photoconductivity as the detection phenomenon, but its spatial variation is read out by an electron beam.

Photon detectors.   *Photon detectors* are those solid-state devices that are operated under the influence of photon effects. Based upon their physical structures, they can be divided into junction type and bulk type. The biased photodiodes, such as the *pin* photodiodes and avalanche photodiodes (APDs), and the unbiased photovoltaic detectors are the junction type; the photoconductive detectors are the bulk type.

Photon detectors have a small size, minimum noise, low biasing voltage, high sensitivity, high reliability, and fast response time. Therefore, they are very useful in optical-fiber communications systems.

Basically, if a photon of sufficient energy excites an electron from a nonconducting state into a conducting state, the photoexcited electron will generate current or voltage in the detector. From the solid-state theory, the electronic excitation requires that the incident photon energy must be equal to or greater than the electronic excitation energy. In other words, the excitation condition is

$$E_{\text{exc}} \leq h\nu \qquad (10\text{-}1\text{-}15)$$

or

$$E_{\text{exc}} \leq \frac{1.24}{\lambda_0} \qquad (10\text{-}1\text{-}16)$$

where E_{exc} = electronic excitation energy in electron volts
      λ_0 = free space wavelength in micrometers

Photoconductive Detectors.   The *photoconductive detectors* are microwave devices that are constructed by a single crystal of semiconductor material, and its two ends are fixed with ohmic contacts as shown in Fig. 10-1-3.

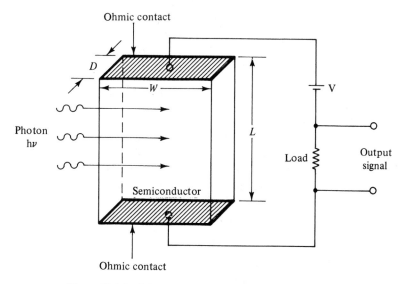

Figure 10-1-3   Schematic diagram of photoconductive detector.

The incident photon energy creates free carriers in the crystal and changes the conductivity of the material as given by

$$\sigma = q(n\mu_n + p\mu_p) \qquad (10\text{-}1\text{-}17)$$

where q = carrier charge
        n = electron concentration
       μ_n = electron mobility
        p = hole concentration
       μ_p = hole mobility

This type of detector can be used for automatic light control in homes and office buildings to turn lights on at dawn and turn them off at dark. Also, they are very useful in optical signaling systems.

There are two types of excitation in a semiconductor, as shown in Fig. 10-1-4. The intrinsic excitation can occur only from the valence band to the conduction band to create an electron-hole pair whereas the extrinsic excitation can take place either from a discrete crystal-defect (dopant) energy level in the forbidden band to conduction band or from the valence band to forbidden band to generate a conduction electron or hole.

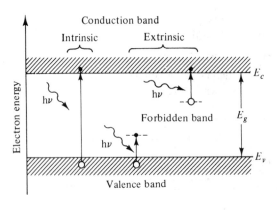

Figure 10-1-4 Processes of photoexcitations. (From S. M. Sze, reprinted by permission of John Wiley & Sons, Inc.).

Intrinsic Photoconductive Detectors.   In the intrinsic mode, the photoconduction is produced by absorption of a photon cross the energy bandgap that creates a free electron and hole simultaneously; the cutoff wavelength is determined by the bandcap energy. The absorption coefficient α is very large due to the large number of available electrons in the valence and conduction bands. Here α is of the order of 10^4 cm^{-1} for photoexcitation near the bandgap energy.

The total current of electrons and holes in an intrinsic photoconductive detector is

$$I = \sigma EWD = qWD(n\mu_n + p\mu_p)E \qquad (10\text{-}1\text{-}18)$$

where W = width of the crystal
D = thickness of the crystal
E = applied electric field across the crystal when the conductivity is measured

The photocurrent is given by

$$I_{\text{ph}} = qWD(\Delta n\mu_n + \Delta p\mu_p)E \qquad (10\text{-}1\text{-}19)$$

where $\Delta n = n - n_0$ is the photoexcited electron concentration
n_0 = electron concentration in equilibrium
$\Delta p = p - p_0$ is the photoexcited hole concentration
p_0 = hole concentration in equilibrium

In general, if there is no trapping, the photoexcited electron and hole concentrations can be expressed as

$$\Delta n = \Delta p = g\tau \qquad (10\text{-}1\text{-}20)$$

where $g = \dfrac{\eta P_{\text{in}}}{h\nu}$ is the generation rate
η = quantum efficiency

P_{in} = incident radiation signal power
τ = carrier lifetime

Extrinsic Photoconductive Detectors.   It is often desirable to reduce the excitation energy level so as to improve the photoconductor performance and minimize the internal noise figure. As a result, the extrinsic photoconductor is used. In this mode, a photon is absorbed at the impurity dopant ionization energy level and then a free electron is created in an *n*-type photoconductor or a free hole in a *p*-type photoconductor. If the incident photon energy is insufficient to induce the intrinsic photoconduction, it still can create an extrinsic photoconduction. Table 10-1-3 lists ionization energies E_i of selected impurities in Ge and Si that correspond to their cutoff wavelengths [1]. The cutoff wavelength is determined by the approximate ionization energy for the impurity. The absorption coefficient α is very small, in the order of 1 to 10 per centimeter.

TABLE 10-1-3. IONIZATION ENERGY AND
WAVELENGTH OF IMPURITY IN GERMANIUM AND
SILICON [3]

Semiconductor	Impurity	Ionization energy in eV	Wavelength λ_0 in μm
Ge	Au	0.15	8.3
Ge	Hg	0.09	14
Ge	Cd	0.06	21
Ge	Cu	0.041	30
Ge	Zn	0.033	38
Ge	B	0.0104	120
Si	In	0.155	8
Si	Ga	0.0723	17
Si	Bi	0.0706	18
Si	Al	0.0685	18
Si	As	0.0537	23
Si	P	0.045	28
Si	B	0.0439	28
Si	Sb	0.043	29

Reprinted by permission of Springer-Verlag, Berlin.

The total currents of the *n*-type and *p*-type modes are given, respectively, by

$$I_n = qWDn\mu_n E = qWDn\mathcal{V}_{dn} \tag{10-1-22}$$

and

$$I_p = qWDp\mu_p E = qWDp\mathcal{V}_{dp} \tag{10-1-23}$$

where $\mathcal{V}_d = \mu E$ is the drift velocity.

The photocurrents are

$$I_{nph} = qWD\Delta n\mu_n E \qquad (10\text{-}1\text{-}24)$$

and

$$I_{pph} = qWD\Delta p\mu_p E \qquad (10\text{-}1\text{-}25)$$

Example 10-1-2:   Photocurrent of a Photoconductive Detector

A typical GaAs photoconductive detector has the following parameters:

Incident radiation power	$P_{in} = 20$ mW
Efficiency	$\eta = 90$ %
Carrier lifetime	$\tau = 1$ ns
Infrared signal frequency	$\nu = 3.45 \times 10^{14}$ Hz
Applied electric field	$E = 400$ V/cm
Crystal width	$W = 100$ cm
Crystal thickness	$D = 10$ cm

  (a) Compute the generation rate.

  (b) Calculate the photoexcited carriers.

  (c) Determine the photocurrent.

Solution:

  (a) From Eq. (10-1-21) the generation rate is

$$g = \frac{\eta P_{in}}{h\nu} = \frac{0.90 \times 20 \times 10^{-3}}{6.625 \times 10^{-34} \times 4.5 \times 10^{14}}$$

$$= 7.88 \times 10^{16} \text{ sec}^{-1}$$

  (b) The photoexcited carriers are

$$\Delta n = \Delta p = g\tau = 7.88 \times 10^{16} \times 10^{-9}$$
$$= 7.88 \times 10^7 \text{ carriers}$$

  (c) Then from Eq. (10-1-19) the photocurrent is

$$I_{ph} = qWD(\Delta n\mu_n + \Delta p\mu_p)E$$
$$= 1.6 \times 10^{-19} \times 10^2 \times 10 \times 7.88 \times 10^7$$
$$\times (8500 + 400) \times 400$$
$$= 44.90 \text{ mA}$$

Photodiodes. *Photodiodes* are usually reverse biased with relatively large biasing voltages in order to reduce the transit time and the diode capacitance for high-speed performance. The photoinduced carriers in the vicinity of the junction modify the current-voltage characteristic so that the radiation level can be measured. Figure 10-1-5 shows a schematic diagram for a *p-n* junction photodiode with its voltage-current characteristic curves.

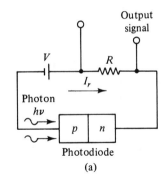

(a)

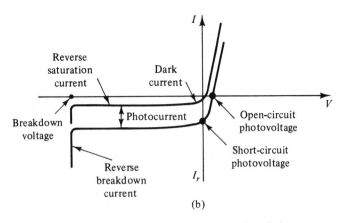

(b)

Figure 10-1-5 Schematic diagram of a photodiode.

When the junction is reverse biased, the observed photosignal is a photocurrent rather than a photovoltage. The dark current due to the thermal generation of electron-hole pairs in the depletion layer is given by

$$I_d = I_{rs}\{\exp[qV/(nkT)] - 1\} \tag{10-1-26}$$

where I_{rs} = reverse-saturation current
  n = constant of the order of unity
  q = charge
  V = applied voltage
  k = Boltzmann's constant
  T = absolute temperature

The average photocurrent due to the optical signal is

$$I_{\text{ph}} = q\eta P_{\text{in}}/(h\nu) \tag{10-1-27}$$

The response time of the detector in optical detection applications is very critical. When a photodiode detector is exposed to a light signal of incident pulses, the photogenerated minority carriers must diffuse to the junction and be swept across the depletion layer to the other side in a time much shorter than the pulse width. Therefore, it is often desirable to increase the width W of the depletion layer, so that most of the photons are absorbed within W rather than in the neutral p and n regions. In addition, a wide W results in a small junction capacitance and eventually reduces the RC time constant of the detector circuit. The p-i-n photodiode and the avalanche photodiode are two of the most common photodetectors with a large, wide depletion layer.

p-i-n Photodiodes.   The p-i-n photodiode has an intrinsic i-region sandwiched between a p-region and an n-region as shown in Fig. 10-1-6. When the diode is reverse biased, the depletion regions occur at the junctions of both p- and n-regions and the effective depletion layer width is increased by the effect of the i-region. Thus, the depletion layer capacitance is much less than that of an ordinary p-n diode and the response time to the modulated light signal is much shorter. This type of photodiode is very useful in optical-fiber communications and star-tracking systems.

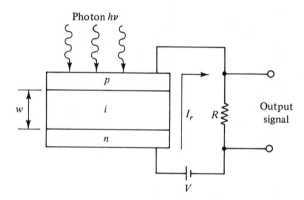

Figure 10-1-6   P-i-n photodiode circuit.

Currents.   When the p-i-n photodiode is reverse biased, the applied voltage appears almost entirely across the i-region. If the carrier lifetime within i-region is long enough compared with the drift time, most of the photogenerated carriers will be collected by the p- and n-regions. When the incident photons are absorbed the electron-hole pairs are released within the i-region. While the electrons drift to the n-region and holes to the p-region under reverse bias, both are added to the reverse current. Under steady-state conditions the total current density through the reverse-biased depletion layer is given by [4]

$$J_{tot} = J_{dr} + J_{diff}$$ (10-1-28)

where J_{dr} = drift current
       J_{diff} = diffused current

The drift current density is due to carriers generated inside the depletion region, and it is expressed as

$$J_{dr} = q\Phi_0(1 - e^{-\alpha W}) \qquad (10\text{-}1\text{-}29)$$

where $\Phi_0 = \dfrac{P_{in}(1 - R)}{Ah\nu}$ is the incident photon flux per unit area per second

P_{in} = incident radiation signal power
R = reflection coefficient
A = detecting area
W = width of i region
α = absorption or ionization coefficient

The diffusion current density is due to the carriers generated outside the depletion layer and diffusing into the reverse-biased junction, and it is given by

$$J_{diff} = q\Phi_0 \frac{\alpha L_p}{1 + \alpha L_p} e^{-\alpha W} + q p_0 \frac{D_p}{L_p} \qquad (10\text{-}1\text{-}30)$$

where L_p = hole diffusion length
D_p = hole diffusion coefficient
p_0 = hole concentration in equilibrium

Then the total current density is

$$J_{tot} = q\Phi_0 \left(1 - \frac{e^{-\alpha W}}{1 + \alpha L_p}\right) + q p_0 \frac{D_p}{L_p} \qquad (10\text{-}1\text{-}31)$$

Under normal operating conditions, the last term is very small and it may be neglected.

Avalanche photodiodes (APD).   If the amplitude of the reverse bias across a photodiode is increased to the breakdown-voltage level, avalanche multiplication occurs, thus resulting in a much larger current gain than the other photodiodes. The *avalanche photodiodes* (APDs) have a low noise figure over a wide bandwidth and their typical gain-bandwidth product is 100 GHz. They are widely used with lasers in optical-fiber communications systems.

Multiplication and Current.   For equal absorption coefficients ($\alpha_n = \alpha_p$), the multiplication factor is expressed by [5]

$$M = 1/(1 - \alpha_n W) \qquad (10\text{-}1\text{-}32)$$

If both leakage and photocurrents are subjected to avalanche multiplication, the combined multiplication factor is given by

$$M = \frac{I}{I_u} = \frac{I_{ph} - I_d}{I_{phu} - I_{du}} = \frac{1}{1 - (V_r/V_b)^n} \qquad (10\text{-}1\text{-}33)$$

where I = total multiplied current
I_u = total unmultiplied current
I_{ph} = multiplied photocurrent
I_d = multiplied dark current
I_{phu} = unmultiplied photocurrent
I_{du} = unmultiplied dark current
V_r = reverse bias voltage
V_b = breakdown voltage
n = a constant

The multiplication factor M can have a value between 1 and 100, depending on V_r. Figure 10-1-7 shows an equivalent circuit for an avalanche photodiode.

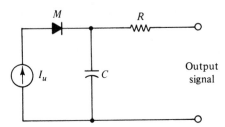

Figure 10-1-7   Equivalent circuit for an avalanche photodiode.

Example 10-1-3:   Photocurrent of an Avalanche Photodiode

A typical avalanche photodiode has the following parameters:

Incident radiation power	P_{in} = 50 mW
Efficiency	η = 9.5 %
Red-light frequency	ν = 4.5 × 10^{14} Hz
Breakdown voltage	V_b = 35 V
Reverse bias voltage	V_r = 34 V
Dark current	I_d = 10 nA
Constant	n = 2

(a) Compute the multiplication factor M.
(b) Calculate the generation rate g.
(c) Determine the average photocurrent I_{ph}.

Solution:

(a) From Eq. (10-1-33), the multiplication factor is

$$M = [1 - (V_r/V_b)^2]^{-1} = [1 - (34/35)^2]^{-1} = 16.67$$

(b) The generation rate is

$$g = \frac{\eta P_{in}}{h} = \frac{0.95 \times 50 \times 10^{-3}}{6.625 \times 10^{-34} \times 4.5 \times 10^{14}}$$
$$= 1.59 \times 10^{17} \text{ sec}^{-1}$$

(c) The average photocurrent is

$$I_{ph} = qg = 1.6 \times 10^{-19} \times 1.59 \times 10^{17}$$
$$= 25.50 \text{ mA}$$

Photovoltaic detectors.   A *photovoltaic detector* is a microwave solid-state device that is operated under the intrinsic photoconductive effect without bias and it requires an internal potential barrier with a built-in electric field to separate the photoexcited electron-hole pairs [6]. The physical structure of a photovoltaic detector consists of a *p-n* junction formed in an intrinsic semiconductor as shown in Fig. 10-1-8.

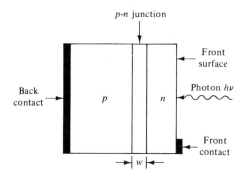

Figure 10-1-8   Diagram of a p-n junction photovoltaic detector.

When the incident photons impinge upon the detector, the induced electron-hole pairs are separated by the junction electric field and a photovoltage is generated as shown at point *A* in Fig. 10-1-9. Figure 10-1-9 shows several different modes of operation for photodiodes. Curve 1 indicates the *V-I* curve for a photodetector without incident radiation, and curve 2 is for detection with incident radiation. The photovoltaic mode operates at point *A* for open-circuit voltage in a high impedance without bias. The photoconductive mode operates at point *C* and the avalanche photodiode mode is at point *D*.

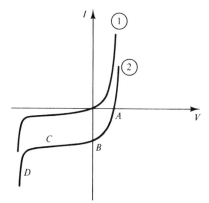

Figure 10-1-9   V-I characteristic curves for photodiodes.

The photovoltaic detector has no bias supply and its circuit is much simpler than the other detectors. Its detectivity is usually 40% greater than a photoconductive mode.

Because the photovoltaic detector operates only in intrinsic photoexcitation, its bandgap energy must be equal to or less than the incident photon energy and its cutoff wavelength is determined by the bandgap energy.

Photoemissive detectors.   *Photoemissive detectors* are microwave solid-state devices that operate under a photoemissive effect. Because of their high gain, low noise, and fast response, they are useful for the detection of low-intensity signals, high-speed signals, and high-resolution spatial information.

Photoemissive Effect.   The *photoemissive effect*, also known as the *external photoeffect,* is a process in which the action of the incident radiation causes the electrons to be emitted from the photocathode surface into a vacuum and be collected by an anode. From modern solid-state theory the ionization energy of a solid, which is also known as the work function ϕ, is the minimum energy required for an electron to escape from the solid at the Fermi level into its surface at rest (not hot). If the incident photon energy is greater than the electron bandgap energy of an intrinsic semiconductor, the electron will have kinetic energy after escaping from the intrinsic semiconductor—that is,

$$h\nu = E_g + \tfrac{1}{2} m\mathcal{V}^2 \tag{10-1-34}$$

where $E_g = \Phi$ is the bandgap energy for an intrinsic semiconductor
     Φ = work function

The electron affinity χ of a semiconductor is defined as the energy level measured from the conduction energy band edge to the vacuum level E_{vac}. That is,

$$\chi = E_{\text{vac}} - E_c = \Phi + E_F - E_g \tag{10-1-35}$$

where E_F = Fermi energy level and the valence-band energy is taken as zero
     The photoemissive threshold energy E_{th} for an electron in a semiconductor is given by

$$E_{\text{th}} = E_g + \chi \tag{10-1-36}$$

This semiconductor optical ionization energy is constant regardless of doping. The photoemission threshold energy can be increased or decreased, however, depending upon whether the electron affinity is positive or negative.

Photoemissive detectors can be divided into two groups. The first group, which is called the classic type, consists of metal photocathodes and semiconductor photocathodes with positive electron affinity (PEA). The second group includes the semiconductor photocathodes that have a negative electron affinity (NEA). The emitting surfaces of both groups are coated with a very thin evaporated metallic layer.

Metal Photocathodes.   Such devices are old types like the vacuum phototubes. Because the metals have a work function of several electron volts, they detect only the radiation at the wavelength range from extra violet to visible light—that is, 0.003 to 0.7 μm.

Semiconductor Photocathode with Positive Electron Affinity (PEA).   Because
the electron affinity is positive, the photoemissive threshold energy is quite high. Even
though their wavelength response to the radiation is better than the metal photocathode,
they detect the radiation from 0.7 to 1.0 μm at the near infrared range. Figure 10-1-10
shows the energy band diagram for a positive electron affinity (PEA) photocathode.

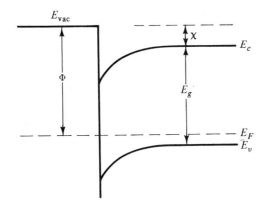

Figure 10-1-10   Energy-band diagram of a
PEA semiconductor photocathode.

Example 10-1-4:   Positive Electron Affinity (PEA) of a Photocathode

A typical cesium antimonide (CsSb) photocathode has the following parameters:

Work function	Φ	= 1.65 eV
Bandgap energy	E_g	= 1.60 eV
Fermi energy level	E_F	= 0.40 eV

Determine the electron affinity of the photocathode.

Solution: From Eq. (10-1-35), the electron affinity is

$$\chi = \Phi + E_F - E_g = 1.65 + 0.40 - 1.60 = 0.45 \text{ eV}$$

It is a PEA photocathode.

Semiconductor Photocathode with Negative Electron Affinity (NEA).   This type
of detector is constructed by overcoating the surface of selected *p*-type semiconductors
with an evaporated layer of low work function material. Figure 10-1-11 shows the energy-
band diagram for a negative electron affinity (NEA) photoemitter.

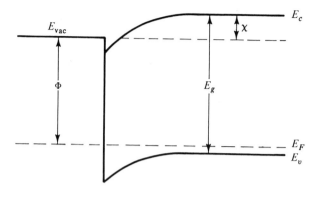

Figure 10-1-11 Energy-band diagram of an NEA semiconductor photocathode.

The incident photon energy for an NEA semiconductor photoemitter must equal or exceed the bandgap energy of the semiconductor in order for photoemission to occur. The operation wavelength range for NEA semiconductor photoemitters is up to 3 μm. NEA photoemitters have five advantages over the classic ones:

1. High quantum efficiency
2. Wide spectral response up to near infrared
3. High resolution
4. High uniform absolute sensitivity and
5. Low dark current at room temperature

However, the NEA photoemitters also have some disadvantages:

1. High cost
2. Small size
3. Low durability
4. Low response
5. Difficult fabrication

Example 10-1-5: Negative Electron Affinity (NEA) of a Photoemitter

A typical GaAs/CsO photoemitter has the following parameters:

Work function	Φ	$= 1.08$ eV
Bandgap energy	E_g	$= 1.43$ eV
Fermi energy level	E_F	$= 0.01$ eV

Determine the electron affinity.

Solution: From Eq. (10-1-35) the electron affinity is

$$\chi = \Phi + E_F - E_g = 1.08 + 0.01 - 1.43 = -0.34 \text{ eV}$$

It is an NEA photoemitter.

10-1-2 Tube Targets

Tube target of a camera tube is the section that detects the incident photon radiation emitted from the scene. The tube target can be constructed in many different ways and is commonly overcoated with a thin layer of detecting material. In general, there are two types: continuous surface and mosaic surface. Most targets have only one floating side where the charge pattern is stored, whereas the other side is held at the same potential with the signal plate. In a bisided target, however, both sides are floating, which gives rise to a charge pattern on both sides of the target. Figure 10-1-12 shows several schematic diagrams of tube targets [1].

Vidicon target.   The Vidicon target is composed of a transparent signal plate overcoated by a photoconductive layer of tin oxide (SnO_2) as shown in Fig. 10-1-12(a). In operation, the signal plate is connected to a positive potential and the scan side of the photoconductor is maintained at the photocathode potential by a low-energy scanning beam. The target may be considered to consist of many elemental capacitors parallel connected with elemental resistors. In the dark the resistance of the elemental resistors is very high, and the elemental capacitors remain fully charged to the signal plate voltage. When a light image is focused on the target, however, photoconduction takes place and the elemental resistance is decreased. As a result, the conduction current discharge the elemental capacitors and a charge image pattern is formed. A video signal is then generated by the capacitive displacement current during the recharging of the elemental capacitors by the scanning electron beam.

Silicon diode-array target.   The silicon diode-array target consists of a two-dimensional array of diodes deposited in an *n*-type silicon wafer as shown in Fig. 10-1-12(b). The diode array is formed by diffusion of a *p*-type dopant through the openings of a silicon oxide film present on the surface. The oxide film separates the diode matrix and prevents the beam from striking the *n*-type material. In operation, the diodes are reverse-biased and each diode remains as an elemental capacitor. When a light image is focused on the diode array target, a charge image pattern is formed. The photoconductions produce a great number of hole-electron pairs for each primary electron by interacting with the silicon layer and discharge the storage capacitor. As a result, a video signal is generated.

Plumbicon target.   The Plumbicon target is composed of a lead oxide (PbO_2) transparent signal plate as shown in Fig. 10-1-12(c). The lead-oxide transparent plate is an *n*-type conductivity signal plate overlaid by an intrinsic (*i*-type) conductivity type of

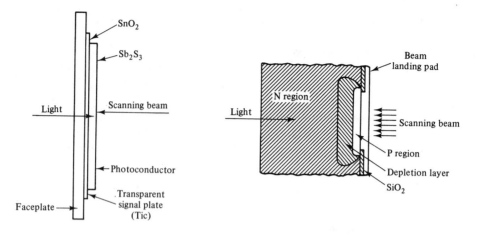

(a) Vidicon target

(b) Silicon target

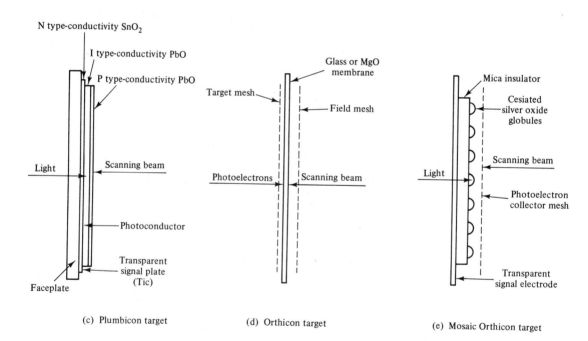

(c) Plumbicon target

(d) Orthicon target

(e) Mosaic Orthicon target

Figure 10-1-12   Schematic diagram of several image tube targets (Reprinted by permission of Howard W. Sams & Co., Inc.).

lead oxide layer that is covered by a thin layer of p-type conductivity lead oxide. In operation, the signal plate is connected to a positive potential of $+50$ V and the p-type layer is maintained at ground potential by a low-energy scanning. The dark current of the photoconductor is extremely low. When a light image is focused on the target, electrons, and holes are produced in the bulk of the i-type photoconductor causing photoconduction. As a result, the conduction current generates a charge image pattern that can be read out as a video signal by a low-energy scanning beam. Therefore, the Plumbicon target is more sensitive than the ordinary Vidicon target.

Orthicon target.   The Orthicon target consists of a thin membrane of semiconducting glass or magnesium oxide as shown in Fig. 10-1-12(d). In operation, both sides of the target are floating, and the potentials of the two sides are determined by the interactions of the scanning and imaging electron beams with the target. On the average, each photon electron produces a multiple number of secondary electrons, which allows the target surface to charge positive. A video signal is then induced as an amplitude modulation of the scanning beam. If there is no positive charge on the target, all primary electrons are reflected to the electron gun. If there are positive charges on the target, the scanning beam deposits a sufficient amount of electrons to neutralize the positive charges and only the remaining electrons are reflected. As a result, the return beam is amplitude modulated by the charge image pattern.

Mosaic Orthicon target.   Figure 10-1-12(e) shows a mosaic Orthicon target. The mosaic surfaces have lower light sensitivity than the continuous surfaces because the void areas cause energy losses. The photoemissive islands are composed of cesium and oxygen-activated minute silver globes in a random pattern.

CCI target.   The charge-coupled imager (CCI) target is a two-dimensional array of metal-oxide semiconductor (MOS) diodes or metal-insulator semiconductor (MIS) diodes. The CCI can move the charge image pattern in the MOS diode along a predetermined path under the control of clock pulses. In operation, the information or signal is stored in the form of an electrical charge pattern in the potential wells created in MOS diodes. Under the control of externally applied voltage (that is, gate voltage), the potential wells and the charge patterns can be shifted rapidly from one well to an adjacent one through the entire CCI structure. Actually, the CCI structure is identical to the charge-coupled device (CCD) that is commonly described in solid-state textbooks. The basic structure is explained below.

Energy-Band of MIS Diode.   A single MIS structure on an n-type semiconductor (or p-type semiconductor) is the basic element of the CCD. Figure 10-1-13 shows the energy band diagrams for a MIS structure [7].

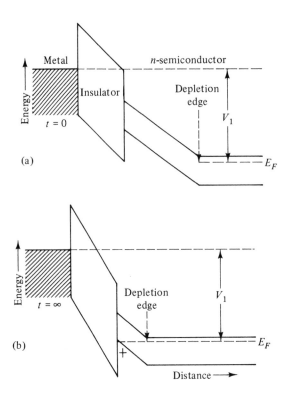

Figure 10-1-13 Energy-band diagram of MIS structure. (After W. S. Boyle and G. E. Smith, reprinted by permission of Bell System, AT & T.).

The voltage applied to the metal electrode is negative with respect to the semiconductor and large enough to cause depletion. When the voltage is first applied at $t = 0$, there are no holes at the insulator-semiconductor interface, as shown in Fig. 10-1-13(a). As holes are introduced into the depletion region, they will accumulate at the interface and cause the surface potential to be more positive, as shown in Fig. 10-1-13(b).

Three-Phase Structure.   The CCD can be constructed in the form of a typical three-phase structure as shown in Fig. 10-1-14 [8]. The CCD consists of a closely spaced array of MIS diodes on an n-type semiconductor substrate with a large negative gate voltage applied. Its basic function is to store and transfer the charge packets from one potential well to an adjacent one. As shown in Fig. 10-1-14(a), $V_1 = V_3$, and V_2 is more negative. In effect, a potential well with stored holes is created at gate electrode 2. The stored charge is temporary because a thermal effect will diffuse the holes out of the wells. Therefore, the switching time of the voltage clock must be fast enough to move all charges out of the occupied well to the next empty one. When the voltage V_3 is pulsed to be more negative than the other two voltages V_1 and V_2, the charge begins to transfer to the potential well at gate electrode 3, as shown in Fig. 10-1-14(b).

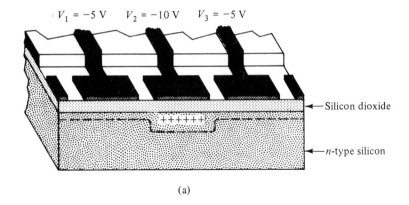

$V_1 = -5$ V $V_2 = -10$ V $V_3 = -5$ V

Silicon dioxide

n-type silicon

(a)

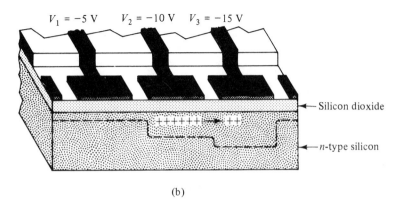

$V_1 = -5$ V $V_2 = -10$ V $V_3 = -15$ V

Silicon dioxide

n-type silicon

(b)

Figure 10-1-14   Cutaway of CCD. (From Boyle and Smith, reprinted by permission of the IEEE, Inc.).

Store and Transfer of Charge Packets.   A linear array of MIS diodes on an n-type semiconductor is shown in Fig. 10-1-15 [7]. For a three-phase CCD, every third gate electrode is connected to a common line as shown in Fig. 10-1-15. At $t = t_1$, a more negative voltage V_1 is applied to gate electrodes 1, 4, 7, and so on, and less negative voltages V_2 and V_3 ($V_2 = V_3$) are applied to the other gate electrodes. It is assumed that the semiconductor substrate is grounded and that the magnitude of V_1 is larger than the threshold voltage V_{th} for the production of inversion under the steady-state conditions. As a result, positive charges are stored in the potential wells under electrodes 1, 4, 7, . . . , as shown in Fig. 10-1-15(a).

At $t = t_2$, when the voltage V_2 at the gate electrode 2, 5, 8, . . . is pulsed to be more negative than V_1 and V_3, the charge packets will be transferred from the potential wells at electrodes 1, 4, 7, . . . to the potential minimum under electrodes 2, 5, 8, . . . , as shown in Fig. 10-1-15(b).

At $t = t_3$, when voltages $V_1 = V_3$ and voltage V_2 remain more negative, the charge packets have been transferred one spatial position and the sequence is ready to be continued, as shown in Fig. 10-1-15(c).

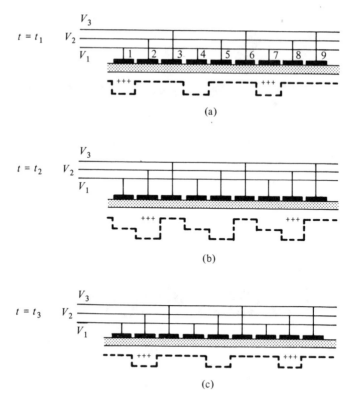

Figure 10-1-15 Store and transfer of charge packets for a three-phase CCD. (From Boyle, courtesy of the Bell System, AT & T.).

10-1-3 Photocathodes

The *photocathode* of an image tube is a semitransparent photoemissive material that deposits on the vacuum side of the tube input faceplate. In general, there are two major types from the principle of operation:

1. *Positive Electron Affinity (PEA) Cathode:* This type consists of polycrystalline materials deposited in high vacuum on the input faceplate such as the alkali antimonide cathode, the silver-oxygen-cesium cathode, and the solar-blind cesium-activated cathode.

2. *Negative Electron Affinity (NEA) Cathode:* This type is composed of epitaxially grown single-crystal materials such as generation 3 GaAs: CsO/AlGaAs photocathode.

Alternatively, according to the spectral sensitivity characteristics, the image tube photocathodes can be classified into three groups:

1. *Color-light Photocathode:* It operates for the wavelength range from 0.40 μm to 0.70 μm. This group includes bialkali cathode and cesium antimony cathode.

2. *Visible Light and Infrared Photocathode:* It operates in the wavelength range from 0.30 μm to 1.20 μm. This group includes multialkali cathode and silver-oxygen-cesium cathode.

3. *Ultraviolet Photocathode:* It operates for the wavelength range from 0.12 μm to 0.32 μm. This group includes solar-blind cathode.

The axis in a lens is a straight line through the geometrical center of the lens and normal to the two faces at the points of intersection as shown in Fig. 10-1-16.

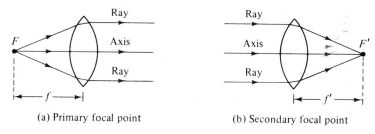

(a) Primary focal point   (b) Secondary focal point

Figure 10-1-16   Focal point and focal length.

The *primary focal point F* lies on the axis and is defined for a positive lens as the point from which diverging rays are refracted by the lens into a parallel beam. The *secondary focal point F'* is defined by applying the principle of reversibility to the same lens. The distance between the center of a lens and either of its focal points is called its *focal length*.

1. *Magnification and Conjugate.* In any optical device the ratio between the transverse dimension of the final image and the corresponding dimension of the original object is defined as *magnification* of the lens. If any object is placed at the position previously occupied by its image, it will be imaged at the position previously occupied by the object. In this condition, the object and image are said to be *conjugate*. The conjugate plane for a collimator is at an infinite distance.

2. *Telescope.* In an image system, the image radiation flux from an object scene is usually collected by a telescope and brought out in a collimated beam with a small pupil at the scan mirror. A terrestrial-type telescope with a two-lens erecting system is commonly used in the image system.

3. *Field-of-View and Stops.* The field-of-view (FOV) in geometrical optics determines how much of the surface of a broad object can be seen through an optical system.

It is often subdivided into narrow field-of-view (NFOV), mediate field-of-view (MFOV), and wide field-of-view (WFOV) depending on the size of the aperture. In order to control the brightness of images, an aperture stop is placed between the lens and the focal plane to limit the incident bundles of rays; the field stop in front of the focal plane determines the extent of the object, or the field, that will be produced in the image.

10-2 SCANNING SECTION AND MULTIPLIER SECTION

The scanning section of an image tube reads out the image charge pattern that is formed in the tube target as a time-based video signal, and the multiplier section raises the signal level that modulates the returned electron beam sufficiently above the noise level for adequate signal display.

10-2-1 Scanning Section

The scanning section is composed of an electron gun and a focusing coil. The electron gun consists of a thermionic cathode, a control grid, and an accelerating grid. The gun produces a low-velocity electron beam and the beam is focused on the target by a magnetic field. The electron beam deposits equal amounts of negative charges on the target to neutralize the positive charge image pattern. Because the target cannot accept more electrons without becoming negative the remainder of the electrons not deposited is repelled as a return beam. As a result, the return beam is amplitude modulated by the absorption of electrons at the target and a video signal is then generated.

10-2-2 Multiplier Section

The multiplier section is used to raise the signal-to-noise ratio of the image tube. The return beam is usually directed to the dynodes of several-stage electrostatically focused multiplier. The dynode material is a silver-magnesium alloy. The multiplier section can amplify the video signal more than 500 times without significant reduction of the signal-to-noise ratio. The response of the image tube is a result of many variables such as spatial frequency, light intensity, and different objects. Mathematically, the image contrast or quality of an image tube can be specified in terms of the modulation transfer function (MTF) as

$$MTF = \frac{B_{max} - B_{min}}{B_{max} + B_{min}} \tag{10-2-1}$$

where B_{max} = maximum intensity of a sine-wave image pattern
B_{min} = minimum intensity of a sine-wave image pattern

Alternatively, the modulation transfer function can be expressed as

$$MTF = \frac{M_i}{M_o} \tag{10-2-2}$$

where M_i = modulation in image
M_o = modulation in object

10-3 VIDICON

The *Vidicon* is a small image tube invented by RCA (Radio Company of America) and is often applied to any of a variety of image tubes that have a photoconductive target. The tube consists of an evacuated tube closed by a flat transparent faceplate at one end and by an electron gun at the opposite end as shown in Fig. 10-3-1 [9].

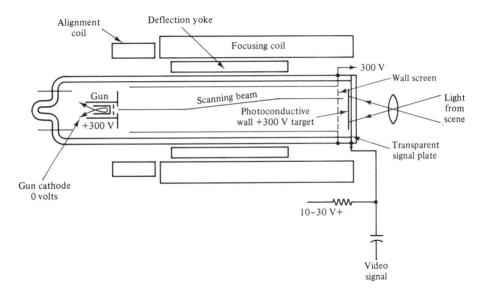

Figure 10-3-1   Schematic diagram of a Vidicon (Courtesy of the Electronics, Inc.)

10-3-1 Physical Structure

The faceplate, which is called the *signal electrode* or *plate,* is overcoated with a thin layer (1 to 2 μm) of photoconductive semiconductor film, such as antimony trisulfide (Sb_2S_3) for image illumination. The electron gun, which is called the *cathode electrode,* emits the electron beam that scans over the photoconductive layer (usually called the *target*). A uniform magnetic field is maintained to focus the beam. The video signal is taken from the target by connecting the amplifier to the transparent signal plate.

A fixed potential of about 20 V positive, relative to the thermionic cathode, is applied to the transparent signal plate. The electron beam deposits electrons on the scanned surface of the photoconductor, charging it down to thermionic cathode potential in the absence of any radiation. In effect, the conductivity of the photoconductor is sufficiently low that very little current flows in the dark.

If a light image is focused on the target, its conductivity is increased to about 10^{-12} mho in the illuminated areas, thus permitting charge to flow. In these areas the scanned surface gradually becomes charged a volt or two positive with respect to the cathode during the frame time (1/30 s) between successive scans.

The electron beam deposits sufficient electrons to neutralize the charge accumulated during the frame time, and in doing so it generates the video signal in the signal plate lead. Because the target is sensitive to light throughout the entire frame time a full storage of charge is achieved.

Many materials, such as tin oxide (SnO_2), a CdSe-based structure, and a multilayer Se(As,Te) structure, are capable of operation with a low-faceplate illumination and a lag of 10% for a 200-nA signal current. The spectral response is up to 13 μm. These three types of the photoconductive materials for the surface structures of Vidicons are shown in Fig. 10-3-2 [9].

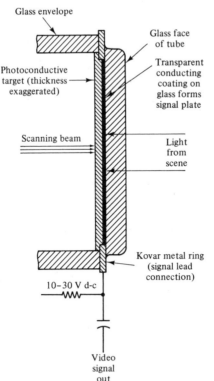

Figure 10-3-2 Surface structure of a Vidicon (Courtesy of the Electronics, Inc.).

The charge-discharge cycle is completed when the positive-charge effect is achieved by photoconduction through the target itself, rather than by photoemission from the scanned surface. This mode of operation requires that the resistivity of the photoconductive target be sufficiently high so that its time constant exceeds the 1/30-s television frame time. A dark resistivity of 10^{12} Ω-cm or greater is satisfactory.

10-3-2 Characteristics and Applications

Photoconductive targets are free from the spurious spots and lag. Sensitivities in excess of 1,000 μA per lumen are obtainable. Resolution is limited only by the electron optics of the beam.

The one-inch diameter Vidicon shown in Fig. 10-3-3 (Orthicon tube at the top) is capable of resolving more than 600 lines. The capability of the target can be made sufficiently large in any size target so that the light signal-to-noise ratio of the output signal can be as high as needed.

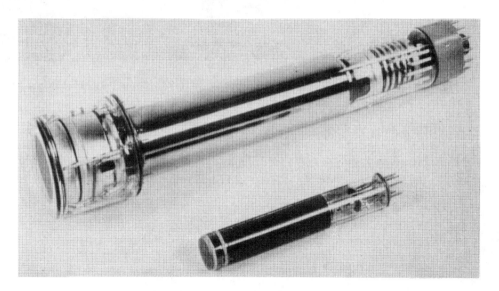

Figure 10-3-3 Photograph of one-inch diagram Vidicon with Orthicon at top. (Courtesy of the Electronics, Inc.).

The one-inch diameter Vidicon has a target sensitivity of 300 μA per lumen and can transmit a noise-free picture with a scene brightness of several foot-lamberts using an $f/2$ lens as shown in Fig. 10-3-4. The entire Vidicon tube and its optical system are scaled down in size in comparison with a standard image Orthicon and the quantity of light in lumens intercepted by the lens is decreased. The Vidicon tubes have many applications in industrial, broadcasting, television, medical as well as military uses.

Figure 10-3-4   Photograph taken by Vidicon in 1950. (Courtesy of the Electronics, Inc.).

10-3-3 IR Vidicon

In a typical *IR (infrared) Vidicon,* the target is mounted on or near the tube's faceplate, and the front surface of the sensor is coated with a good electric conductor as shown in Fig. 10-3-5.

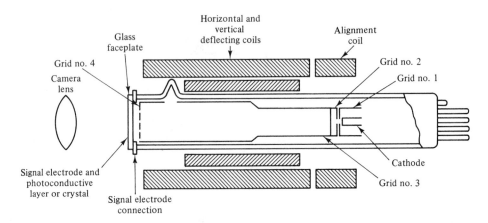

Figure 10-3-5   Schematic diagram of an IR Vidicon (Courtesy of RCA).

An electron beam scans the back surface of the photosensor line by line. In the process, the electron beam deposits enough of an electronic charge on each resolution element to raise the potential to that of the thermionic cathode. Between successive passes of the electron beam over a resolution element, the photocathode alters its resistivity from point to point in response to the imaged scene. The charge pattern on the back surface is changed accordingly by differential leakage to the electrical coating on the front surface. The electron beam current, in returning the surface-to-thermionic cathode potential, then varies from point to point and becomes the video signal current. An electron multiplier amplifier within the tube structure itself often is used to increase both signal and electron beam energy levels above the noise level of the video amplifier.

The contrast of an object for an IR Vidicon is a function of the spectral bandpass, the differential temperature between object and its surroundings, and the absolute temperature of the background. For small differential temperature ΔT above a background temperature T, the contrast of an object can be expressed by

$$C = 1.44 \times 10^4 \times \frac{\Delta T}{\lambda T^2} \qquad (10\text{-}3\text{-}1)$$

where ΔT = differential temperature in degrees Kelvin
$\quad T$ = absolute temperature in degrees Kelvin
$\quad \lambda$ = bandpass wavelength in micrometers

To predict the performance of an IR Vidicon, two spectral bands must be considered: the middle IR from 1 to 5 μm for commercial uses and the far IR from 8 to 13 μm for military use. These two bands correspond roughly to the IR window in the atmosphere. Figure 10-3-6 shows the spectral response of two RCA IR Vidicons.

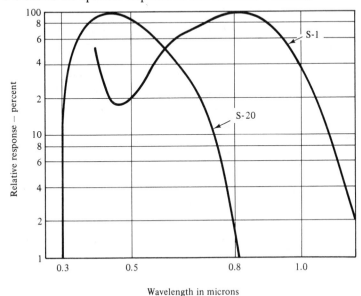

Figure 10-3-6   Spectral response of two RCA IR Vidicons (Courtesy of RCA).

For satellites, an IR Vidicon horizon sensor with no moving parts, high resolution, and fast response has many advantages. Applications in surveillance and terminal guidance become more attractive as aircraft reach supersonic speeds and in view of the high radiation levels of missile plumes and reentry vehicles.

Example 10-3-1: Contrast of An IR Vidicon

An IR Vidicon has the following parameters:

$$\begin{array}{lll}
\text{Differential temperature} & \Delta T = 2\ {}^{\circ}\text{K} \\
\text{Background temperature} & T\ \ = 300\ {}^{\circ}\text{K} \\
\text{Bandpass wavelength} & \lambda\ \ = 8\ \mu\text{m}
\end{array}$$

Calculate the contrast of its object.

Solution: The contrast is

$$\begin{aligned}
C &= 1.44 \times 10^4 \times \frac{2}{8 \times (300)^2} \\
&= 0.04 \\
&= 4\ \%
\end{aligned}$$

10-3-4 Silicon-Target Vidicon

The *silicon-target Vidicon* has a photoconductive layer that consists of a two-dimensional array of silicon photodiodes processed in an n-type silicon wafer. It is one of the most sensitive image tubes in the Vidicon family and it has extremely broad spectral response and high quantum efficiency. In operation, the target is reverse biased by connecting the n-type wafer to a positive potential at $+8$ V, and maintaining the p side of the diode array at ground potential by the scanning electron beam. The target provides a very low dark current at 5 to 10 nA. The peak quantum efficiency reaches 85% at a wavelength of 0.52 μm in light response. The magnetic focus and electrostatic deflection in the FPS (focus projection and scanning) Vidicon provides very high resolution. Figure 10-3-7 shows a schematic diagram of the FPS Vidicon invented by RCA.

Some silicon Vidicon tubes made by Westinghouse Electric offer very high sensitivity and have a very broad spectral response that extends beyond 1 μm. Resistance

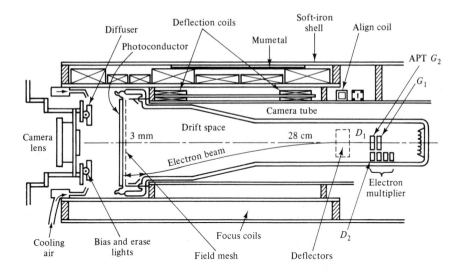

Figure 10-3-7   Schematic diagram of silicon-target FPS Vidicon (Reprinted by permission of Howard W. Sams & Co., Inc.).

to burn makes them ideally suited to surveillance applications. Their rugged construction and compact size (4.5-in. length) are ideal characteristics for military applications.

10-3-5 Plumbicon

The *Plumbicon* image tube is a lead-oxide target Vidicon and has advantage over the conventional Vidicon because it can be used for black-and-white and color broadcast as well. The target is composed of a transparent n-type lead-oxide conduction signal plate (SnO_2 layer) overlaid by an intrinsic and a p-type conduction lead-oxide layer as shown in Fig. 10-1-1(c). The thickness of the photoconductor is in the range of 10 to 20 μm for high target resolution.

In operation, the photoconductor is reverse biased by maintaining the signal plate at a positive potential and the p-type layer on the scanning side at ground potential. Because the n-i and i-p interfaces are separate the dark current is less than 5 nA. As a result, the black level uniformity is extremely good. Figure 10-3-8 shows the signal current as a function of the target potential [1].

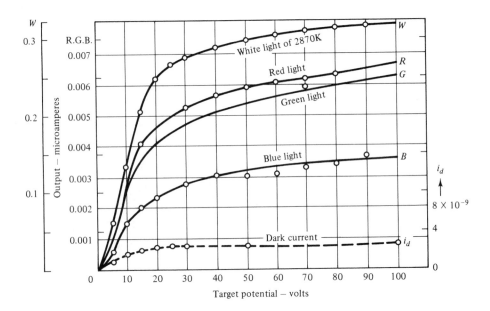

Figure 10-3-8 Photocurrent curves of Plumbicon (Reprinted by permission of Howard W. Sams & Co., Inc.).

10-4 THERMICON

Photothermic image tube is a device that is based on the thermal variation of a photo-emissive effect (or photoelectric effect). When a light image is focused on the target of a Thermicon, a charge flow is induced. After the electron beam scans the target, a video signal is generated. There are two well-known types: *Vidicon* and *Thermicon*. These devices are commonly used for commercial television (TV) in the broadcast industry and for military detection in Walleye and Redeye missiles.

10-4-1 Physical Structure

Thermicon is the trade name for one of the photothermionic image converters operated under the thermal variation of photoemission. The schematic diagram of a Thermicon is shown in Fig. 10-4-1 [10].

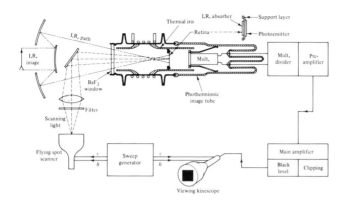

Figure 10-4-1 Schematic diagram of a Thermicon (Courtesy of the Optical Society of America).

The target, which is called the *retina* in this tube, is a multilayer film (0.5 to 0.7 μm). An evacuated tube envelope, equipped with a sealed window of barium fluoride or sodium chloride, surrounds the retina.

The retina consists of several layers. The first serves the purpose of absorbing infrared but transmitting visible light, the second layer provides structural support, and the third is the temperature-sensitive photocathode. When a thermal radiation is incident on the retina, a temperature distribution is produced and then probed by a suitably filtered light beam from a flying spot scanner. As the light spot moves across the face of the retina, it triggers a current from the photolayer. The magnitude of the current is then modulated by the temperature distribution. As the current is received by the anode, it is amplified and impressed on the intensity control grid of a viewing monitor. If the electron beam of the viewing monitor is swept in synchronism with the flying spot scanner, it will produce, point by point, a visible reproduction of the thermal image.

10-4-2 Operational Principle

The basic principle of operation for a Thermicon is the thermal variation of photoemission. As shown in Fig. 10-4-2, the differential temperature ΔT of the retina from its mean temperature T results as an image of a scene with differential temperature $\Delta \theta$ from its mean value θ.

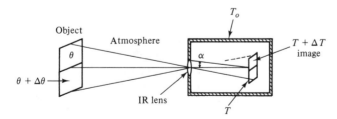

Figure 10-4-2 Schematic diagram of temperature image process (Courtesy of the Optical Society of America).

If the retina is freely supported in a shielded vacuum envelope at an ambient temperature T_0, the relationship between differential temperatures in image and object is given by [10]

$$\Delta T = \tau \Gamma [A/(\epsilon_1 + \epsilon_2)](\Delta \theta) \left(\frac{\theta}{T}\right)^3 \tag{10-4-1}$$

where T = retina mean temperature in degrees Kelvin
   ΔT = retina differential temperature in degrees Kelvin
   θ = object mean temperature in degrees Kelvin
   $\Delta \theta$ = object differential temperature in degrees Kelvin
   τ = atmospheric transmittance
   Γ = optical conversion factor
   A = mean absorptivity of the retina at temperature T
   $\epsilon_1 = \epsilon_2$ is the mean emissivity of the two retina sides

For typical values: $\tau = 0.37$, $\Gamma = 0.059$ (for $f_n = 2.0$), and $A = \epsilon_1 = \epsilon_2$. Eq. (10-4-1) is simplified to

$$\Delta T = 0.011 \left(\frac{\theta}{T}\right)^3 (\Delta \theta) \tag{10-4-2}$$

If the walls surrounding the retina are cooled to $T_0 = 0$ °K, and if the effect of the atmosphere can be made to vanish by such means as matching the atmospheric transmission profile with spectral filter characteristics of the cooled lens, the relative temperature contrast $\Delta T/T$ will be independent of the apparatus constants and becomes

$$\Delta T/T = \Delta \theta/\theta \tag{10-4-3}$$

As a result, the relative temperature contrasts of the image are then equal to those of the object, regardless of distance, aperture, and other parameters. Alternatively, the differential temperature of the image can be expressed as

$$\Delta T = \{[\tau \Gamma A/(\epsilon_1 + \epsilon_2)] \sin^2 \alpha\}^{1/4} (\Delta \theta) \tag{10-4-4}$$

where α = subtended angle of the image.

10-4-3 Characteristics

The sensitivity of a Thermicon is defined as the minimum detectable differential temperature in the object scene at a stated signal-to-noise ratio. The resolution is the number of picture elements in the frame area that is discernible with a stated signal-to-noise ratio. The performance of a standard television tube, for instance, requires about 100,000 elements. The frame repetition rate is the frequency with which the image field is recurring. A standard television tube needs a rate of 30 frames per second. The signal produced by the temperature variations ΔT of a picture element consists of a fractional change in the photocurrent triggered by the electron shot noise. By increasing the bandwidth—that is, decreasing the dwell time of the probing light spot on the element—the random fluctuations in the number of emerging photoelectrons will increasingly compete with the signal. Limitations to the performance of thermal imaging tubes arise not only from the existence of competing noise, but also from the deterioration of the signal. The cooling unit for the photosurface is necessary to utilize thermal photoeffects and Fig. 10-4-3 shows a cross-section view of a Thermicon [10].

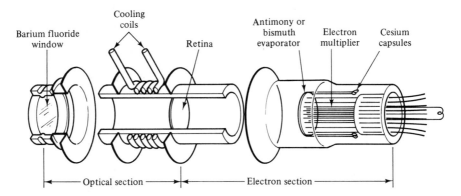

Figure 10-4-3   Cross-section view of a Thermicon (Courtesy of the Optical Society of America).

Example 10-4-1: Differential Temperature of A Thermicon

A Thermicon has the following operating parameters:

Atmospheric transmittance	τ	$= 0.37$
Optical conversion factor	Γ	$= 0.059$
Absorptivity and emissivity	A	$= \epsilon_1 = \epsilon_2$
Object mean temperature	θ	$= 400\ °K$
Object differential temperature	$\Delta\theta$	$= 300\ °K$
Retina mean temperature	T	$= 350\ °K$

Determine the retina differential temperature.

Solution: From Eq. (10-4-2), the retina differential temperature is

$$\Delta T = 0.011 \times \left(\frac{400}{350}\right)^3 (300)$$
$$= 0.011 \times 1.49 \times 300$$
$$= 4.93\ °K$$

10-4-4 Image Intensifier

The WL-30677 is a distortion-free electrostatically focused *image intensifier*. Both the input and output windows are fiber optics with usable image diameters of 40 and 25 mm, respectively, as shown in Fig. 10-4-4.

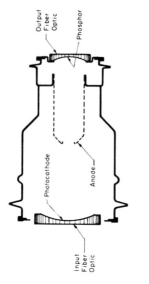

(a) WL-30677 (b) Equivalent cross section

Figure 10-4-4 Schematic diagram of WL-30677 image intensifier (Courtesy of We Electric.).

Light collected from the scene is formed into an optical image on the flat surface of the input fiber optic. This optical image is then transferred by the optical fiber to the interior concave surface where a high-sensitivity photocathode transforms the photon image into an electron image; that is, photoelectrons are released in direct proportion to the light intensity at each spatial point of the image.

The photoelectrons are then accelerated and focused onto the output phosphor layer by an electrostatic lens formed by the potential applied between the anode cone and the photocathode. As the electrons strike the phosphor layer, kinetic energy acquired from the accelerating voltage is transformed into radiant energy. An aluminum film evaporated onto the phosphor layer serves as a reflector causing all of the light emitted from the phosphor to leave through the output optic. The output optical fiber transmits this amplified image from the curved phosphor to the flat output surface, ready for direct viewing at the phosphor, and thus the optical image is reduced to approximately 0.64 its size at the photocathode. The image is also inverted.

Due to its high brightness gain, the WL-30677 can be used in many applications. Scientific applications include image intensification in electron microscopy, astronomy, nuclear physics, and television camera tubes. The sensitivity for most practical scenes is limited only by photon-shot noise, so the WL-30677 image intensifier is effective in many low-light-level applications such as passive night surveillance.

10-5 ORTHICON

The image *Orthicon* is a high-sensitivity, storage-type camera tube used in television pickup and studio cameras. It is especially suitable for televising optical image with a wide range of light levels. The image Orthicon operates on the charge integration principle as shown in Fig. 10-5-1 [1].

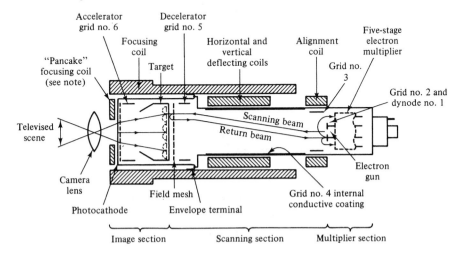

Figure 10-5-1 Schematic diagram of Orthicon image tube (Reprinted by permission of Howard W. Sams & Co., Inc.).

10-5-1 Physical Structure

The image Orthicon tube consists of three main sections: the image section, the scanning section, and the multiplier section.

1. *Image Section:* The image section is composed of a semitransparent photocathode deposited on the inside of the faceplate, a target, and an accelerating grid (grid no. 6). A magnetic field is used to focus the electron beam. When a light image is focused on the photocathode by an optical system, the photoelectrons generated by the light image are accelerated to about 500-eV energy by the electromagnetic lens.

   The first target material used in an image Orthicon was an optical-quality soda lime glass film about 5 μm thick, but its lifetime is very short. The disadvantage of the magnesium oxide target is high secondary emission coefficient and high lateral resistivity. The electronically conducting glass target is the best target because it provides high-quality pictures, long life, and low secondary emission coefficient. The meshes on both sides of the target is designed to intercept the secondary electrons.

2. *Scanning Section:* The scanning electron beam is produced by a low-velocity electron gun consisting of a thermionic cathode, a control grid (grid no. 1), and an accelerating grid (grid no. 2). The electron beam is brought to sharp focus on the target by the magnetic field (grid no. 4) and deposits the same amount of negative charges to neutralize the positive charges on the target. A video signal is generated as an amplitude modulation of the scanning electron beam. The reflected electrons from the target are returned to the electron gun.

3. *Multiplier Section:* The purpose of electron multiplier section in the image Orthicon is to increase the small signal level above the noise level of the video amplifier first stage. For this reason, the return electron beam is directed to the first dynode of a five-stage electrostatically focused multiplier. The secondary electrons are directed from the first dynode to the second dynode, and so on. The amplification of the multiplier section is extremely high.

10-5-2 Operational Principle

The image Orthicon tube operates on the charge integration principle. The charge integration is performed on both the continuous surfaces of the target: one side with the imaging electron beam and on the other side with the scanning beam as shown in Fig.

10-5-2 [1]. The light energy or photon energy is transferred to electrons by the photocathode. The image electron beam from the scene deposits a positive charge pattern on the target corresponding to the imaged scene. The scanning electron beam deposits an equal amount of negative charges to neutralize the positive charge pattern. The output charge pattern is read out as an amplitude modulation of the return beam.

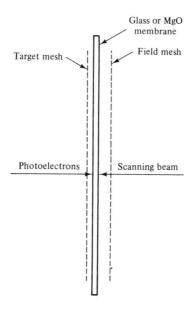

Figure 10-5-2   Schematic diagram of Orthicon continuous surface (Reprinted by permission of Howard W. Sams & Co., Inc.).

10-5-3 Characteristics and Applications

The signal-to-noise ratio of different image Orthicons is in the range of 30 to 40 dB, depending on the mesh-to-target capacitance and target voltage. The light transfer characteristic of the image Orthicon tube is shown in Fig. 10-5-3 [1]. The knee of the transfer characteristics corresponds to a target potential charged up to the mesh potential. As light level approaches the knee and is beyond the knee, the target potential approaches the mesh potential and the capture of the secondary electrons by the mesh becomes inefficient. As a result, the output signal remains almost constant.

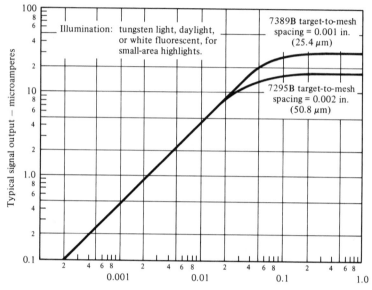

Figure 10-5-3 Light transfer characteristic of Orthicon tube (Reprinted by permission of Howard W. Sams & Co., Inc.).

10-6 ISOCON

The *Isocon* image tube was developed by isolating the return electrons from the scattered electrons in order to improve the signal-to-noise ratio of the Orthicon tube. The main differences between the Orthicon and Isocon are the beam separation techniques employed in the Isocon tube. Figure 10-6-1 shows the cross section of an Isocon image tube [1].

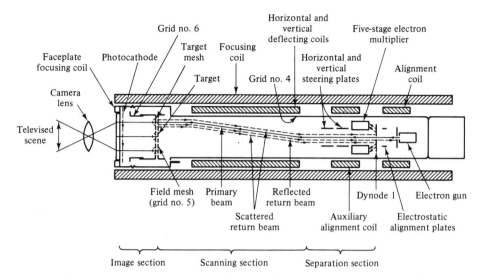

Figure 10-6-1 Cross section of an Isocon tube (Reprinted by permission of Howard W. Sams & Co., Inc.),

The main noise source of the Orthicon tube is the shot noise coming with the current in the return electron beam. The return beam contains scattered and reflected electrons as shown in Fig. 10-6-2.

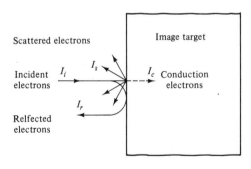

Figure 10-6-2 Components of return electron beam.

In the Orthicon tube, the entire return beam is directed into the electron multiplier. In the Isocon tube, however, the reflected electrons are isolated and only the scattered electrons are received by the electron multiplier. The scattered portion of the return beam is proportional to the charges on the target. Several electrons are scattered for each electron entering the target to neutralize the positive charges. The higher is the secondary emission, the greater the scatter gain.

10-7 CCI TUBE

The *charge-coupled imager* (CCI) tube is an MIS or MOS diode-array sensor in which the charge image pattern to be televised is stored in potential wells created at the diode surface by clock pulses. The stored charge patterns are transported to the output location of the imager by moving the potential wells analogously to electronic scanning.

There are several ways to achieve the operation of a CCI camera tube. Figure 10-7-1 shows two schematic diagrams of a frame transfer CCI either with a temporary storage array or with separated photosensors. In a frame transfer CCI with a temporary storage array, the optical image is detected by a two-dimensional array of photosensors and integrated during a TV frame time. The integrated charge image patterns are then transferred into the temporary storage array by clock pulses *A* and *B* during the vertical blanking time. Next, the charge image patterns are transferred down, one horizontal line at a time, into the output register by clock *B* during the horizontal blanking time. The charge image patterns of a horizontal line is then read out from the output register by high-speed clock *C* during the horizontal line period.

The frame transfer CCI with separated photosensors consists of a two-dimensional array of photosensors and vertically nonphotosensitive charge-coupled-device (CCD) channels. The charges produced by the optical image on the photosensors are integrated during a TV frame time. During the vertical blanking time, the integrated charge patterns are transferred from the photosensors into the vertical CCD channels. Finally, the charge image pattern is transferred down from the CCD channels one horizontal line at a time into the output register by clock *A* and is read out by high-speed clock *B* during the horizontal line period.

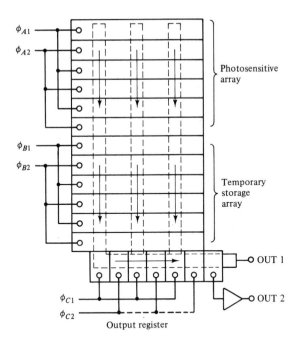

(a) Frame transfer CCI with a temporary storage array

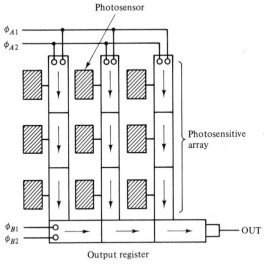

(b) Frame transfer with separated photosensors

Figure 10-7-1   Operational diagram of CCI tube (Reprinted by permission of Howard W. Sams & Co., Inc.).

10-8 SEC TUBE

The *secondary-electron conduction* (SEC) tube is a high-sensitivity, wide-dynamic range, and storage-type image tube used for outdoor broadcasting at night and for scientific and industrial applications. The SEC tube has three major sections: the image section, the SEC target section, and the scanning section, as shown in Fig. 10-8-1 [1].

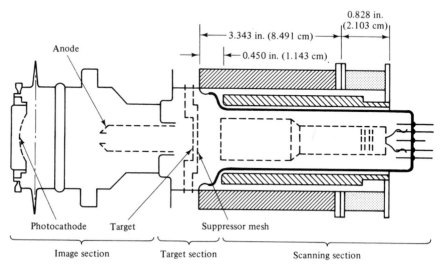

Figure 10-8-1 Schematic diagram of SEC tube (Reprinted by permission of Howard W. Sams & Co., Inc.)

    The image section is an electrostatically focused image tube diode that has the SEC target in its image plane. The components of the image section consist of a semitransparent photocathode deposited on the inside of the optical-fiber faceplate and an electrostatic image tube lens for focusing the photocathode current on the SEC target. The secondary-electron conduction target is formed of a low-density layer of potassium chloride (KCl) supported by an aluminum oxide and a conductive aluminum layer. The conductive aluminum layer serves as a signal plate. The scanning section is composed of an electron gun and magnetic focusing components. The electron image of the cathode is inverted and it has a less-than-unity magnification (about 0.64).

    In operation, a positive voltage is applied to the signal plate and the scan side of the target is charged to the gun cathode potential by a low-velocity scanning beam. Thus, the target capacitor between the signal plate and the scan side of the target is charged to the signal plate voltage. When a light image is focused on the target, many low-energy secondary electrons are emitted. The secondary-electron conduction discharges the target capacitor and produces a charge image pattern corresponding to the local intensity of the photoelectron image. The charge pattern is then read out by scanning the target with a low-velocity electron beam that returns the potential of the charged area to the cathode potential.

The SEC tube can provide a typical high-light current of 150 nA under normal operating conditions and its signal-to-noise ratio is about 150 or 43.5 dB if a suitable video amplifier is used.

10-9 SIT TUBE

The *silicon intensifier target* (SIT) image tube is a very high-sensitivity and storage-type camera tube used for very-low-light levels. The SIT tube is composed of three main sections: the image section, the scanning section, and the electron-gun section, as shown in Fig. 10-9-1 [1].

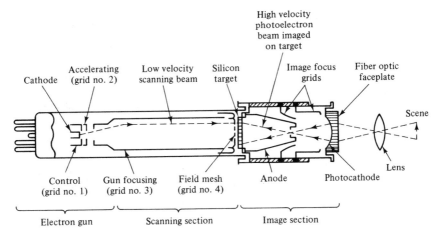

Figure 10-9-1   Schematic diagram of SIT tube (Reprinted by permission of Howard W. Sams & Co., Inc.).

The SIT image tube consists of a two-dimensional diode array formed in an n-type silicon wafer as described in Section 10-1-2(2). The silicon diode array provides the SIT tube with a very high target gain on the order of 2000 to 2500. Also the SIT image tube is composed of an electrostatic image inverting diode and of a low-velocity scanning beam similar to the SEC tube.

In operation, when a light image being televised is focused on the semitransparent photocathode by an optical system, a large number of hole-electron pairs are produced by the primary photoelectrons due to the interaction with the silicon layer. An electric field is generated between a small positive potential of the back plate (signal plate) of the target and a negative potential of the scan side of the target. This electric field diffuses the hole to the scan side of the target and discharges the elemental diode capacitors correspondingly to the local intensity of the primary photoelectron image. As a result, a charge image pattern is produced and this charge pattern is read out on the back plate by neutralizing the positive charges with the scanning electron beam.

The dark current of the SIT tube is extremely low at 5 to 10 nA at normal target voltage of 8 V. The high target gain enables the SIT tube to operate at lower light levels better than any other image tubes.

REFERENCES

[1] CSORBA, ILLES P. *Image Tubes*. Indianapolis, Ind., Howard W. Sams & Co., Inc. 1985, pp. 314, 326, 330, 339, 344, 348, 355, 359, 367.

[2] WILSON, JOHN, and HAWKES, J. F. B., *Optoelectronics: An Introduction*. Englewood Cliffs, N.J., Prentice-Hall, Inc., 1983, pp. 325, 326, 328.

[3] KEYES, R. J., *Optical and Infrared Detectors*. Berlin, Springer-Verlag, 1980, p. 12.

[4] GARTNER, W. W., Depletion-Layer Photo-Effects in Semiconductors. *Phys. Rev.*, 116 (1959), p. 84.

[5] SZE, S. M., *Physics of Semiconductor Devices,* 2nd ed., New York, John Wiley & Sons, Inc., 1981, pp. 769, 800.

[6] LONG, D., Photovoltaic and Photoconductive Infrared Detectors. In R. J. Keyes, ed., *Optical and Infrared Detectors*. Berlin: Springer-Verlag, 1980.

[7] BOYLE, W. S., and SMITH, G. E., Charge-Coupled Semiconductor Devices. *Bell Syst. Tech. J.*, 49, April 1970, pp. 587–93.

[8] BOYLE, W. S., and SMITH, G. E. Charge-Coupled Devices—A New Approach to MIS Device Structures. *IEEE Spectrum*, Vol. 8, No. 7, July 1971, pp. 18–19.

[9] WEIMER, PAUL K., and others, The Vidicon Photoconductive Camera Tube. *Electronics*, May 1950, pp. 70–73.

[10] GARBUNY, M., et al., Image Converter for Thermal Radiation. *J. Opt. Soc. Am.*, Vol. 51, No. 3, Mar. 1961, pp. 261–73.

PROBLEMS

Radiation Detectors

10-1. A radiation detector is used to detect the infrared energy at a wavelength of 10 μm. What is the Debye temperature at that frequency?

10-2. A radiation detector is used to detect the infrared at a wavelength of 3 μm.
   (a) Find the frequency for the infrared.
   (b) Determine the bandgap energy in eV.
   (c) Compute the photon energy $h\nu$ in watt-sec (or joules).
   (d) Calculate the quantum condition temperature.

Detector Figures of Merit

10-3. A detector has a sensitive area of 2 cm^2 and an output signal voltage of 10 V (rms) with an irradiance of 4 W/cm^2. Determine the responsivity of the detector.

10-4. A detector has a sensitive area of 3 cm^2, an output signal voltage of 8 V (rms), and a signal-to-noise ratio of 8. The irradiance is 4 W/cm^2.
(a) Determine the noise equivalent power (NEP) of the detector.
(b) Find the detectivity D of the detector.

10-5. A detector has the following parameters:

$$V_s = 10 \text{ V(rms)} \qquad V_n = 0.50 \text{ V(rms)}$$
$$H = 4 \text{ W/cm}^2 \qquad A_d = 5 \text{ cm}^2$$
$$B = 10^{15} \text{ Hz}$$

(a) Determine the noise equivalent power (NEP) of the detector.
(b) Find the detectivity D.
(c) Calculate the specific detectivity D^*.

10-6. The current responsivity of a radiation detector is expressed by Eq. (10-1-6). Verify the equation.

10-7. A certain photon detector has the following parameters:

Detecting area	$A_d = 1.5 \text{ cm}^2$
Wavelength	$\lambda_0 = 2 \ \mu\text{m}$
Quantum efficiency	$\eta_d = 30 \ \%$
Bandwidth	$B = 1 \text{ Hz}$
Noise current	$I_n = 10 \text{ pA}$

(a) Find the responsivity $R(f)$.
(b) Compute the noise equivalent power (NEP).
(c) Calculate the detectivity D.
(d) Determine the specific detectivity D^*.

Modes of Operation

10-8. Describe the principles of operation for the photon-effect, thermal-effect, and the photo-thermic-effect detectors.

10-9. Explain the characteristics and applications of the photon, thermal, and photothermic detectors.

Photon Detectors

10-10. A GaAs photoconductive detector is used to detect infrared signal. Its electron concentration n is 10^{13} cm^{-3} and its hole concentration p is 10^7 cm^{-3}.
(a) Determine the conductivity of the detector.
(b) Find the current density if the applied electric field is 100 V/cm.

10-11. A certain *p-i-n* photodiode has an absorption coefficient α of 5×10^3 cm^{-1} at $\lambda = 2$ μm. Its reflection coefficient R is 0.10 and its *i*-region width W is 10 μm. The incident signal power P_{in} is assumed to be 4 W and the detecting area A is 4 cm^2.
 (a) Determine the incident photon flux density.
 (b) Calculate the drift current density.

10-12. A certain avalanche photodiode has an *i*-region W of 14 μm and an absorption coefficient α ($\alpha_n = \alpha_p$) of 7×10^3 cm^{-1}.
 (a) Determine the multiplication factor M.
 (b) If both the leakage and photocurrents are considered, calculate the combined multiplication factor M for $V_b = 40$ V, $V_r = 39$ V, and $n = 2$.

10-13. A certain PEA (positive-electron-affinity) semiconductor photocathode has a bandgap energy E_g of 1.20 eV and a work function Φ of 1.30 eV. Its Fermi level is 0.30 eV above the maximum valence band level E_v.
 (a) Determine the electron affinity χ of the material.
 (b) Find the photoemissive threshold energy E_{th} for electrons in the semiconductor.

10-14. A certain NEA (negative-electron-affinity) semiconductor photocathode has a bandgap energy E_g of 1.60 eV, a Fermi level 0.20 eV above the maximum valence band E_v, and a work function Φ of 1.0 eV.
 (a) Determine the electron affinity χ of the material.
 (b) Find the photoemissive threshold energy E_{th}.

10-15. A photocathode is illuminated with radiation of wavelength 0.60 μm. The cathode has a work function Φ of 1.10 eV. Determine the anode voltage required to produce zero anode current.

10-16. If the anode voltage for a photocathode is 40 V with respect to the cathode and the cathode is illuminated with a radiation of wavelength 0.30 μm, determine the electron velocity at the anode.

10-17. Describe the advantages of the NEA photocathode over the classic metal photocathode.

10-18. An IR Vidicon has the following parameters:

Differential temperature	$\Delta T = 2.5$ °K
Background temperature	$T = 300$ °K
Bandpass wavelength	$\lambda = 9$ μm

 Compute the contrast of its object.

10-19. A Thermicon has the following operating parameters:

Atmospheric transmittance	$\tau = 0.37$
Optical conversion factor	$\Gamma = 0.059$
Absorptivity and emissivity	$A = \epsilon_1 = \epsilon_2$
Object mean temperature	$\theta = 450$ °K
Object differential temperature	$\Delta\theta = 300$ °K
Retina mean temperature	$T = 320$ °K

 Find the retina differential temperature ΔT.

10-20. Describe the operational principle of the Plumbicon.

10-21. Describe the operational principle of the Orthicon.

10-22. Describe the operational principle of the Isocon.

10-23. Describe the operational principle of the CCI, SEC, and SIT image tubes.

CHAPTER 11

Infrared Image Tubes

11-0 INTRODUCTION

The infrared image tubes use thermal effect at the infrared frequency region from wavelength 0.70 μm to wavelength 15 μm for vision image display. Infrared (IR) radiation is an electromagnetic radiation generated by vibration and rotation of the atoms and molecules within any material at temperature above absolute zero—that is, 0°K or −273°C. In recent years there has been an increasing emphasis on the research, design, development, and deployment of various infrared devices and systems for military applications at night or during the day when vision is diminished by fog, haze, smoke, or dust. Infrared systems are defined as those that sense the passive infrared radiation emitted by some target or source and process it to the point that a visual image of that target or source is formed.

The history of infrared discovery is an interesting one. In 1800 Sir William Herschel discovered infrared radiation when he worked for the British Royal Navy [1, 2, 3], but at that time he did not use the term *infrared*. Herschel referred to the new portion of the spectrum by such names as "invisible rays," "radiant heat," "dark heat," and "the rays that occasion heat." Sir Herschel found that the heating effect increased as he moved the thermometer toward the red from the blue end of the spectrum. In 1829 Nobili made the first thermocouple, which was an improved thermometer based on the thermoelectric effect discovered by Seebeck in 1821 [4]. In 1833, Melloni invented a thermopile that was made by a number of thermocouples connected in series. The thermopile was more sensitive than the thermocouple and it could detect the radiant heat from a person at a distance of 10 m [5]. In 1901 Langley and Abbot developed an improved bolometer that could detect the radiant heat from a cow at a distance of 400 m [6]. During World War

I, an infrared search system could detect aircraft at a distance of 1.6 km and a person at a distance of 300 m. In 1917 Case constructed the first photoconductive sensor by using thallous sulfide [7]. Many sensitive infrared detectors, such as photon detectors and image converters, were developed during World War II. The sniperscope, consisting of an image converter and an illuminator mounted on a carbine, could enable a soldier to fire accurately at targets at night as far away as 60 m. In the late 1950s the Sidewinder and Falcon heat-seeking infrared-guided missiles were developed. Subsequently infrared devices and systems were installed in the Walleye, Redeye, and Chaparral missiles and the A-6E aircraft. Furthermore, infrared techniques became applicable to the altitude stabilization of space vehicles, measurement of planetary temperatures, earth mapping, and the early detection of cancer [8]. The fundamental work on infrared thermal imaging system was contributed by many dedicated scientists and engineers, such as Hudson [4, 9], Jones [10], and Johnson [11].

All bodies radiate energy throughout the infrared spectrum when their temperature is above absolute zero. If the radiating source is hot enough (above 1000 °K), some of the emitted energy may be visible to the human eye in the 0.4 to 0.7 micrometer (μm) range. Energy emitted at wavelengths between 0.75 and 1000 μm is defined as *infrared radiation*. It should be noted that wavelengths between 100 and 1000 μm are ultra-microwaves or millimeter waves. Table 11-0-1 shows the infrared spectrum.

TABLE 11-0-1. INFRARED SPECTRUM

Division	Wavelength (μm)	Frequency (Hz)
Near infrared (NIR)	0.7–3	4.29×10^{14}–1×10^{14}
Middle infrared (MIR)	3–6	1×10^{14}–5×10^{13}
Far infrared (FIR)	6–15	5×10^{13}–2×10^{12}
Extreme infrared (XIR)	15–1000	2×10^{12}–3×10^{11}

The near, middle, and far infrared divisions include spectral intervals in which the earth's atmosphere is relatively transparent, the so-called *atmospheric windows*. It is these windows that will be utilized by any infrared sensor that must look through the earth's atmosphere. The extreme infrared, often called *ultramicrowave,* is usually used only for laboratory applications where the instrument can be evacuated because the atmosphere is essentially opaque [9]. Typical targets of interest have peak emittance at the wavelength of about 10 μm; a good atmospheric window exists between 8- and 14 μm; and scattering is much lower at the wavelength of 10 μm. It is obvious, then, why the 8- to 12-μm band has been chosen consistently for thermal imaging at ranges longer than 900 m. Shorter wavelengths of 3 to 5 μm, for instance, can readily be applied to thermal imaging at ranges shorter than 900 m. Figure 11-0-1 shows the transmittance of the atmosphere for a 1.8-km horizontal path at sea level in infrared range [12].

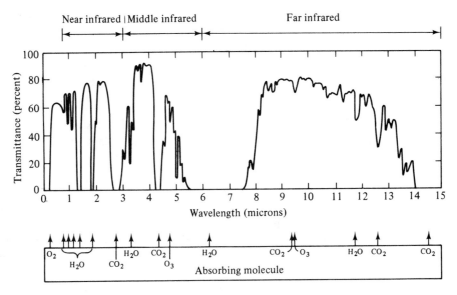

Figure 11-0-1 Transmittance of atmosphere for 1.8-km horizontal path at sea level (From R. D. Hudson, reprinted by permission of John Wiley & Sons, Inc.).

11-1 INFRARED RADIATION

A fundamental law of physics states that all bodies at temperatures above absolute zero in degrees Kelvin ($T > 0$ °K or -273°C) emit radiation. The amount of the infrared energy emitted depends upon the absolute temperature, the nature of the body, and the wavelength of the radiation. The emitting body may be a blackbody, which absorbs all incident radiation, or a graybody, which absorbs a portion of the incident radiation.

Spectral radiant emittance. The spectral radiant emittance of a blackbody is given by Planck's law [13] as

$$W(\lambda) = C_1 \lambda^{-5} [e^{C_2/(\lambda T)} - 1]^{-1} = \frac{2\pi hc^2}{\lambda^5} \left[\frac{1}{\exp[hc/(\lambda kT)] - 1} \right] \quad \text{W/cm}^2/\mu\text{m} \quad (11\text{-}1\text{-}1)$$

where $C_1 = 2\pi hc^2 = 3.7415 \times 10^4$ W-$(\mu\text{m})^4/\text{cm}^2$
     $h = 6.6256 \times 10^{-34}$ W-s^2 is Planck's constant
     $c = 3 \times 10^{10}$ cm/s is the velocity of light in vacuum
     $C_2 = ch/k = 1.4388 \times 10^4$ μm-°K
     $k = 1.38 \times 10^{-23}$ watt-s/°K is Boltzmann's constant
     λ = wavelength in micrometers
     T = absolute temperature in degrees Kelvin

Radiant emittance.   The total radiant emittance into a hemisphere from a black-body at a given absolute temperature can be obtained by integrating Eq. (11-1-1) over wavelength limits extending from zero to infinity. That is,

$$W = \int_0^\infty W(\lambda)d\lambda = \frac{2\pi^5 k^4}{15c^2 h^3} T^4 = \sigma T^4 \qquad \text{W/cm}^2 \qquad (11\text{-}1\text{-}2)$$

where $\sigma = 5.6697 \times 10^{-12}$ W/cm^2/°K^4. Equation (11-1-2) is well known as the Stefan-Boltzmann's or Boltzmann's law, and it indicates the radiant flux in watts per unit area of a source.

Irradiance.   When the radiant flux from a source arrives at a detector, the irradiance H incident upon the surface of a detector is less than the radiant flux at the source due to the atmospheric absorption and molecular scattering in the transmission path. The irradiance H is expressed in watts per centimeter square, and it can be measured at the detector site.

Maximum spectral radiant emittance.   Differentiating Eq. (11-1-1) with respect to wavelength λ and equating the derivative equal to zero result in Wien's displacement law. That is,

$$\lambda_m T = a \qquad \mu\text{m-°K} \qquad (11\text{-}1\text{-}3)$$

where $a = 2898$ μm-°K
           λ_m = wavelength in micrometers for maximum spectral radiant emittance
           T = absolute temperature in degrees Kelvin

For example, the maximum wavelength for peak radiation from a 300 °K source can be determined from Wien's displacement law as about 10 μm.

Example 11-1-1:   Computations of Infrared Radiations

The jet engine of an aircraft emits infrared radiation at a wavelength of 10 μm and at a temperature of 200 °C.

(a) Calculate the spectral radiant emittance of the jet engine in W/cm^2/μm and W/m^2/m.

(b) Compute the radiant emittance in W/cm^2.

(c) Find the temperature in degrees Kelvin for maximum radiant emittance if the wavelength is fixed to be 10 μm and determine the maximum radiant emittance in W/cm^2.

(d) Determine the wavelength for maximum spectral radiant emittance if the temperature is fixed to be 200 °C and find the maximum spectral radiant emittance in watts/cm^2/μm.

Solution:

(a) Given: $\lambda = 10~\mu m$ and $T = 273 + 200 = 473~°K$. From Eq. (11-1-1),

$$W(\lambda) = \frac{3.7415 \times 10^4}{10^5} \left[\exp\left(\frac{1.4388 \times 10^4}{10 \times 473}\right) - 1 \right]^{-1}$$

$$= 18.70 \times 10^{-3}~W/cm^2/\mu m$$

$$W(\lambda) = \frac{2\pi hc^2}{\lambda^5} \left[\exp\left(\frac{ch}{\lambda kT}\right) - 1 \right]^{-1}$$

$$= \frac{2 \times 3.1416 \times 6.625 \times 10^{-34} \times (3 \times 10^8)^2}{(10 \times 10^{-6})^5}$$

$$\times \left[\exp\left(\frac{3 \times 10^8 \times 6.625 \times 10^{-34}}{10 \times 10^{-6} \times 1.38 \times 10^{-23} \times 473}\right) - 1 \right]^{-1}$$

$$= 18.72 \times 10^7~W/m^2/m$$

(b) The radiant emittance is

$$W = \sigma T^4 = 5.6697 \times 10^{-12} \times (473)^4 = 0.2838~W/cm^2$$

(c) From Wien's displacement law,

$$\lambda T_m = a = 2898$$

$$T_m = \frac{2898}{10} = 289.8~°K$$

$$W = \sigma T^4 = 5.6697 \times 10^{-12} \times (289.8)^4 = 0.04~W/cm^2$$

(d) The wavelength for maximum spectral radiant emittance is

$$\lambda_m = \frac{a}{T} = \frac{2898}{473} = 6.13~\mu m$$

The maximum spectral radiant emittance is

$$W(\lambda) = \frac{3.7415 \times 10^4}{(6.13)^5} \left[\exp\left(\frac{1.4388 \times 10^4}{6.13 \times 473}\right) - 1 \right]^{-1}$$

$$= 0.036~W/cm^2/\mu m$$

Radiant intensity and radiance.   The radiant intensity J and radiance N of a blackbody in a hemisphere are expressed as

$$J = \frac{WA}{\pi} - NA \qquad W/sr \qquad (11\text{-}1\text{-}4)$$

and

$$N = \frac{W}{\pi} = \frac{J}{A} \qquad W/sr/cm^2$$

where A is the area of a radiating surface in cm^2.

It should be noted that a hemisphere has a solid angle of 2π. The unit of a solid angle is expressed in steradians that is often written as sr. However, here it is assumed that the detector is a Lambertian receiver, which has an effective integrated solid angle of 1π for a hemisphere. If the source is an isotropic radiator, the radiation emits uniformly throughout the entire solid angle of 4π. If the source is small compared with the field of view of the infrared system—that is, a point source—the irradiance varies with the distance but not the angle about the radiator. If the source is large compared with the system field of view—that is, an extended source—the irradiance is constant. This situation can be explained by Lambert's cosine laws, which states that the radiant intensity in any direction propagating from any point on a surface is a function of the cosine of the angle θ between the said direction and the normal line to the surface at that point. In other words, the maximum radiation is in the direction normal to the surface and zero radiation in the tangential direction. This is why a scanning detector always receives the same amount of radiation regardless of the change of the angle θ between the detector's line of sight and the normal line to the radiating surface. As the radiating area viewed by the detector is increased, the angle θ is also increased and the value of cosin θ is decreased. Consequently, the total irradiance is constant.

Emissivity.   The emissivity of a thermal radiator is a measure of its radiation efficiency. It is defined as

$$\text{Emissivity} = \frac{\text{total radiant emittance of a graybody}}{\text{total radiant emittance of a blackbody at the same temperature}} \qquad (11\text{-}1\text{-}5)$$

$$\epsilon = \frac{W'}{W} = \frac{\displaystyle\int_0^\infty \epsilon(\lambda)\, W(\lambda)\, d\lambda}{\displaystyle\int_0^\infty W(\lambda)\, d\lambda} = \frac{1}{\sigma T^4} \int_0^\infty \epsilon(\lambda) W(\lambda)\, d\lambda$$

The emissivity ϵ is therefore an indication of the graybody of the thermal radiator. In other words, the lower the emissivity, the grayer the radiator, the higher the emissivity, the blacker the body. The blackbody has an emissivity of unity. Table 11-1-1 lists the emissivity of commonly used materials in total normal radiation.

TABLE 11-1-1. EMISSIVITY OF COMMONLY USED MATERIALS [4]

Material	Temperature (°C)	Emissivity ϵ
Metals and Their Oxides		
Aluminum		
polished sheet	100	0.05
sheet as received	100	0.09
anodized sheet, chromic acid process	100	0.55
vacuum deposited	20	0.04
Brass		
highly polished	100	0.03
rubbed with 80-grit emery	20	0.20
oxidized	100	0.61

TABLE 11-1-1. EMISSIVITY OF COMMONLY USED MATERIALS [4] (CONTINUED)

Material	Temperature (°C)	Emissivity ϵ
Copper		
polished	100	0.05
heavily oxidized	20	0.78
Gold: highly polished	100	0.02
Iron		
cast, polished	40	0.21
cast, oxidized	100	0.64
sheet, heavily rusted	20	0.69
Magnesium: polished	20	0.07
Nickel		
electroplated, polished	20	0.05
electroplated, no polish	20	0.11
oxidized	200	0.37
Silver: polished	100	0.03
Stainless steel		
type 18-8, buffed	20	0.16
type 18-8, oxidized at 800 °C	60	0.85
Steel		
polished	100	0.07
oxidized	200	0.79
Tin: commercial tin-plated sheet iron	100	0.07
Other Materials		
Brick: red common	20	0.93
Carbon		
candle soot	20	0.95
graphite, filed surface	20	0.98
Concrete	20	0.92
Glass: polished plate	20	0.94
Lacquer		
white	100	0.92
matte black	100	0.97
Oil, lubricating (thin film on nickel base)		
nickel base alone	20	0.05
film thickness of 0.001, 0.002, 0.005 in.	20	0.27, 0.46, 0.72
thick coating	20	0.82
Paint, oil: average of 16 colors	100	0.94
Paper: white bond	20	0.93
Plaster: rough coat	20	0.91
Sand	20	0.90
Skin, human	32	0.98
Soil		
dry	20	0.92
saturated with water	20	0.95
Water		
distilled	20	0.96
ice, smooth	−10	0.96
frost crystals	−10	0.98
snow	−10	0.85
Wood: planed oak	20	0.90

After R. D. Hudson, Jr., reprinted by permission of John Wiley & Sons, Inc.

11-2 INFRARED RADIATION SOURCES

A *blackbody* is defined as any object that completely absorbs all incident radiation. Conversely, the radiation emitted by a blackbody at a given temperature is maximum. Therefore a blackbody is an ideal radiator and absorber of radiation for all wavelengths at all temperatures, and its emissivity is equal to unity. Since the blackbody is a theoretical thermal radiator, it is commonly used as a standard source to calibrate all other infrared devices. Objects with emissivity less than unity are called *graybodies,* and the great majority of radiating objects belong to this group.

In general, the sources of infrared radiation can be classified into two groups, natural and artificial, as shown in Table 11-2-1. In addition to the radiation from a target or object, a certain amount of background radiation will be present. The background radiation appears in the infrared detection system as unwanted noise and must be filtered out for proper defection.

TABLE 11-2-1. INFRARED RADIATION SOURCES

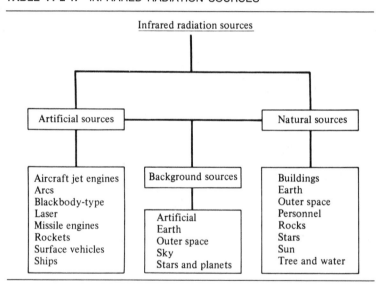

Artificial sources.   This type of infrared radiation source includes the controlled sources, such as blackbody type, and active sources, such as aircraft and rocket engines.

Most blackbodies used as a standard source for the calibration of infrared devices are of the cavity type with an opening of 1.27 cm or less and operate in the temperature range of 400 to 1300 °K. The spectral radiant emittances of a blackbody as calculated from Planck's law in Eq. (11-1-1) in terms of temperature and wavelength are shown in Fig. 11-2-1 [4]. The dashed curve in Fig. 11-2-1 indicates the locus of these maxima as computed from Wien's displacement law in Eq. (11-1-3).

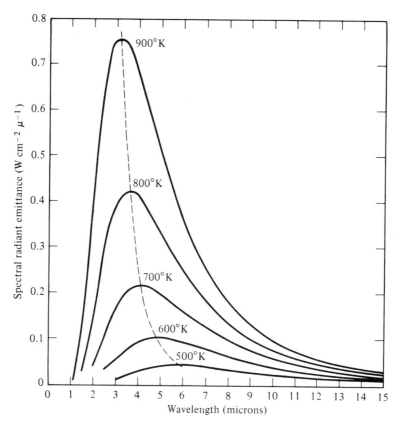

Figure 11-2-1   Spectral radiant emittance of a blackbody (After R. D. Hudson, reprinted by permission of John Wiley & Sons, Inc.)

Example 11-2-1:   Spectral Radiant Emittance of a Blackbody

A blackbody emits radiation at 900 °K.

(a) Determine the wavelength in μm for maximum spectral radiant emittance.

(b) Find the peak spectral radiation in W/cm^2/μm.

Solution:

(a) From Eq. (11-1-3), the wavelength for peak radiation is

$$\lambda_m = \frac{a}{T} = \frac{2898}{900} = 3.22 \ \mu m$$

(b) From Eq. (11-1-1), the peak spectral radiation is

$$W(\lambda) = \frac{3.7415 \times 10^4}{(3.22)^5} \left[\exp\left(\frac{1.4388 \times 10^4}{3.22 \times 900} \right) - 1 \right]^{-1}$$

$$= 0.76 \ \text{W/cm}^2/\mu m$$

Aircraft jet engines, rockets, and missiles are powerful active sources of infrared radiation. The prime sources of infrared radiation on aircraft and missiles are the hot metal of a jet tailpipe or an engine exhaust manifold and the jet plume. Figure 11-2-2 shows the exhaust temperature contours of the turbojet engine JT4A used on the Boeing 707 [4]. At supersonic speeds, the skin of an aircraft or missile becomes an infrared radiation source because of aerodynamic heating. Figure 11-2-3 shows the equilibrium surface temperature caused by aerodynamic heating in altitudes above 11.28 km and laminar flow for the Boeing 733 (a supersonic transport–SST) and the British-French Concorde [4]. A large aircraft may emit several kilowatts of infrared energy, but the human body emits only about 2 W.

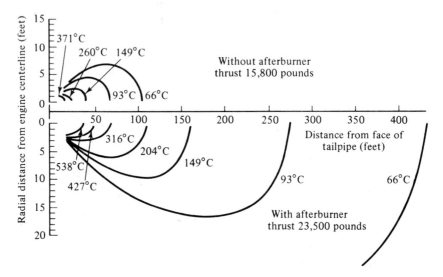

Figure 11-2-2 Exhaust temperature contours of Boeing 707 (From R. D. Hudson, reprinted by permission of John Wiley & Sons, Inc.).

The laser beam provides coherent sources of extremely high radiance in the portion of the spectrum extending from the ultraviolet to microwaves. The first application of the laser in the infrared portion of the spectrum was for military operations and communication systems.

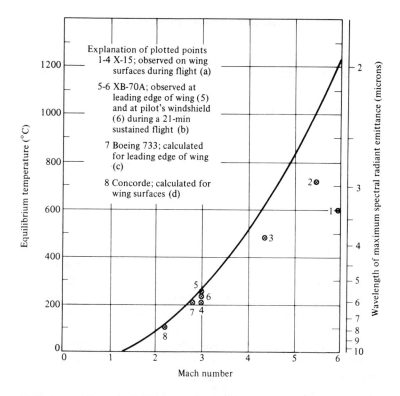

Explanation of plotted points

1-4 X-15; observed on wing
    surfaces during flight (a)

5-6 XB-70A; observed at
    leading edge of wing (5)
    and at pilot's windshield
    (6) during a 21-min
    sustained flight (b)

7 Boeing 733; calculated
    for leading edge of wing
    (c)

8 Concorde; calculated for
    wing surfaces (d)

(a) Various articles on the X–15 program, *Aviation Week*: **75**, 52 (November 20, 1961); **75**, 60 (November 27, 1961); **77**, 35 (August 13, 1962); **78**, 38 (June 10, 1963).

(b) C.M. Plattner, XB–70A flight research – part 2. *Aviation Week* **84**, 60 (June 13, 1966).

(c) C.M. Plattner, Variable-sweep wing keynotes Boeing 733 SST proposal. *Aviation Week*, **80**, 36 (May 4, 1964).

(d) Size, Speed, Safety of SST are debated. *Aviation Week*, **80**, 29 (June 1, 1964).

Figure 11-2-3   Surface temperature of SST and Concorde (From R. D. Hudson, reprinted by permission of John Wiley & Sons, Inc.).

Surface vehicles may radiate sufficient infrared energy to be considered as targets. The paint used on such vehicles usually has an emissivity of 0.85 or greater. The exhaust pipes and mufflers may radiate several times as much energy as the rest of the vehicle because of their high temperature.

Example 11-2-2:   Radiant Emittance of Jet Engine

The jet engine of an aircraft radiates infrared energy at a temperature of 100 °C at a distance of 100 m. Determine the radiant emittance of the jet engine at that location.

Solution: From Eq. (11-1-2), the radiant emittance is

$$W = \sigma T^4 = 5.67 \times 10^{-12} \times (373)^4$$
$$= 109.75 \text{ mW/cm}^2$$
$$= 1.098 \text{ kW/m}^2$$

Natural radiation sources.   This type of infrared radiation sources includes the terrestrial sources such as rocks, trees, earth, and water; the celestial sources such as the sun, sky, stars, and planets; and the buildings on the ground.

The sun radiates a total radiant emittance of the order of 1 kW/m² normal to the earth's surface at sea level. In other words, the sun radiates as a 5900 °K blackbody. Of this solar radiation, approximately 596 mW/m² is absorbed by the earth's surface from the sun at wavelength longer than 3 μm.

During the day, the infrared radiation from the surface of the earth is a combination of reflected and scattered solar energy and thermal emission from the earth itself. At night, when the sun has disappeared, the spectral distribution becomes that of a graybody at the ambient temperature of the earth. Figure 11-2-4 illustrates the spectral radiance of typical terrain objects as observed during the daytime [4].

Example 11-2-3:   Spectral Radiance of Soil

A certain soil is at temperature of 32 °C.

(a) Calculate its emitting wavelength for maximum spectral radiant emittance.
(b) Compute its peak spectral radiance in W/cm²/μm.

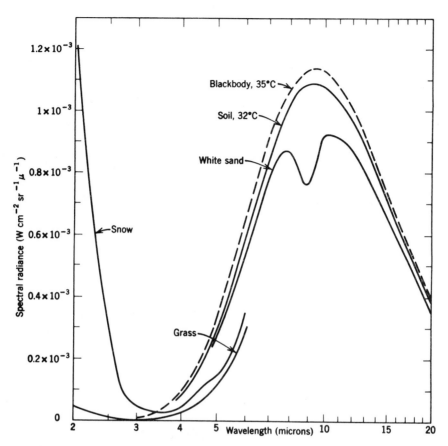

Figure 11-2-4   Spectral radiance of terrain objects (From R. D. Hudson, reprinted by permission of John Wiley & Sons, Inc.).

Solution:

(a) From Eq. (11-1-3), the wavelength for peak radiance is

$$\lambda_m = \frac{2898}{305} = 9.50 \ \mu m$$

(b) From Eq. (11-1-1), the peak spectral radiance is

$$W(\lambda) = \frac{3.7415 \times 10^4}{(9.50)^5} \left[\exp\left(\frac{1.4388 \times 10^4}{9.5 \times 305} \right) - 1 \right]^{-1}$$
$$= 0.484 \times 7.02 \times 10^{-3}$$
$$= 3.40 \times 10^{-3} \ W/cm^2/\mu m$$

11-3 INFRARED DETECTORS

An infrared detector is a key to the infrared imaging system for the purpose of collecting irradiance energy from a target and converting it into some other measurable form, such as electrical current or photographic image. At wavelengths of 8 to 12 μm and at video frequencies on the order of 0.01 to 5 MHz, there are only a few detectors with sufficient sensitivity for high-performance infrared imaging systems. These detectors are mercury-doped germanium (Ge: Hg), mercury cadmium telluride (Hg: Cd, Te), and lead tin telluride (Pb: Sn, Te). The quality and characteristic of a detector are usually described by three parameters as described earlier:

    a. Responsivity (R)

    b. Specific detectivity (D*)

    c. Time constant (τ)

    a. Responsivity (R).   Responsivity (R) is a measure of merit for thermal detectors, and it is defined as the ratio of the detector output over input. That is,

$$R(f) = \frac{V_s}{HA_d} \quad \text{V/W} \tag{11-3-1}$$

where V_s = signal voltage in volts (rms) is the fundamental component of the signal
     H = incident radiant flux in watts per centimeters squared in the rms value of the fundamental component of the irradiance on the detector
    A_d = sensitive area of the detector in centimeters squared

    b. Specific detectivity (D*).   The responsivity shown in Eq. (11-3-1) indicates only the behavior of the detector signal and does not give any information on the amount of noise in the output of the detector that will ultimately obscure the signal. In order to know the signal-to-noise ratio at the output of the detector, it is necessary to define another parameter: detectivity. Before doing so, it is desirable to define the noise equivalent power (NEP) at the detector output as

$$\text{NEP} = \frac{HA_d}{V_s/V_n} = \frac{HA_dV_n}{V_s} = \frac{V_n}{R(f)} \quad \text{watts} \tag{11-3-2}$$

where V_n = noise voltage in rms volts at the detector output.

    The specific detectivity ($D*$) is defined as directly proportional to the square root of the product of the detector area and the noise bandwidth and inversely proportional to the noise equivalent power [10]. That is,

$$D^*(f) = \frac{(A_d \cdot \Delta f)^{1/2}}{NEP} \qquad cm \cdot Hz^{1/2}/W \qquad (11\text{-}3\text{-}3)$$

where Δf = noise bandwidth in hertz.

The detectivity (D) of a detector is related to the specific detectivity (D^*) by the following equation [10]:

$$D = D^* (A_d \cdot \Delta f)^{-1/2} = \frac{1}{NEP} \qquad watt^{-1} \qquad (11\text{-}3\text{-}4)$$

This relationship was verified by Jones [10] because $DA_d^{1/2}$ = constant is well known from extensive theoretical and experimental studies. It should be noted that the quantity of the specific detectivity (D^*) refers to an electrical noise bandwidth of 1 Hz and a detector area of 1 cm^2. It is customary to indicate D^* by two numbers in parentheses. The first number shows the temperature of the blackbody and the second indicates the spatial frequency. For instance, D^* (400 °K, 800) means a value of D^* measured with a 400 °K blackbody at a spatial frequency of 800 Hz.

c. Time constant (τ).   The time constant of a detector is defined as the time required for the detector output to reach 63% of its final value after a sudden change in the irradiance. The responsive time constant (τ_r) of a detector [10] is defined as

$$\tau_r = \frac{\frac{1}{4}R_m^2}{\int_0^\infty [R(f)]^2 \, df} \qquad seconds \qquad (11\text{-}3\text{-}5)$$

where R_m = maximum value of $R(f)$ with respect to frequency.

Similarly, the detective time constant (τ_d) of a detector is expressed [10] as

$$\tau_d = \frac{\frac{1}{4}D_m^{*2}}{\int_0^\infty [D^*(f)]^2 \, df} \qquad seconds \qquad (11\text{-}3\text{-}6)$$

Mercury-doped germanium (Ge: Hg).   As described previously, all infrared thermal imaging systems are designed to operate at the wavelengths of 8 to 12 μm. The response of the detector is then limited to that wavelength band for high performance. A variety of detectors have been constructed with germanium (Ge) as a host lattice. Impurities such as gold (Au) yielded a response to about 10 μm, copper (Cu) to 30 μm, and mercury (Hg) to 14 μm. For the band range of 8 to 14 μm, Ge: Hg was found more effective than the others [14] because it has a detectivity of 4×10^{10} cm-Hz$^{1/2}$/W and a detective time constant of 1 ns. The disadvantage of the Ge: Hg detector is that it requires cooling to 25 to 30 °K. The cooling requirement and cost have prevented Ge: Hg from being continuously used in the infrared imaging system.

Mercury cadmium telluride (Hg: Cd,Te).   The mercury cadmium telluride (Hg: Cd,Te) is an alloy consisting of a mixture of the compounds HgTe and CdTe. Its spectral response varies from 9.5 to 12 μm and its impedance ranges from 20 to 110 Ω. The detectivity is about 1×10^{10} cm-Hz$^{1/2}$/W and the detective time constant is about 0.05 ns [14].

Lead tin telluride (Pb: Sn,Te).   Lead-tin-telluride detectors are available only in the photovoltaic mode. This device has a detectivity of 1×10^{10} cm-Hz$^{1/2}$/W and a detective time constant of 0.1 ns. Its operating wavelengths vary from 8 to 12 μm. Figure 11-3-1 shows the characteristic properties of the three detectors just discussed [14]. In addition, many other infrared detectors are available:

1. Thermal detectors—thermocouple, thermopile, and bolometer,
2. Photon or quantum detectors—photoelectric detector, photoconductive (PC) detector, and photovoltaic (PV) or *p-n* junction detector, and
3. Imaging detectors—infrared film, Vidicon, and photothermionic image converter (Thermicon).

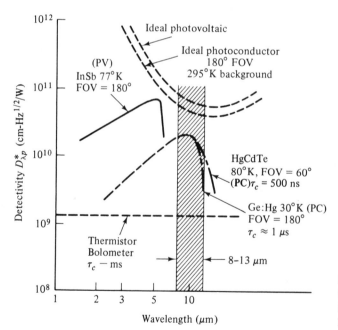

Figure 11-3-1 Specific detectivity (D*) of leading detectors (After J. J. Richter, reprinted by permission of the IEEE, Inc.).

The Vidicon is a small television-type camera tube in which an electron beam scans a photoconductive target. The Thermicon is based on the thermal variation of photoemission and produces the scene image on its retina. These two image tubes were studied in Chapter 10. Table 11-3-1 lists the properties of several infrared photoconductive detectors [8]. Figure 11-3-2 illustrates the infrared detectors against the infrared spectrum [9].

TABLE 11-3-1. INFRARED PHOTOCONDUCTIVE DETECTORS

Material	Maximum temperature for background limited operation	Long wavelength cutoff (50%) (μm)	Peak wavelength (μm)	Absorption coefficient (cm⁻¹)	Quantum efficiency	Resistance (Ω)	D^* peak (cm · Hz$^{1/2}$/W)	Approximate response time (s)
InAs	110	3.6	3.3	$\sim 3 \times 10^3$	0.5–0.8	10^3–10^4	3×10^{11}	5×10^{-7}
InSb		5.6	5.3	$\sim 3 \times 10^3$			6×10^{10}	5×10^{-6}
							-1×10^{11}	
Ge: Au	60	9	6	~ 2	0.2–0.3	4×10^5	3×10^9–10^{10}	3×10^{-8}
Ge: Au(Sb)	60	9	6			10^4	6×10^9	1.6×10^{-9}
							7×10^9	
							-4×10^{10}	
Ge: Hg	35	14	11	~ 3	0.2–0.6	1.4×10^4	4×10^{10}	3×10^{-8}
		14	10.5	~ 4	0.62	1.2×10^5		-10^{-9}
Ge: Hg(Sb)	35	14	11			5×10^9	1.8×10^{10}	3×10^{-10}–2×10^{-9}
								3×10^{-10}–3×10^{-9}
Ge: Cu	17	27	23	~ 4	0.2–0.6	2×10^4	2.4×10^{10}	3×10^{-4}–10^{-8}
								4×10^{-9}–1.3×10^{-7}
Ge: Cu(Sb)	17	27	23		0.05–0.3	2×10^5	2×10^{10}	$<2.2 \times 10^{-9}$
Hg: Cd, Te $x = 0.2$		14	12	$\sim 10^3$		60–400	10^{10}	$< 10^{-8}$
						20–200	6×10^{10}	$<4 \times 10^{-4}$
Pb: Sn, Te $x = 0.17$–0.2		11	10	$\sim 10^4$		42	3×10^8	1.5×10^{-8}
		15	14			52	1.7×10^{10}	1.2×10^{-4}

After Levinstein and Mudar; reprinted by permission of the IEEE, Inc.

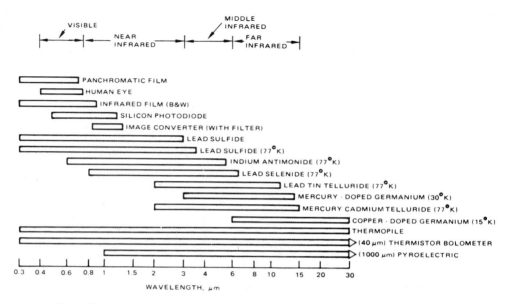

Figure 11-3-2 Infrared detectors versus spectral ranges (From R. D. Hudson, reprinted by permission of the IEEE, Inc.).

Example 11-3-1:   Merit Figures of Infrared Detectors

A certain infrared detector has the following parameters:

Infrared wavelength	$\lambda_0 = 8\ \mu m$
Detecting area	$A_d = 100 \times 10^{-4}\ cm^2$
Bandwidth	$\Delta f = 1\ Hz$
Incident radiant flux	$H = 1\ W/cm^2$
Signal voltage	$V_s = 1\ mV$
Noise voltage	$V_n = 0.25\ pV$

(a) Determine the responsivity $R(f)$.

(b) Find the noise equivalent power (NEP).

(c) Calculate the detectivity D.

(d) Compute the specific detectivity D^*.

Solution:

(a) From Eq. (10-1-5), the responsivity is

$$R(f) = \frac{V_s}{HA_d} = \frac{10^{-3}}{1 \times 10^{-2}} = 0.1 \text{ V/W}$$

(b) From Eq. (10-1-9), the noise equivalent power is

$$\text{NEP} = \frac{V_n}{R(f)} = \frac{0.25 \times 10^{-12}}{0.1} = 2.5 \text{ pW}$$

(c) From Eq. (10-1-10), the detectivity is

$$D = \frac{1}{\text{NEP}} = \frac{1}{2.5 \times 10^{-12}} = 4 \times 10^{11} \text{ W}^{-1}$$

(d) From Eq. (10-1-11), the specific detectivity is

$$D^* = D(A_d \times \Delta f)^{1/2} = 4 \times 10^{11} \times (100 \times 10^{-4} \times 1)^{1/2}$$
$$= 4 \times 10^{10} \text{ cm-Hz}^{1/2}/\text{W}$$

11-4 FLIR TUBE

The forward-looking infrared (FLIR) tube is a device used in night vision and heat-sensing (or thermal-effect) applications. When the scene or object is illuminated by an external light source, such as the sun, night sky, or the lamp, the brightness and color radiation is determined by the reflectivity of the objects of the scene. In thermal imaging, the thermal radiation of the scenery is used for imaging. The FLIR tube consists of a two-dimensional array of thermal detectors and of optical components to produce the scene image at night or during the day when the vision is diminished by fog, haze, smoke, or dust. Figure 11-4-1 shows the photograph of the FLIR tube installed in the radar system of the naval A-6E airplane.

Figure 11-4-1 Photograph of FLIR tube (Courtesy of Hughes Aircraft Co.).

The FLIR tube uses the newest detection techniques based on a two-dimensional detector array to scan the object space and generate a video signal for visual display. The two-dimensional detector array is used to produce a wide field of view and high resolution. The quality of an infrared thermal imaging system is usually described by three parameters such as the modulation transfer function (MTF), the minimum resolvable temperature (MRT), and the noise equivalent differential temperature (NEΔT). The merit of a detector is commonly characterized by two parameters: detectivity (D^*) and responsivity (R).

11-4-1 FLIR Detector Array

The charge-coupled device (CCD) is self-scanning system and is used in the FLIR image tube. There are three types of CCD imaging systems: the line imager, the interline transfer, and the frame transfer as shown in Fig. 11-4-2.

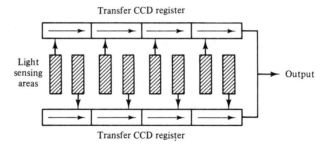

(1) Line imager

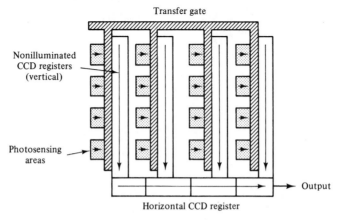

(2) Interline transfer

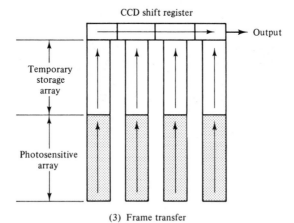

(3) Frame transfer

Figure 11-4-2   CCD-area imagers for FLIR tube.

1. *Line Imager:* The shaded areas of the CCD line imager represent photosensing elements with potential wells for optical charge integration. After charge patterns (or charge packets) are accumulated, they are first transferred into two parallel CCD shift registers; they are then shifted to the output following the directions of the arrows. Line imagers with 1,500 elements or more have been reported in use.

2. *Interline Transfer:* The interline-transfer CCD can be viewed as stacking line imagers in parallel. The charge packets are transferred into parallel shift-register lines and sequentially shifted down to the output register, which is read by transferring the charge patterns horizontally.

3. *Frame Transfer:* The frame-transfer system has an opaque temporary storage array with the same number of elements as the photosensing array. The charge patterns under the photosensing array are transferred over to the temporary storage array as a frame of an image. As a result, the information in the temporary storage array is shifted down one by one to the output shift register and transferred out horizontally.

The FLIR tube views an object with variable field of view as selected by the operator, and it collects the radiation emitted by the object and delivers the energy to the detectors. The FLIR sensor then converts the information radiation into an electrical signal and provides the image signals in a real-time form. Finally, the signal data converter (SDC) processes the signals generated by the detectors for display.

11-4-2 Range-Finder Tube

The FLIR image tube is a device for image detection, recognition, and identification of a scene or object and provides no information on the range (or distance) and location of the object. The forward-air-controller (FAC) receiver tube (or laser range-finder tube), as shown in Fig. 11-4-3, is an important part of the FLIR image system.

The FAC receiver tube has the following functions: (1) the capability to locate the properly coded laser sources of an object and indicate the angular position of the object with respect to the optical reference line, (2) the sensitivity to detect the returned object energy at a required rate, and (3) the ability to detect a required minimum false acquisition rate.

Figure 11-4-3   Photograph of FAC receiver tube (Courtesy of Hughes Aircraft Company).

The FAC receiver consists of four major components:

1. A telescope
2. A detector with four preamplifiers
3. A pulse amplifier for each channel
4. A peak detector with a dump pulse generator

Telescope.   The telescope is a wide-field optical system consisting of an air-spaced doublet. An optical band-pass filter is located on the front of the objective lens as shown in Fig. 11-4-4. The function of the telescope is to collect the laser energy reflected by a designated target and bring it to the detector for signal processing.

Detector and preamplifier.   The detector is sectioned into four quadrants, each with a preamplifier for channels A, B, C, and D, respectively. The amount of laser energy detected by each quadrant is determined by the angular position of the designated target relative to the optical reference line (ORL) as shown in Fig. 11-4-5.

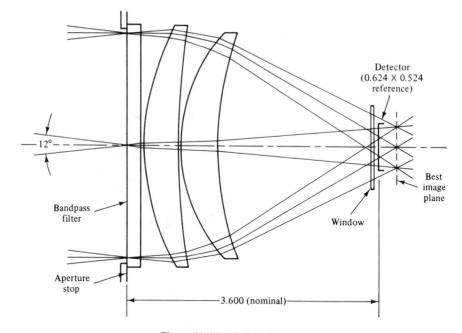

Figure 11-4-4   Optical structure.

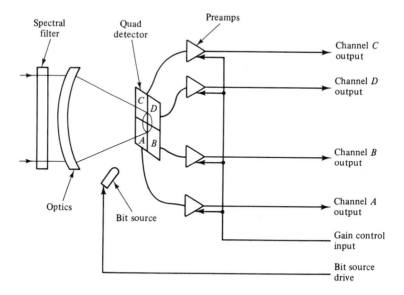

Figure 11-4-5   FAC receiver detector.

The detector is a photovoltaic silicon *p-i-n* type. The laser energy reflected from a designated target is absorbed by the detector and changed into a current output. This current output is then amplified by a preamplifier to be a pulse voltage output as shown in Fig. 11-4-6.

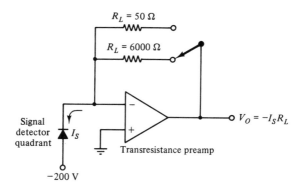

$R_L = 50\ \Omega$

$R_L = 6000\ \Omega$

Signal detector quadrant I_S

$V_O = -I_S R_L$

Transresistance preamp

$-200\ V$

· Wide bandwidth
· Dual gain
· Packaged 2/hybrid

Figure 11-4-6   FAC receiver detector with preamplifier.

Pulse amplifier.   The voltage outputs of the four signal channels A, B, C, and D from the detector preamplifiers are fed to four pulse amplifiers, respectively. Two pulse amplifiers amplify channel A and channel C, separately, and the other two amplify channel B and channel D. The pulse amplifiers shall have linear amplification of about 55.4 dB (55.4 dB = 588). The gain of all four channels are equally controlled by the automatic gain control (AGC) in order to preserve the line-of-sight error information that is amplitude modulated. The outputs of the four amplified channels are then fed to the peak detector and dump pulse generator for further signal processing as shown in Fig. 11-4-7.

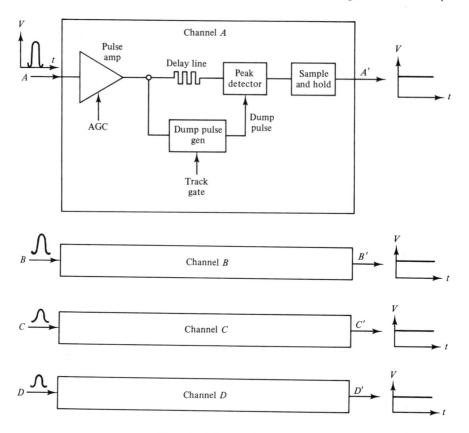

Figure 11-4-7 Pulse amplification.

Peak detector and dump pulse generator.   The dump pulse generator and peak detector are two critical components in the FAC receiver tube. The dump pulse generator operates in two modes: search mode, where the tube sensitivity is maximum, and track mode, where the tube sensitivity is maximum for a period when pulses are expected and the sensitivity is low for the calibration period to decrease false triggers. When the sum of the four pulse trigger signals exceeds the preset threshold level, the dump pulse generator is triggered and a raw alarm pulse is generated. All raw alarms, except those that occur within the track gate, are ignored by the track logic as shown in Fig. 11-4-8.

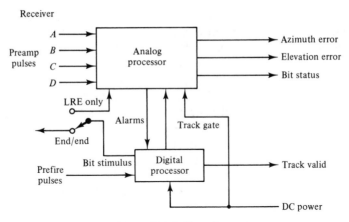

Figure 11-4-8   Raw alarm.

The missing rate of the raw alarm pulses indicates the sensitivity of the FAC receiver tube. The peak detector operates in conjunction with the dump pulse generator. Just before a new signal pulse arrives at the peak detector, the dump pulse discharges (dumps) the voltage across a capacitor. The sum channel integrator generates a sample pulse that holds the outputs of the other two summing components.

11-5 FLIR IMAGE SYSTEM

The FLIR image system consists of the optical unit, the detector array, the signal processors, and the display indicators, as shown in Figure 11-5-1.

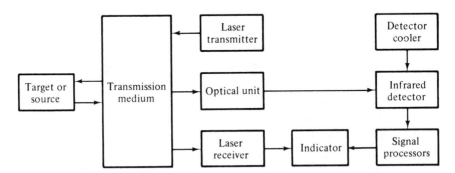

Figure 11-5-1   Block diagram of an infrared system.

Since a simple infrared system is passive, it can be used for the detection, recognition, and identification of a target by sensing the radiation emitted by the target, and it provides no information on the distance to the target. If an infrared system has some illuminating devices, such as a built-in laser transmitter and receiver, the entire infrared system may become enabled to determine the range to the target. The target is the object of interest for which the infrared system is designed and built. The radiated energy varies with the temperature of the target, its emissivity, and the viewing angle. The atmosphere is not a very favorable transmission medium. If the operating wavelength is chosen in the range of the atmospheric window as shown in Fig. 11-0-1, however, the atmospheric absorptance may reduce to a minimum. The optical unit collects the radiation emitted by the target and delivers it to the detector just like an antenna in the radar system. Before reaching the detector, the information radiation from the target may be recovered by an optical demodulator from the unwanted emission in the background. The detector then converts the information radiation into an electrical signal. Since a wide field of view (WFOV) and high resolution require a large number of detector elements, the *forward-looking infrared* (FLIR) unit uses the newest detection techniques based on discrete detector arrays to scan object space and generate video signal. Finally, the signal processor amplifies the electrical signal and sends the coded target information to the indicator for display. In addition, the infrared detector must be cooled down by a cryogenic cooler to a specified operating level of temperature—30 °K for Ge: Hg detector—for high performance. This cooling is necessary because the internal radiation emitted by the detector itself must be reduced to a minimum. Otherwise it is difficult, if not impossible, to separate the two radiations with equal magnitude from the target and the detector itself.

The performance of an infrared imaging system is described by the following three parameters:

1. The noise equivalent differential temperature ($NE\Delta T$)
2. The minimum resolvable temperature (MRT)
3. The modulation transfer function (MTF)

Noise equivalent differential temperature ($NE\Delta T$).   The signals from the detectors in an infrared imaging system respond to the variation in the irradiance at the entrance pupil of the optical system as the detectors are scanned across the target scene. Small differences in irradiance or temperature can be detected as an ac video signal while the high ambient irradiance or temperature is presented as a dc level or white noise. Then the noise signal is subtracted by a single RC low-pass filter. The noise equivalent differential temperature ($NE\Delta T$) is defined as the temperature difference required at the input of the detector to produce a peak signal-to-noise (rms) ratio of unity at the detector preamplifier output [10]. That is,

$$NE\Delta T = \frac{V_n}{V_s} \Delta T \qquad °\text{C} \qquad\qquad (11\text{-}5\text{-}1)$$

where V_s = ac signal voltage in volts (peak to peak)
$\quad\quad\ V_n$ = noise voltage in volts (rms)
$\quad\quad\ \Delta T$ = target temperature difference in °C

In Eq. (11-5-1), it is assumed that the reflectance of a collimator is 100%. The peak signal voltage from a target may be measured by an oscilloscope. The noise voltage may be recorded by turning off the scanner, covering the optical aperture with a flat black cover or opaque one, and reading the noise on an rms voltmeter.

Minimum resolvable temperature (*MRT*).   The minimum resolvable temperature (MRT) is a measure that is sensitive to the thermal imaging system/human observer combination. Its output is a function describing the minimum temperature difference necessary to resolve various spatial frequency patterns projected into the input port of the infrared system under test. A square 4-bar pattern, 7-to-1 aspect ratio, is currently the standard pattern used for such measurements.

Johnson's Imaging Model.   One of the earliest experimental attempts to relate threshold resolution to the visual discrimination of images of real scenes is attributed to Johnson [11]. The basic experimental scheme was to move a real scene object, such as a car, out of range until it could just barely be discerned on a detector's display at a given discrimination level, such as detection, recognition, or identification. Then the real scene object was replaced by a bar pattern of contrast similar to that of the scene object. The number of bars per minimum object dimension in the pattern was then increased until the bars could barely be individually resolved [15]. Figure 11-5-2 shows the Johnson's imaging model.

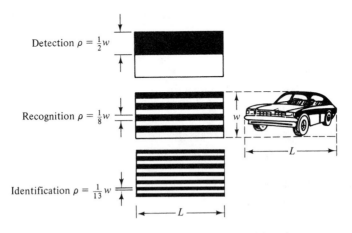

Figure 11-5-2   Johnson's imaging model.

In Figure 11-5-2 the car is replaced by a bar pattern. According to Johnson's hypothesis, if the car is to be barely detected, the bar pattern width ρ should be equal to one-half the car's minimum dimension. For simple recognition, the bar width should be one-eighth whereas for identification the bar width should be one-thirteenth the car's minimum dimension. The length of the bars was unspecified, but it was assumed that the length is to be equal to the car's longest dimension.

In the laboratory measurement, the bars are normally oriented vertically, but 45° and horizontal orientations are sometimes used also. The observer is allowed to adjust the display brightness and contrast controls for the most resolvable pictures at some minimum temperature ΔT. This process is repeated for other spatial frequencies. The ambient temperature during the measurement must not be changed by more than ± 0.20°C from its initial values. The smaller the resolvable temperature, the better the infrared system.

The characteristic spatial frequency f of a 4-bar pattern, 7-to-1 aspect ratio, is defined as

$$f = \frac{1}{2\theta} = \frac{L}{2\rho} \qquad \text{cycles/mrd} \tag{11-5-2}$$

where L = focal length of a collimator in inches

ρ = bar width in mils is the smallest resolution dimension of the target that is to be viewed

θ = arctan (ρ/L) is the angle subtended by target (or bar) width in milliradians

Figure 11-5-3 schematically shows the relationship between the bar width and focal length of a collimator.

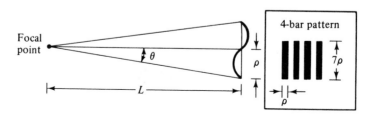

Figure 11-5-3   Relationship between bar width and focal length.

Example 11-5-1:   Spatial Angle of a 4-Bar Pattern or a Target

A 4-bar pattern has a bar width of 10 mils and the focal length of the collimator is 100 in.

(a) Determine the angle subtended by the bar width in milliradians (mrd).

(b) Find the characteristic spatial frequency in cycles per milliradian (cycles/mrd).

Solution:

(a) From Fig. 11-5-3, the angle subtended by the 4-bar pattern is

$$\theta = \arctan \left(\frac{\rho}{L}\right) \simeq \frac{\rho}{L} \qquad \text{for } \rho \ll L$$

$$= \frac{10 \times 10^{-3}}{100} = 0.1 \text{ mrd}$$

(b) From Eq. (11-5-2), the characteristic spatial frequency is

$$f = \frac{100}{2 \times 10 \times 10^{-3}} = 5 \text{ cycles/mrd}$$

In practice, the minimum resolvable temperature (*MRT*) is usually plotted against some normalized spatial frequency f_0 instead of the characteristic spatial frequency f.

Modulation transfer function (MTF).   From electrical system theory the total system transfer function is the product of the individual system element transfer functions. An infrared thermal imaging system is to collect the thermal energy emitted by a target and convert it to an electrical current or voltage for display. Since an infrared imaging system is composed of optical, physical, chemical, mechanical, electrical, and electronic elements, the system is more complicated than any ordinary electrical or electronic system. Therefore it is extremely difficult, if not impossible, to calculate analytically the infrared system transfer function.

From Fourier theory, any analytical function can be transferred to a Fourier series. The modulation transfer function (*MTF*) of an infrared system may be defined as the modulus of the Fourier transform of the one-dimensional spatial impulse response of the system. For a 4-bar pattern, 7-to-1 aspect ratio, the object consists of alternate light and dark bands that vary sinusoidally. The distribution of brightness, which is a spatial function, may be resolved into a Fourier transform as

$$B(x) = B_0 + B_1 \cos (2\pi f x) \qquad (11\text{-}5\text{-}3)$$

where f = spatial frequency of the brightness variation in cycles per milliradian
$\quad B_0$ = dc or average level of brightness
$\quad B_1$ = magnitude of the variable brightness
$\quad X$ = spatial coordinate in milliradians

Figure 11-5-4 shows the energy response of a 4-bar pattern. For simplicity, the third and higher harmonic terms are neglected. The modulation of the object pattern at the input port of the system is given by Smith [16] as

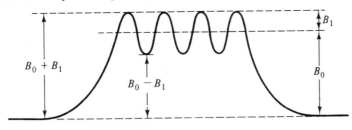

Figure 11-5-4   Object energy response of 4-bar pattern.

$$M_0 = \frac{(B_0 + B_1) - (B_0 - B_1)}{(B_0 + B_1) + (B_0 - B_1)} = \frac{B_1}{B_0} \qquad (11\text{-}5\text{-}4)$$

where $(B_0 + B_1)$ = maximum brightness

$(B_0 - B_1)$ = minimum brightness

When the object brightness pattern passes through the infrared system, the image pattern is affected by the system transfer function as shown in Fig. 11-5-5.

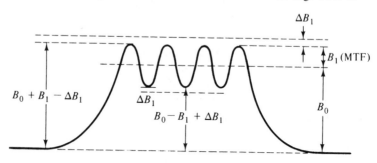

Figure 11-5-5 Image energy response of 4-bar pattern.

The modulation of the image pattern at the output port of the system is given by Smith [16] as

$$M_i = \frac{[B_0 + B_1(MTF)] - [B_0 - B_1(MTF)]}{[B_0 + B_1(MTF)] + [B_0 - B_1(MTF)]} = \frac{B_1}{B_0}(MTF) = M_0(MTF) \qquad (11\text{-}5\text{-}5)$$

Therefore the modulation transfer function (MTF) of an infrared thermal imaging system is expressed as

$$MTF = \frac{M_i}{M_0} \qquad (11\text{-}5\text{-}6)$$

In practice, the quantities of the image modulation M_i and the object modulation M_0 are unknown. In the infrared system laboratory, the maximum and minimum image pattern can be measured from an oscilloscope as shown in Fig. 11-5-5. Then, Eq. (11-5-5) can be expressed as

$$MTF = \frac{M_i}{M_0} = \frac{B_0}{B_i} M_i$$

$$= \frac{B_0}{B_i} \cdot \frac{(B_0 + B_1 - \Delta B_1) - (B_0 - B_1 + \Delta B_1)}{(B_0 + B_1 - \Delta B_1) + (B_0 - B_1 + \Delta B_1)} = \frac{B_1 - \Delta B_1}{B_1} \qquad (11\text{-}5\text{-}7)$$

where ΔB_1 is the differential portion decreased by the effect of the system transfer function, B_1 is the magnitude of the variable object brightness, $(B_0 + B_1 - \Delta B_1)$ is the maximum image pattern, and $(B_0 - B_1 + \Delta B_1)$ is the minimum image pattern. In conclusion, the higher the modulation transfer function, the better the infrared imaging system.

System measurement and applications. The infrared thermal imaging system is currently based on discrete detectors and optical scanning mechanism for image display. The basic principles of operation and measurement are shown in Fig. 11-5-6.

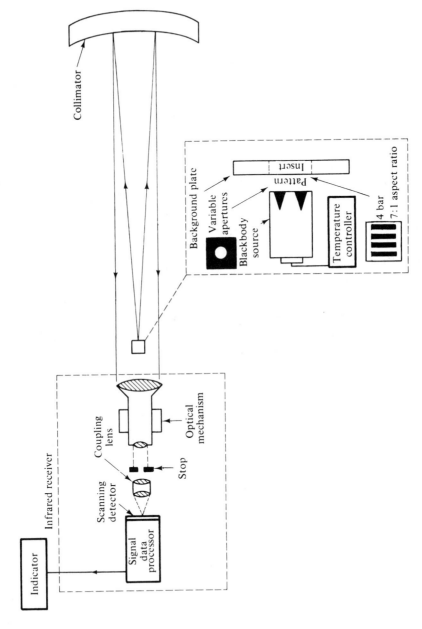

Figure 11-5-6 Block diagram for FLIR image system measurement.

Collimator

Background plate

Variable apertures

Insert

Pattern

Blackbody source

Temperature controller

4 bar

7:1 aspect ratio

Infrared receiver

Coupling lens

Optical mechanism

Scanning detector

Stop

Signal data processor

Indicator

423

System Operation.   The infrared thermal energy emitted by a target is collected by the telescope of the infrared system and brought out in a collimated beam with a small pupil at the scan mirror. The scan mirror provides the scan motion to generate one angular dimension of the sensor field of view. The scanned energy at the smaller aperture is collected by the detector lens and focused on the infrared detectors. The signals from the detectors and preamplifier are processed through the data processors for image display.

System Measurements.   A square 4-bar pattern, 7-to-1 aspect ratio, is inserted into the plate slot. The temperature of the source is reduced until a match is obtained between the background and target. When a differential temperature ΔT is increased, the 4-bar image is just distinguished. The differential temperature ΔT value is the target temperature difference for the tested pattern. The same procedure is repeated for various spatial frequencies. When the best image of each spatial frequency appears on the oscilloscope, the maximum, minimum, and average levels of brightness are recorded. The noise voltage or power can be measured by covering the aperture of the infrared receiver with an opaque filter [17, 18, 19].

FLIR image system applications.   The infrared thermal imaging system has unlimited application potential in many areas. Some current applications are listed below.

Military Applications.   Military fire control at night or during the day when the vision is diminished due to fog, smoke, or haze. Detection and tracking of ships, aircraft, missiles, surface vehicles, and personnel; submarine detection and range finding. It has been demonstrated that a missile can be fired accurately at a target in the night by using an infrared device as far away as 2 kilometers.

Medical Applications.   Early detection and identification of cancer, obstacle detection for the blind, location of blockage in a vein, and early diagnosis of incipient stroke.

Scientific Applications.   Satellite and space communication, environmental survey and control, detection of life and vegetation on other planets, and measurement of lunar and planetary temperature.

Industrial Applications.   Aircraft landing aid and traffic counting, forest fire detection, and natural source detection.

11-6 HIGH-SPEED SHUTTER TUBES

In microwave modulation, it often is desirable to utilize a switching tube to switch on a crystal at a very fast rise time, stay on for a predetermined period, and turn off at a very fast fall time. The time period is in the nanosecond range and the tube should be able to hold off high voltage at the kilovolt level. There are three types of microwave switching tubes available for laser light modulation: the ground-cathode switching vacuum planar triode, the hot-cathode gas-filled thyratron, and the cold-cathode gas-filled krytron [20].

11-6-1 Planar Triode

General description.   Most vacuum triodes are constructed cylindrically—that is, the filament or cathode structure, the grid, and the anode are cylindrical in shape and are mounted with the axis of each cylinder along the center of the tube. However, some vacuum triodes are manufactured with the cathode, grid, and anode in the shape of a flat surface. The triodes so constructed are called *planar triodes*. Figure 11-6-1 illustrates cross section of a planar triode, the Varian EIMAC-8941.

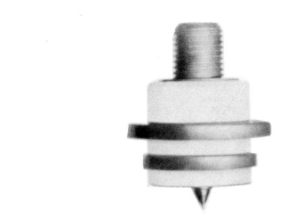

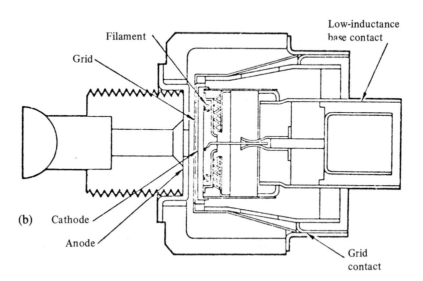

Figure 11-6-1   Picture and cross section of planar triode Varian 8941. (Courtesy of Varian Associates, Inc. EIMAC Division).

For microwave applications, it is necessary to provide very small spacings between the tube elements and to achieve very short lead lengths with the triode. The close spacings reduce the electron transit time and thus allow the tube to operate at microwave frequencies. The short leads increase the operating frequency by reducing lead inductance.

In microwave circuit design, it often is necessary to choose a planar triode of high amplification and zero bias. The zero-bias triode is an excellent choice for grounded-grid microwave amplifiers or high-speed laser modulators. The main advantages of the zero-bias planar triodes are power gain and circuit simplicity. Since no bias supply is required, protection circuits for loss of bias or drive are not needed. Figure 11-6-2 shows the constant plate and grid current characteristics of the Varian EIMAC-8941 planar triode.

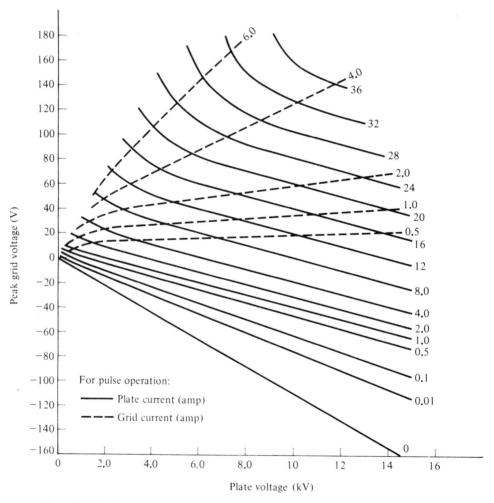

Figure 11-6-2   Constant plate and grid current characteristics of Varian EIMAC-8941 planar triode (Courtesy of Varian Associates, Inc., EIMAC Division).

Microwave applications.   The planar triode is normally used for advanced airborne, ground, and space applications. The tube can also be used in grid or plate pulser as laser modulator or series regulator at microwave frequencies. The plate voltage may reach 25 kV and the pulse width may be within tens of nanoseconds. The lifetime of the tube may last several thousand hours in proper operation. Currently the planar triode is commonly used to switch on and off the Pockels cell for laser modulation. When the Varian EIMAC-8941 planar triode is used for pulse operation as shown in Fig. 11-6-3, the pulse across the Pockels cell has a peak voltage of 4 kV, a pulse width of 8 ns, and a rise or fall time of about 3 ns [21].

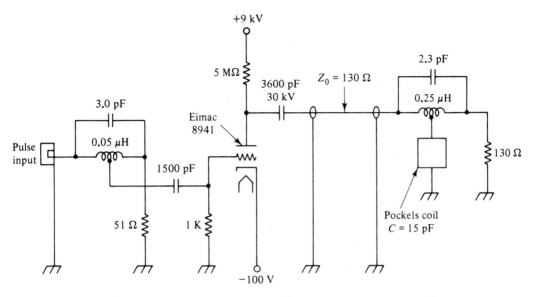

Figure 11-6-3   Circuit diagram with the Varian EIMAC-8941 planar triode for laser modulation. (Courtesy M. M. Howland at Lawrence Laboratory).

11-6-2 Thyratron

Conventional thyratrons.   In a conventional thyratron, the grid usually consists of a cylindrical structure which surrounds both the plate and the cathode, and a baffle or series of baffles containing small holes is inserted between the cathode and the plate. Due to the almost complete shielding between the cathode and the plate, the application of a small grid voltage before conduction occurs is sufficient to overcome the electric field at the cathode due to the high plate voltage. Once the arc glow has begun, the grid loses complete control over the arc, since the grid is now immersed in a plasma and is separated from the arc by the positive ion sheath.

There are two types of grid control: one in which the grid potential is always positive, and the other in which the grid control is generally negative except for a very low plate voltage. The physical distinction between these two types of thyratron lies essentially in the more complete shielding by the grid of the cathode from the plate in the positive-control thyratron. The major features of these two types are as follows:

Positive-Control-Grid Thyratrons.   The grid potential of the thyratron tube is made positive for conduction to begin—that is, in order for the electrons emitting from the cathode to acquire sufficient velocity for ionization of the gas molecules by collision, the grid potential must be made positive to accelerate the electrons. The positive grid is necessary because the electric field between the cathode and plate is practically zero due to the complete shielding of a plasma sheath.

Negative-control-grid thyratrons.

When the shielding between the cathode and plate is less complete at a very high plate potential, the grid potential must be made negative in order to prevent conduction from taking place momentarily. When the plate potential is very low and the grid potential is sufficiently positive, the conduction occurs between the cathode and the grid, and the tube may be destroyed. In this situation, the negative-control grid is needed.

Grounded-grid thyratrons.

The internal spacings of a grounded-grid thyratron are basically the same as those of a conventional thyratron. The major differences are in the treatment of the grid and anode structures, which are made of refractory materials. The mounting flange also is attached at the grid seal. The hot cathodes used are identical to those of conventional thyratrons. In the grounded-grid mode, the grid is used as the negative electrode for the main discharge; the normal cathode is used as the negative electrode only for triggering. In this way, some of the desirable triggering and operating characteristics of thyratrons can be combined with high peak current capability and low-inductance structure. The main current pulse is carried by a metal vapor arc between the grid and anode. Figure 11-6-4 illustrates the circuit of grounded-grid thyratrons.

In operation, the hot cathode is negatively pulsed to form a low-density plasma in the grid-cathode region, just as in a conventional thyratron. This plasma supplies electrons to the anode field penetrating the grid slots and thus initiate anode-grid voltage breakdown. The arrangement of the external circuit then forces transition to a grid-anode metal vapor arc.

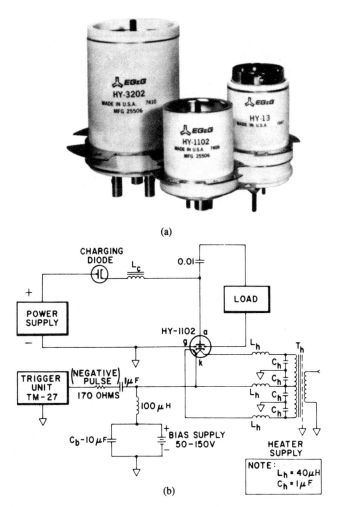

Figure 11-6-4 Picture and current diagram of grounded-grid thyratron. (Courtesy of EG & G Company, Inc., Electro-Optics Division).

The grounded-grid thyratrons manufactured by EG and G, such as HY-1002, HY-1102, and HY-3202, are used for fast switching applications in laser modulation. The main characteristics of a grounded-grid thyratron are: low load impedance less than 1 Ω, operating voltage from 10 to 34 kV, peak current from several to 10 kA, and fast rise time of 7 ns.

11-6-3 Krytron

General description.   The krytron is a four-element (grid, anode, cathode, and keep-alive), cold-cathode, and gas-filled switching tube as shown in Fig. 11-6-5.

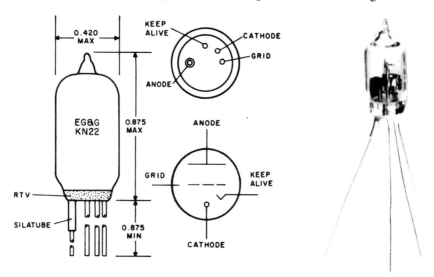

Figure 11-6-5   Schematic diagram of a krytron (Courtesy of EG & G Company Inc., Electro-Optics Division).

The control-grid structure encloses the anode except for a small opening at the top. It is through this small opening in the grid that conduction current must pass. The sole function of the gas in the tube is to provide ions for the neutralization of space charge and permit a high current to be obtained at much lower voltage than is required by the three-half-power law in a vacuum tube. Commutation is normally initiated by a positive pulse applied to the high-impedance grid, and a high peak current is conducting in an arc discharge-glow mode. This unique design allows the krytron to hold off high voltages and still have a low-voltage drop during conduction. A column of ionized gas is maintained by a keep-alive current to provide an initial source of plasma for short delay time.

Characteristics.   In general, when a positive pulse is applied to the grid and raises the grid voltage above the breakdown level a glow discharge occurs between the grid and the cathode. If the grid current is sufficient high, the main-gap voltage between the anode and the cathode is broken down and the glow discharge is transferred to the anode electrode. In this way, a small input current at the grid electrode controls the output current in the load circuit. Once conduction between the cathode and the anode has begun, the glow discharge cannot be extinguished by means of the control grid. The anode-cathode voltage must be lowered below the sustaining value long enough to permit deionization to take place. The minimum time of this process is called the *deionization time* or *recovery time*. Table 11-6-1 shows the characteristics of several commonly used krytrons.

TABLE 11-6-1. CHARACTERISTICS OF KRYTRONS

| Tube type | Anode voltage (KVDC) | | | Trigger voltage (V) | Current (A) | | | Pulse | | | | |
	Min	Typical	Max		Max peak	Typical peak	Keep alive (μA)	Duration (μS)	Delay time (μS)	Jitter (μS)	Repetition (PPM)	Life data ($\times 10^7$)
KN-2	0.3	2.0	4.0	200	500	40	50	5	0.20	0.02	6	1
KN-4	0.4	1.2	5.0	250	2500	270	150	10	0.30	0.03	1	0.0025
KN-6	0.7	2.6	5.0	250	3000	715	50	10	0.25	0.03	1	0.0035
KN-6B	0.7	2.8	8.0	250	3000	715	50	10	0.50	0.05		0.0035
KN-9	0.3	1.5	4.0	200	500	1	50	5	0.20	0.02	24×10^3	1.50
KN-22	0.4	4.0	5.0	750	100	80	300	0.04	0.04	0.005	3×10^3	2

Note: Delay time is the time required for the pulse to reach 50% of its final value.

Microwave applications.   The krytrons manufactured by EG and G such as KN-22, KN-6B, and KN-9 are commonly used to switch the Pockels cell crystals for laser modulation. In general, the features of krytrons are: no warm up needed, ability to hold off high voltage up to 10 kV, fast rise and fall times of 7 ns, and short delay time of about 40 ns. Hence the krytrons are suitable for use in very fast switching devices. Figure 11-6-6 shows a circuit diagram for krytron Pockels pulses.

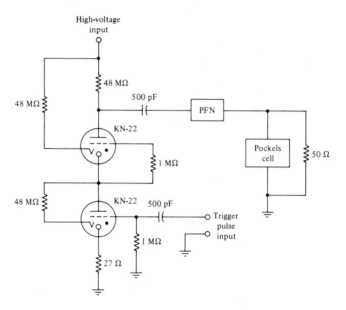

Figure 11-6-6   Circuit diagram of krytron Pockels pulses.

REFERENCES

[1] HERSCHEL, W., Investigation of the powers of the prismatic colours to heat and illuminate objects: with remarks that prove the different refrangibility of radiant heat. *Phil. Trans. Roy. Soc.*, London, pt. II, Vol. 90, 1800, p. 255.

[2] HERSCHEL, W., Experiments on the refrangibility of the invisible rays of sun. *Phil. Trans. Royal Soc.*, London, pt. II, Vol. 90, 1800, p. 284.

[3] HERSCHEL, W., Experiments on the solar, and on the terrestrial rays that occasion heat. *Phil. Trans. Royal Soc.,* London, pt. II, Vol. 90, 1800, p. 293, 437.

[4] HUDSON, R. D., Jr., *Infrared System Engineering.* John Wiley & Sons, New York, 1968, pp. 6, 21, 36, 40, 43, 45, 92, 102, 107, 115.

[5] BARR, E. S., The infrared pioneers—II. Macedonia Mellori. *Infrared Phys.*, 1962, p. 67.

[6] BARR, E. S., The infrared pioneers—III. Piermont Langley. *Infrared Phys.*, Vol. 3, 1963, p. 195.

[7] CASE, T. W., Notes on the change of resistance of certain substances in light. *Phys. Rev.*, Vol. 9, 1917, pp. 305–10.

[8] LEVINSTEIN, H., and J. MUDAR, Infrared detectors in remote sensing. *Proc. IEEE,* Vol. 63, No. 1, Jan. 1975, p. 6–14.

[9] HUDSON, R. D., Jr., and HUDSON, J. W., The military applications of remote sensing by infrared. *Proc. IEEE,* Vol. 63, No. 1, Jan. 1975, pp. 104–28.

[10] JONES, R. C., Phenomenological description of the response and detecting ability of radiation detectors. *Proc. IRE,* Vol. 47, No. 9, Sept. 1959.

[11] JOHNSON, J., Analysis of image forming systems. *Proc. of Image Intensifier Symposium,* AD220160, Ft. Belvoir, Va., Oct. 1958.

[12] GEBBIE, H. A., et al., Atmospheric transmission in the 1- to 14-μm region. *Proc. Roy. Soc.,* A206, 87 (1951).

[13] PLANCK, M., *Theory of Heat Radiation.* Dover, New York (reprint of 1910 edition).

[14] RICHTER, J. J., Infrared thermal imaging. *Proc. IEEE Southeast Region 3 Conference,* Orlando, Fla., April 29, 1974.

[15] ROSELL, F. A., Levels of visual discrimination for real scene objects vs. bar pattern resolution for aperture and noise limited imagery. *Proc. of the National Aerospace and Electronics Conference,* Dayton, Ohio, January 10–12, 1975, pp. 327–34.

[16] SMITH, F. D., Optical image evaluation and the transfer function. *Applied Optics,* Vol. 2, No. 4, April 1963, pp. 335–350.

[17] LIAO, SAMUEL Y. System performance of the detecting and ranging set (DRS) for the Navy A-6E TRAM aircraft. *Report for Hughes Aircraft Company,* El Segundo, Calif., August 1978.

[18] WOOD, J. T., Test and evaluation of thermal imaging systems. *Northeast Electronics Research and Engineering Meeting,* record part 3, November 1973.

[19] LIAO, SAMUEL Y. Capability and reliability of forward-air-controller (FAC) receiver subsystem for the Navy A-6E TRAM aircraft. Hughes Aircraft Company, El Segundo, Calif., August 1981.

[20] LIAO, SAMUEL Y., Pockels cell subsystem for laser modulation. *Report for Hughes Aircraft Company,* El Segundo, Calif., August 1980.

[21] HOWLAND, M. M., et al., Very fast, high peak-power, planar triode amplifiers for driving optical gates. *Lawrence Laboratory,* UCRL 82538, Livermore, Calif., June 12, 1979.

SUGGESTED READINGS

1. HUDSON, RICHARD D., Jr., *Infrared System Engineering.* John Wiley & Sons, New York, 1969.

2. *IEEE Proc.,* Vol. 63, No. 1, January 1975. Special issue on infrared technology for remote sensing.

3. LIAO, SAMUEL Y., *Microwave Devices and Circuits* (Chap. 6). Prentice-Hall, Englewood Cliffs, N.J., 1980.

PROBLEMS

11-1 Infrared Radiation

11-1-1. A military aircraft emits the infrared radiation in a wavelength of 4 μm at a temperature of 1100 °K.
   (a) Calculate the spectral radiant emittance of the aircraft in W/cm^2/μm.
   (b) Find the radiant emittance in W/cm^2.
   (c) Estimate the temperature in degrees Kelvin for the maximum spectral radiant emittance if the wavelength is fixed to be 4 μm and determine the maximum spectral radiant emittance in W/cm^2/μm.
   (d) Determine the wavelength for maximum spectral radiant emittance if the temperature is fixed to be 1100 °K and compute the maximum radiation in W/cm^2/μm.

11-1-2. The irradiance H is the radiation incident upon the surface of a detector and it is usually less than the radiant flux at the source due to the atmospheric absorption and the molecular scattering in the transmission path. If the loss is considered 30% and the radiant flux from the source is 20 W/cm^2, determine the irradiance H at a detector.

11-2 Infrared Radiation Sources

11-2-1. The blackbody source is usually used as the standard one for the calibration of infrared devices. It emits radiation into a hemisphere at 600°C with an emitting area of 150 cm^2.
   (a) Compute the radiant emittance of the blackbody.
   (b) Calculate the irradiance of the blackbody.
   (c) Find the radiant intensity.
   (d) Determine the radiance.
   (e) Determine the wavelength for maximum spectral radiant emittance.

11-3 Infrared Detectors

11-3-1. The quality and characteristics of an infrared detector are usually described by three parameters: responsivity (R), specific detectivity (D^*), and time constant (τ). Describe the quality and characteristics of the three commonly used infrared detectors.
   (a) Mercury doped-germanium, Ge: Hg (photoconductive detector)
   (b) Mercury cadmium-telluride, Hg: Cd,Te (photoconductive detector)
   (c) Lead tin-telluride, Pb: Sn,Te (photovoltaic detector)

11-3-2. A certain infrared detector has the following parameters:

Detecting area	$A_d = 200 \times 10^{-4}$ cm^2
Signal voltage	$V_s = 2$ mV
Noise voltage	$V_n = 0.5$ pV
Incident radiant flux	$H = 1.5$ W/cm^2
Bandwidth	$\Delta f = 1$ Hz
Infrared wavelength	$\lambda_0 = 10$ μm

(a) Determine the responsivity $R(f)$.

(b) Find the noise equivalent power (NEP).

(c) Calculate the detectivity D.

(d) Compute the specific detectivity D^*.

11-5 FLIR Image System

11-5-1. A FLIR imaging system is usually described by three parameters:

(a) The modulation transfer function (MTF)

(b) The minimum resolvable temperature (MRT)

(c) The noise equivalent differential temperature ($NE\Delta T$)

Describe the three parameters in detail.

11-5-2. One of the earliest experimental attempts to relate threshold resolution to the visual discrimination of images of real scenes is attributed to Johnson. Explain Johnson's imaging model for thermal imaging measurements.

11-5-3. A square 4-bar pattern, 7 to 1 aspect ratio, is currently used as the standard pattern for measurement of the minimum resolvable temperature (MRT). The bar width of the 4-bar pattern is 5 mils and the focal length of the collimator is 50 in.

(a) Find the angle subtended by the bar width in milliradians.

(b) Determine the characteristic spatial frequency f_0 in cycles per milliradians.

11-5-4. The modulation transfer function (MTF), minimum resolvable temperature (MRT), and noise equivalent differential temperature ($NE\Delta T$) are the three most important parameters for measuring an infrared imaging system. Explain how to measure them in detail.

11-5-5. The modulation transfer function (MTF) of an infrared thermal imaging system is the product of the individual system element transfer functions. In an infrared imaging system laboratory, the differential brightness ΔB_1 decreased by the effect of the system transfer function, the magnitude of the variable object brightness B_1 and the dc or average level of the brightness B_0 responding to the characteristic spatial frequency f_0 are 0.1, 0.4, and 1, respectively. (Refer to Fig. 11-5-5.)

(a) Determine the modulation of the object pattern at the input port of the system.

(b) Find the modulation transfer function of the system at the characteristic spatial frequency f_0.

11-5-6. In an infrared imaging system laboratory, the temperature of the source is reduced to such a level so that a match is obtained between the background and target. The differential temperature ΔT, the noise voltage, and the ac signal voltage were measured to be 2°C, 2 V (rms), and 10 V (peak to peak), respectively.

(a) Determine the noise-to-signal ratio for the system.

(b) Calculate the noise equivalent differential temperature ($NE\Delta T$).

11-6 High-Speed Shutter Tubes

11-6-1. The Krytron-22 has a pulse rate of 10 pulses per second with a duration of 0.04 μs. Estimate the life time of the Krytron-22 for a satisfactory operation of 2×10^7 pulses.

11-6-2. Describe the major characteristics and applications of the grounded-grid thyratron.

11-6-3. As shown in Fig. 11-8-5, a pulse-forming network (*PFN*) is needed to shape the output voltage from the Krytron switch to be a specific waveform for the Pockels cell actions. The pulse waveform requires a fast rise and fall time of 8 to 15 ns and a pulse width of 27 ns at 50% of the pulse height. Design a pulse-forming network (*PFN*) to meet the requirements and accomplish the function.

11-6-4. The planar triode EIMAC-8941 has an amplification factor of 200 and a transconductance of 75 mmho. Determine the plate resistance of the planar triode. Explain the major differences between a conventional triode and a planar triode and describe the advantages of the planar triode over the conventional triode.

where the reduction factor Γ is known as the *space-charge noise-reduction factor* and generally has a value between 0.10 and 0.316. The approximate value for the reduction factor is expressed as

$$\Gamma = 1.22 \times 10^{-2} \sqrt{\frac{T_c}{V_p}} \qquad (12\text{-}1\text{-}3)$$

where T_c = cathode temperature in degrees Kelvin
$\quad V_p$ = cathode-plate potential difference in volts

For example, if the oxide cathode of an electron tube emits electrons at 1000 °K and the cathode-plate potential difference is 25 V, the value of the reduction factor Γ is 0.100.

If an electron tube has several electrodes the general expression for noise current in the nth electrode is

$$\overline{i_n^2} = 2eI_N B \left[1 - (1 - \Gamma^2) \frac{I_N}{I_s} \right] \qquad (12\text{-}1\text{-}4)$$

where I_N = dc current to the nth electrode
$\quad I_s$ = total dc current in the tube

It can be seen that if the reduction factor Γ^2 is zero, the noise current in the nth electrode is the full shot noise, and if I_N/I_s is unity, the noise current has the minimum value.

In vacuum electron tubes the random emission and fluctuation of space-charge-limited currents contribute a tube noise. In temperature-limited tubes this noise is called *shot noise* and is due to random emission. In space-charge-limited emission tubes the noise is very small and is called *reduced shot noise*. Both types of noise are characterized by a uniform distribution of energy over the frequency spectrum.

In a velocity-modulation klystron, an electron beam passes the two grids of buncher and catcher cavity through a shunt resistance appeared between the grids. Noise is generated, but there is evidently very little space-charge smoothing action in the tube. This is expected from the fact that there is no virtual cathode between the grids of the cavity. As a result, the noise is nearly the full shot noise of the electron beam, and therefore the noise power delivered to the cavity is given by

$$P_n = 2eI_0 B R_{\text{sh}} \qquad (12\text{-}1\text{-}5)$$

where $\quad e = 1.6 \times 10^{-19}$ coulomb is the electron charge
$\quad I_0$ = electron beam current in amperes
$\quad B$ = frequency bandwidth in hertz
$\quad R_{\text{sh}}$ = shunt resistance of the cavity in ohms

In many cases, the electron transit time across the grids is an appreciable fraction of a cycle, and the transfer of energy from the electrons to the cavity is not perfect. As a result, the delivered noise power is reduced by the beam-coupling coefficient factor.

12-1-2 Thermal Noise

Thermal noise is also called *Johnson noise* or *Nyquist noise* and it is caused by the vibrational kinetic energy of conduction electrons and holes at a finite temperature. If some particles are charged (or ionized), vibrational kinetic energy can be coupled electrically to another device if a suitable transmission path is provided. The noise level of an electronic system is a measure with respect to the theoretical thermal noise level. The thermal noise power at certain temperature T in a bandwidth B is expressed by

$$N = kTB \qquad \text{watts} \qquad (12\text{-}1\text{-}6)$$

where $k = 1.38 \times 10^{-23}$ joules/°K is Boltzmann's constant
$\quad T =$ absolute temperature in degrees Kelvin
$\quad B =$ bandwidth in hertz

If a vacuum-tube amplifier is operating with a bandwidth of 1 Hz at room temperature its noise power is calculated as

$$N = 1.38 \times 10^{-23} \times 290 \times 1 = 4.00 \times 10^{-21} \text{ watt}$$
$$= -204 \text{ dBW}$$
$$= -174 \text{ dBm}$$

If the amplifier bandwidth is 1 MHz (60 dB) and its noise figure is 10 dB the amplifier internal noise level is

$$N = -174 + 60 + 10 = -104 \text{ dBm}$$

In addition, the minimum discernible signal must be at least 3 dB higher than the noise level. This means the smallest signal that can be amplified by this amplifier is -101 dBm.

Thermal noise of resistor.   The mean-squared value of the noise voltage generated by a matched resistor R is given by

$$\overline{V_n^2} = 4kTBR \qquad (12\text{-}1\text{-}7)$$

where $R =$ matched resistor in ohms.

Example 12-1-2:   Noise Voltage of a Resistor

If the matched resistor R has 200 Ω at room temperature and the bandwidth is 100 Hz.

(a) Calculate the output noise voltage.

(b) Determine the output noise voltage if the tube amplifier has a voltage gain of 40 dB.

Solution:

(a) The noise voltage is

$$V_n = (4 \times 1.38 \times 10^{-23} \times 290 \times 100 \times 200)^{1/2}$$
$$= 1.79 \times 10^{-8} \text{ V}$$
$$= -155 \text{ dBV}$$

(b) The output noise voltage is

$$V_n = 1.79 \times 10^{-8} \times 100 = 1.79 \times 10^{-6} \text{ V}$$
$$= -115 \text{ dBV}$$

12-1-3 Signal-to-Noise Ratio and Noise Figure

Signal-to-noise ratio (SNR).   The signal-to-noise ratio for a tube amplifier is defined as

$$SNR = \frac{S}{N} = \frac{\text{available signal power}}{\text{available noise power}} \qquad (12\text{-}1\text{-}7)$$

Noise figure.   The noise figure is defined as the ratio of the signal-to-noise ratio at the amplifier input S_i/N_i to the SNR at the amplifier output S_0/N_0. That is,

$$F = \frac{S_i/N_i}{S_0/N_0} = \frac{N_0}{N_i G} = \frac{N_0}{kTBG} \qquad (12\text{-}1\text{-}8)$$

where G = tube amplifier gain. When the load resistance is matched to the output resistance of a vacuum tube the available input signal power to the load is

$$S_i = \frac{V_i^2}{4R} \qquad \text{watts} \qquad (12\text{-}1\text{-}9)$$

where V_i = input signal voltage in volts. Assuming only thermal noise present $(\overline{V_n^2} = 4kTBR)$ the available noise power is

$$N_i = \frac{\overline{V_n^2}}{4R} = kTB \qquad (12\text{-}1\text{-}10)$$

Then the SNR at the input port of a load system is given by

$$SNR = \frac{S_i}{N_i} = \frac{V_i^2}{4kTBR} \qquad (12\text{-}1\text{-}11)$$

Noise threshold.   Noise threshold is defined as the lower limit of the SNR for signal detection. In general, 6-dB SNR is required for acquisition. However, if the frequency of the incoming signal is well known acquisition at 3-dB SNR is practical.

Noise floor.   The noise floor of the noise test set is a measure of its limit of noise sensitivity. The limit is set by the noise levels of both the detector and signal amplifier.

There are many noise sources in a vacuum tube. Thermal noise arises from the attenuation resistors. Random emission current (shot noise) and random emission velocities result in initial noise current and noise velocities on the electron beam. Partition noise may arise from interception on the electrodes of the electron. Secondary electrons can also contribute noise. Most of these noise sources can be minimized to a low-noise level by proper tube design techniques except of shot and thermal noises.

Example 12-1-3:

An electron-tube amplifier has the following parameters:

$$
\begin{array}{lll}
\text{Input signal voltage} & V_i & = 10^{-6} \text{ V} \\
\text{Bandwidth} & B & = 2 \text{ kHz} \\
\text{Matched resistance} & R & = 100 \text{ } \Omega \\
\text{Operating temperature} & T & = 290 \text{ °K}
\end{array}
$$

Compute the signal-to-noise ratio of the amplifier.

Solution: From Eq. (12-1-11), the *SNR* is

$$
SNR = \frac{V_i^2}{4kTBR} = \frac{(10^{-6})^2}{4 \times 1.38 \times 10^{-23} \times 290 \times 10^3 \times 100}
$$

$$
= 625
$$

12-2 AMPLITUDE NOISE AND PHASE NOISE

All electron tubes have some different types of noises and those noises will ultimately cause amplitude modulation (AM) noise and phase modulation (PM) noise in the electronic systems.

12-2-1 Amplitude Noise

Amplitude noise refers to the variations of amplitude modulation sidebands caused by the tube noise. This means that the amplitudes of the suppressed-carrier sidebands contain the noise levels of the tested electron tubes.

12-2-2 Phase Noise

In general, phase noise is defined as any variations from the ideal phase of a typical sinewave. The phase fluctuations created by the tube noises will eventually cause the frequency instability of the signal. Today's microwave engineering demands signal sources with better frequency stability for electronic systems such as Doppler radars, missile receivers, and space communication links. The phase noise on either the transmitter oscillator or the receiver local oscillator (or amplifier) caused by the tube noises can limit range resolution and sensitivity of an electronic system.

12-3 NOISE MEASUREMENT CIRCUITS

In general, there are three basic circuits for the tube noise measurements:

1. Noise reference circuit
2. Noise detection circuit
3. Noise amplification circuit

12-3-1 Noise Reference Circuit

The noise level of an electron tube is often too low to be measured and detected. Figure 12-3-1 shows a block diagram of a noise reference circuit.

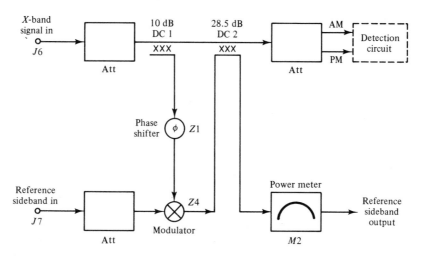

Figure 12-3-1   Block diagram of a noise reference circuit.

The primary function of the noise reference circuit is to generate sidebands 60 dB below the carrier on a double-sideband (DSB) basis in order to determine the level of the detected tube noise with respect to the carrier. The low-level modulator is to provide an AM or PM tone of known level with respect to the carrier. The tone can be used as a standard reference level when it is displayed on a spectrum analyzer. In operation, after having been adjusted for an AM or PM characteristic and after having its level noted on the analyzer, it is switched off and the level of the carrier noise can be noted with respect to the tone. When the carrier power and the sidebands power are matched in the secondary arm DC2, each power is down 28.5 dB from the carrier level in the primary arm. The sidebands power coupled to the primary arm is down another 28.5 dB. This power is double-sideband and it is 57 dB total below the carrier. Each sideband is 3 dB below the center carrier; then as a result, the reference noise level is 60 dB below the carrier. In tube noise measurement, if the reference level is set at 60 dB below the carrier and the noise reading is 40 dB below the reference level, then the tube noise is 100 dB below the carrier.

12-3-2 Noise Detection Circuit

The noise detection circuit performs the demodulation process of the tube noise measurement and it contains detectors, low-pass filters, and delay line component. There are two different circuits for either AM or PM detection.

1. *AM Noise Detection:* The amplitude noise detection circuit is designed to demodulate the amplitude modulated signal. Figure 12-3-2 shows a block diagram for AM noise detection.

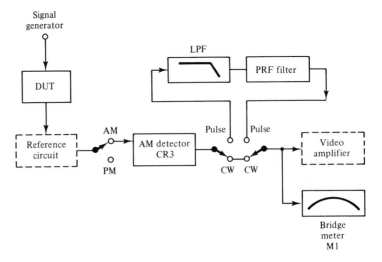

Figure 12-3-2   Block diagram for AM noise detection.

   The crystal detector CR3 demodulates the AM noise and the detected amplitude is then passed through the CW switches to the video amplifier for amplification. If the pulse mode is selected, the detected amplitude must pass through an extra PRF filter and a low-pass filter (LPF) to avoid overloading for the test set.

2. *PM Noise Detection:* The phase noise detection for an electron tube is more complicated than that of AM noise detection circuit. There are two main circuit components for the tube PM noise detection: a double-balanced mixer and a delay line. Figure 12-3-3 shows a block diagram for PM noise detection.

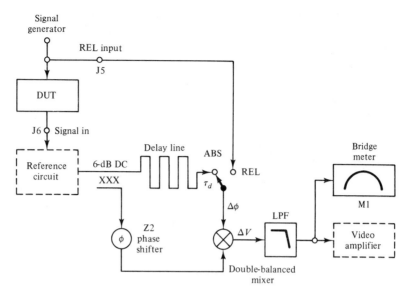

Figure 12-3-3   Block diagram for PM noise detection.

Double-balanced mixer.   The double-balanced mixer consists of a microwave magic-tee and a matched-reversed pair of microwave diodes as shown in Fig. 12-3-4.

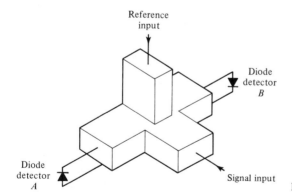

Figure 12-3-4   Double-balanced mixer.

The double-balanced mixer, acting as a phase detector, transforms the instantaneous phase fluctuations into voltage fluctuations ($\Delta\phi \rightarrow \Delta V$). With two input signals 90° out of phase (phase quadrature), the voltage output is proportional to the input phase fluctuations. The voltage fluctuation can then be measured by a baseband spectrum analyzer and converted to phase noise units.

Delay line.   The delay line converts the short-term frequency fluctuations of a signal into phase fluctuations ($\Delta f \rightarrow \Delta \phi$). The nominal frequency arrives at the input port of the delay line at a particular phase. As the frequency changes slightly, the phase shift incurred in the fixed delay time changes proportionally. The delay line converts the frequency change at the line input to a phase change at the line output when compared to the undelayed signal arriving at the mixer in the second port.

12-3-3 Noise Amplification Circuit

The low-noise video amplifier receives and amplifies the detected tube noise and then presents the noise to the spectrum analyzer for display and evaluation. Figure 12-3-5 shows a block diagram for the noise amplification circuit.

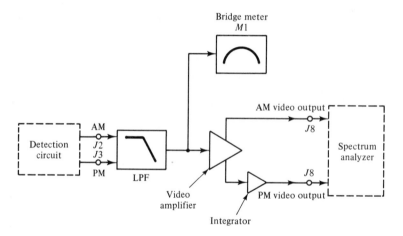

Figure 12-3-5   Block diagram of video amplifier.

After the tube noise has been detected, a switch routes either AM or PM noise to the video amplifier ($Q1$, $Q2$) for amplification. The video amplifier is a stable low-noise feedback amplifier having 40-dB gain from its input to output. At the amplifier input, a 500-kHz low-pass filter prevents the amplifier being overloaded by high-amplitude wideband noise from the PM detector. The integrator has two time constants and is used to give an overall PM response characteristic that is constant amplitude with frequency. The PM detector preceding the video amplifier has a sensitivity that increases as a direct function of video frequency while the integrator has a gain characteristic that is an inverse function of frequency.

12-4 NOISE MEASUREMENT TECHNIQUES

As described previously, all tube noises occur as results of various effects such as shot noise, thermal noise, vibrations, ionizations, and electron-beam fluctuations. In a high-power traveling-wave tube (TWT), the ripples of the power supply voltage and current can also create the electron-beam voltage variations. Consequently, these voltage variations can cause phase-shift fluctuation in the tube output, or phase noise. Noises in a 1-kHz bandwidth from 2 to 200 kHz away from the carrier frequency present serious problems in Doppler radar because this spread represents the Doppler shift of typical target.

12-4-1 AM Noise Measurement

When making AM noise measurement the mode switch is set to AM ABS position. The RF input level is adjusted for a midscale reading. At this point the noise output provides a 60-dB single-sideband reference noise below the carrier at a frequency equal to the frequency of the reference input signal. For making tube noise measurement, it is necessary to remove the reference input signal and then perform the spectrum analysis.

12-4-2 PM Noise Measurement

PM noise is almost always found to predominate over AM noise and the measurement of PM noise is considerably more complicated than that of AM noise. There are two commonly used methods of measuring PM noise of an electron tube in the frequency domain: Frequency discriminator and quadrature phase measurement.

1. *Frequency Discriminator:* PM noise measurement with a frequency discriminator for an electron tube is the easiest way to perform. Figure 12-4-1 shows a block diagram for frequency discriminator measurement.

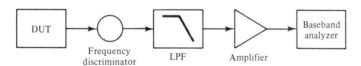

Figure 12-4-1   Frequency discriminator measurement.

Any variations in the carrier frequency or phase are changed into voltage variations by the discriminator. These voltage variations are then analyzed by a low-frequency spectrum analyzer as a function of frequency away from the carrier. The frequency discriminator must have an extremely high quality factor Q in order to have sufficient sensitivity for measurements close to the carrier.

2. *Quadrature Phase Method:* The phase noise of an electron tube can be measured by using a double-balanced mixer as shown in Fig. 12-4-2. If two signals of the same frequency and 90° out of phase are applied to the double-balanced mixer, the output will contain a low-frequency signal whose amplitude represents the sum of the phase noise of the sources. This noise power can be amplified by a video amplifier and then analyzed by a spectrum analyzer to display the tube noise as a function of frequency away from the carrier frequency.

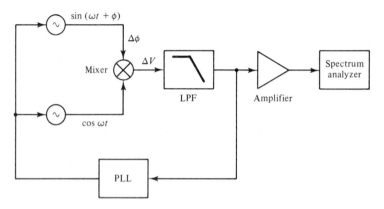

Figure 12-4-2   Quadrature phase noise measurement.

Example 12-4-1:   Phase Noise Measurement

If the reference sideband noise level is set at 60 dB below the center line of the carrier, the reference sideband voltage on the analyzer reads 10 mV and the tube noise voltage indicates 0.1 mV.

(a) Determine the tube noise in dB.

(b) Find the tube noise power with respect to the carrier.

Solution:

(a) The tube noise is

$$\text{Noise} = 20 \log(V_n/V_t) = 20 \log(10/0.1)$$
$$= 40 \text{ dB below the reference level}$$

(b) The measured noise power of this tube is

$$\text{Noise power} = -40 - 60 = -100 \text{ dB with respect to the carrier}$$

REFERENCES

[1] LIAO, SAMUEL Y. Analysis of microwave noise test set. *Report to Hughes Aircraft Company.* El Segundo, Calif., August 1986.

PROBLEMS

12-1. An electron tube has a temperature-limited emission current I_0 of 200 mA and the bandwidth is 2 kHz.

    (a) Compute the mean-squared shot noise current.

    (b) Calculate the noise power delivered to a shunt resistor of 40 kΩ.

12-2. If the matched resistor R has 400 Ω at room temperature and the bandwidth is 1 kHz.

    (a) Compute the output thermal noise voltage.

    (b) Calculate the output thermal noise voltage if the tube amplifier has a voltage gain of 50 dB.

12-3. A tube amplifier has the following operating parameters:

Input signal voltage	$V_i = 2 \times 10^{-6}$ V
Bandwidth	$B = 1$ kHz
Matched resistance	$R = 50$ Ω
Operating temperature	$T = 290°$K

Calculate the signal-to-noise ratio of the amplifier.

12-4. If the reference sideband noise level is set at 60 dB below the center, the reference sideband voltage on the analyzer reads 5 mV and the tube noise voltage indicates 0.2 mV.

    (a) Compute the tube noise in dB.

    (b) Calculate the tube noise level with respect to the carrier.

BIBLIOGRAPHY

BOOKS

1. CHODOROW, MARVIN, and CHARLES SUSSKIND, *Fundamentals of Microwave Electronics,* McGraw-Hill Book Company, New York, 1964.

2. COLEMAN, JAMES T., *Microwave Devices,* Reston Publishing Company, Inc., Reston, Va., 1982.

3. COLLIN, ROBERT E., *Foundations for Microwave Engineering,* McGraw-Hill Book Company, New York, 1966.

4. COLLINS, GEORGE B., *Microwave Magnetrons,* McGraw-Hill Book Company, New York, 1984.

5. CSORBA, ILLES P., *Image Tubes,* Howard W. Sams & Co., Inc. Indianapolis, Indiana, 1985.

6. GANDHI, OM P., *Microwave: Engineering and Applications,* Pergamon Press, New York, 1981.

7. GEWARTOWSKY, J. W., and H. A. WATSON, *Principles of Electron Tubes,* D. Van Nostrand Company, New York, 1965.

8. GILMOUR, A. S., JR., *Microwave Tubes,* Artech House, Dedham, Massachusetts, 1986.

9. GITTINS, J. F., *Power Traveling-Wave Tubes,* American Elsevier, Inc., New York, 1965.

10. GRAY, TRUMAN S., *Applied Electronics,* Massachusetts Institute of Technology Press, Cambridge, 1954.

11. HAMILTON, DONALD ROSS, et al., *Klystrons and Microwave Triodes*. MIT Radiation Laboratory Series. Vol. 7, Boston Technical Lithographers, Inc., 1963.

12. HARMAN, WILLIS W., *Fundamentals of Electron Motion,* McGraw-Hill Book Company, New York, 1953.

13. ISHII, T. KORYU, *Microwave Engineering,* Ronald Press Company, New York, 1966.

14. LIAO, SAMUEL Y., *Microwave Devices and Circuits,* 2nd ed., Prentice-Hall, Inc., Englewood Cliffs, N. J., 1985.

15. MILLMAN, JACOB, *Vacuum-Tube and Semiconductor Electronics.* McGraw-Hill Book Company, New York, 1958.

16. PIERCE, J. R., *Theory and Design of Electron Beam,* 2nd ed., D. Van Nostrand Company, New York, 1954.

17. REICH, HERBERT J., et al., *Microwave Principles,* D. Van Nostrand Company, New York, 1957.

18. REICH, HERBERT J., et al., *Microwave Theory and Techniques,* D. Van Nostrand Company, New York, 1953.

19. SMULLIN, LOUIS D., and HERMANN A. HAUS, *Noise in Electron Devices.* Massachusetts Institute of Technology Press, Cambridge, 1959.

20. SOOHOO, RONALD F., *Microwave Electronics.* Addison-Wesley Publishing Company, Reading, Mass., 1971.

21. SPANGENBERG, KARL R., *Vacuum Tubes,* McGraw-Hill Book Company, New York, 1948.

22. TSUI, JAMES BAO-YEN, *Microwave Receivers with Electronic Warfare Applications,* John Wiley & Sons, Inc. New York, 1986.

JOURNALS

1. BAIRD, J. M., Survey of fast wave tube development. *Technical Digest, 1979 Int. Electron Devices Meeting,* pp. 156–63.

2. BARNETT, LARRY R., et al., An experimental wide-band gyrotron traveling-wave amplifier. *IEEE Trans. Electron Devices,* Vol. ED-28, July 1981, pp. 872–75.

3. BOOT, H. A. H., et al., Historical notes on the cavity magnetron. *IEEE Trans. Electron Devices,* Vol. ED-23, July 1976, pp. 724–729.

4. BROWN, WILLIAM C., The microwave magnetron and its derivatives. *IEEE Trans. Electron Devices,* Vol. ED-31, Nov. 1984, pp. 1595–1605.

5. CHU, KWO RAY, et al., Characteristics and optimum operating parameters of a gyrotron traveling-wave amplifier. *IEEE Trans. Microwave Theory and Techniques,* Vol. MTT-27, No. 2, Feb. 1979, pp. 178–87.

6. CHU, KWO RAY, and J. L. HIRSHFIELD, Comparative study of the axial and azimuthal bunching mechanisms in electromagnetic cyclotron instabilities. *Phys. Fluids,* Vol. 21, No. 3, Mar. 1978, pp. 461–66.

7. DOHLER, GUNTER, et al. Peniotron oscillator operating performance. *Technical Digest, 1981 Int. Electron Devices Meeting,* pp. 328–31.

8. DOHLER, GUNTER, et al., The Peniotron: a fast wave device for efficient high-power mm-wave generation. *Technical Digest, 1978 Int. Electron Devices Meeting,* pp. 400–03.

9. ENDERLEY, C. E., et al., The Ubitron amplifier—a high-power millimeter-wave TWT. *Proc. IEEE,* Vol. 53, Oct. 1965, p. 1648.

10. FERGUSON, P. E., and R. S. SYMONS, A gyro-TWT with a space-charge limited gun. *Technical Digest, 1981 Int. Electron Devices Meeting,* pp. 198–201.

11. GARBUNY, M., et al., Image converter for thermal radiation. *Journal, Optical Society of America,* Vol. 51, No. 3, Mar. 1961, pp. 261–73.

12. HOWLAND, M. M., et al., Very fast, high peak-power, planar triode amplifier for driving optical gates. *Lawrence Laboratory,* UCRL 82538, Livermore, Calif., June 1979.

13. JORY, H. R., Gyro-device developments and applications. *Technical Digest, 1981 Int. Electron Devices Meeting,* pp. 182–85.

14. KILGORE, G. R., Recollections of pre–World War II magnetrons and their applications. *IEEE Trans. Electron Devices,* Vol. ED-31, Nov. 1984, pp. 1593–95.

15. KOSMAHL, H. G., Modern multistage depressed collectors—a Review. *Proc. IEEE,* Vol. 70, Nov. 1982, pp. 1325–1334.

16. LIAO, SAMUEL Y., The effect of collector voltage overdepression on tube performance of the gridded traveling-wave tubes. *Report to Hughes Aircraft Company,* El Segundo, Calif., August 1977.

17. MENDEL, J. T., Helix and coupled-cavity traveling-wave tubes. *Proc. IEEE,* Vol. 61, Mar. 1973, pp. 280–98.

18. PANTELL, R. H., Backward-wave oscillations in an unloaded guide. *Proc. IRE,* Vol. 47, 1959, p. 1146.

19. PHILLIPS, R. M., The Ubitron, a high-power traveling-wave tube based on a periodic beam interaction in unloaded waveguide. *IRE Trans. Electron Devices,* Vol. ED-7, Oct. 1960, pp. 231–41.

20. PREIST, D. H., and M. B. SHRADER, The klystrons—an unusual transmitting tube with potential for UHF-TV. *Proc. IEEE,* Vol. 70, Nov. 1982, pp. 1318–25.

21. SKOWRON, JOHN F., The continuous-cathode (emitting-sole) crossed-field amplifier. *Proc. IEEE,* Vol. 61, Mar. 1973, pp. 330–36.

22. STAPRANS, ARMAND, et al., High-power linear-beam tubes. *Proc. IEEE,* Vol. 61, Mar. 1973, pp. 299–330.

23. SYMONS, ROBERT S., et al., An experimental gyro-TWT. *IEEE Trans. Microwave Theory and Tech.,* Vol. MTT-29, No. 3, Mar. 1981, pp. 181–184.

24. VARIAN ASSOCIATES, *Microwave Tube Manual,* Air Force publication number T.0.00-25-251, October 1979.

25. WEIMER, PAUL K., et al., The Vidicon: Photoconductive camera tube. *Electronics,* May 1950, pp. 70–73.

26. YAMANOUCHI, K., et al., Cyclotron fast wave tube. The double ridge traveling-wave Peniotron. *Int. Conference on Microwave Tubes,* Vol. 5, 1964, pp. 96–102.

27. YINGST, T. E., et al., High-power gridded tubes. *Proc. IEEE,* Vol. 61, 1973, pp. 357–81, 1973.

APPENDIX A

Constants of Materials

1. CONDUCTIVITY σ IN mhos PER METER

Conductor	σ	Insulator	σ
Silver	6.17×10^7	Quartz	10^{-17}
Copper	5.80×10^7	Polystyrene	10^{-16}
Gold	4.10×10^7	Rubber (hard)	10^{-15}
Aluminum	3.82×10^7	Mica	10^{-14}
Tungsten	1.82×10^7	Porcelain	10^{-13}
Zinc	1.67×10^7	Diamond	10^{-13}
Brass	1.50×10^7	Glass	10^{-12}
Nickel	1.45×10^7	Bakelite	10^{-9}
Iron	1.03×10^7	Marble	10^{-8}
Bronze	1.00×10^7	Soil (sandy)	10^{-5}
Solder	0.70×10^7	Sands (dry)	2×10^{-4}
Steel (stainless)	0.11×10^7	Clay	10^{-4}
Nichrome	0.10×10^7	Ground (dry)	$10^{-4} - 10^{-5}$
Graphite	7.00×10^4	Ground (wet)	$10^{-2} - 10^{-3}$
Silicon	1.20×10^3	Water (distilled)	2×10^{-4}
Water (sea)	3 to 5	Water (fresh)	10^{-3}
		Ferrite (typical)	10^{-2}

2. DIELECTRIC CONSTANT—RELATIVE PERMITTIVITY ϵ_r

Material	ϵ_r	Material	ϵ_r
Air	1	Rubber	2.5–4
Alcohol (ethyl)	25	Sands (dry)	4
Bakelite	4.8	Silica (fused)	3.8
Glass	4–7	Snow	3.3
Ice	4.2	Sodium chloride	5.9
Mica (ruby)	5.4	Soil (dry)	2.8
Nylon	4	Styrofoam	1.03
Paper	2–4	Teflon	2.1
Plexiglass	2.6–3.5	Water (distilled)	80
Polyethylene	2.25	Water (sea)	20
Polystyrene	2.55	Water (dehydrated)	1
Porcelain (dry process)	6	Wood (dry)	1.5–4
Quartz (fused)	3.80	Ground (wet)	5–30
		Ground (dry)	2–5
		Water (fresh)	80

3. RELATIVE PERMEABILITY μ_r

Diamagnetic material	μ_r	Ferromagnetic material	μ_r
Bismuth	0.99999860	Nickel	50
Paraffin	0.99999942	Cast iron	60
Wood	0.99999950	Cobalt	60
Silver	0.99999981	Machine steel	300
		Ferrite (typical)	1,000
Paramagnetic material	μ_r	Transformer iron	3,000
		Silicon iron	4,000
Aluminum	1.00000065	Iron (pure)	4,000
Beryllinum	1.00000079	Mumetal	20,000
Nickel chloride	1.00004	Supermalloy	100,000
Manganese sulphate	1.0001		

4. PROPERTIES OF FREE SPACE

Velocity of light in vacuum c	2.997925×10^8	meters per second
Permittivity ϵ_0	8.854×10^{-12}	farad per meter
Permeability μ_0	$4\pi \times 10^{-7}$	henry per meter
Intrinsic impedance η_0	377 or 120π	ohms

5. PHYSICAL CONSTANTS

Boltzmann constant	k	1.381×10^{-23}	Joules per degrees Kelvin
Charge of electron	e	1.602×10^{-19}	coulomb
Mass of electron	m	9.109×10^{-31}	kilogram
Charge-to-mass ratio of electron	$\dfrac{e}{m}$	1.759×10^{11}	coulombs per kilogram
Planck's constant	h	6.626×10^{-34}	Joules-second

APPENDIX B

First-Order Bessel

Function Values

x	$J_1(x)$	x	$J_1(x)$	x	$J_1(x)$	x	$J_1(x)$	x	$J_1(x)$
0.00	0.000	0.92	0.413	1.86	0.582	2.86	0.389	3.84	-0.003
0.02	+0.010	0.94	0.420	1.88	0.5815	2.88	0.3825	3.86	0.011
0.04	0.020	0.96	0.427	1.90	0.581	2.90	0.375	3.88	0.019
0.06	0.030	0.98	0.4335	1.92	0.5805	2.92	0.368	3.90	0.027
0.08	0.040	1.00	0.440	1.94	0.580	2.94	0.361	3.92	0.035
0.10	0.050	1.02	0.4465	1.96	0.579	2.96	0.354	3.94	0.043
0.12	0.060	1.04	0.453	1.98	0.578	2.98	0.3465	3.96	0.051
0.14	0.070	1.06	0.459	2.00	0.577	3.00	0.339	3.98	0.058
0.16	0.080	1.08	0.465	2.02	0.575	3.02	0.3315	4.00	0.066
0.18	0.090	1.10	0.471	2.04	0.574	3.04	0.324	4.10	0.103
0.20	0.0995	1.12	0.477	2.06	0.572	3.06	0.316	4.20	0.139
0.22	0.109	1.14	0.482	2.08	0.570	3.08	0.309	4.30	0.172
0.24	0.119	1.16	0.488	2.10	0.568	3.10	0.301	4.40	0.203
0.26	0.129	1.18	0.493	2.12	0.566	3.12	0.293	4.50	0.231
0.28	0.139	1.20	0.498	2.14	0.564	3.14	0.285	4.60	0.2565
0.30	0.148	1.22	0.503	2.16	0.561	3.16	0.277	4.70	0.279
0.32	0.158	1.24	0.508	2.18	0.559	3.18	0.269	4.80	0.2985
0.34	0.1675	1.26	0.513	2.20	0.556	3.20	0.261	4.90	0.315
0.36	0.177	1.28	0.5175	2.22	0.553	3.22	0.253	5.00	0.3275
0.38	0.187	1.30	0.522	2.24	0.550	3.24	0.245	5.05	0.334
0.40	0.196	1.32	0.526	2.26	0.547	3.26	0.237	5.10	0.337
0.42	0.205	1.34	0.5305	2.28	0.543	3.28	0.229	5.16	0.341
0.44	0.215	1.36	0.534	2.30	0.540	3.30	0.221	5.20	0.343
0.46	0.224	1.38	0.538	2.32	0.536	3.32	0.212	5.26	0.345
0.48	0.233	1.40	0.542	2.34	0.532	3.34	0.204	5.30	0.346
0.50	0.242	1.42	0.5455	2.36	0.5285	3.36	0.196	5.32	0.346

x	$J_1(x)$	x	$J_1(x)$	x	$J_1(x)$	x	$J_1(x)$	x	$J_1(x)$
0.52	0.251	1.44	0.549	2.38	0.524	3.38	0.1865	5.34	0.346
0.54	0.260	1.46	0.552	2.40	0.520	3.40	0.179	5.36	0.346
0.56	0.269	1.48	0.555	2.42	0.516	3.42	0.171	5.38	0.346
0.58	0.278	1.50	0.558	2.44	0.511	3.44	0.1625	5.40	0.345
0.60	0.287	1.52	0.561	2.46	0.507	3.46	0.154	5.47	0.343
0.62	0.295	1.54	0.563	2.48	0.502	3.48	0.146	5.50	0.341
0.64	0.304	1.56	0.566	2.50	0.497	3.50	0.137	5.56	0.3375
0.66	0.312	1.58	0.568	2.52	0.492	3.52	0.129	5.60	0.334
0.68	0.321	1.60	0.570	2.54	0.487	3.54	0.121	5.66	0.3285
0.70	0.329	1.62	0.572	2.56	0.482	3.56	0.112	5.70	0.324
0.72	0.337	1.64	0.5735	2.58	0.476	3.58	0.104	5.80	0.311
0.74	0.345	1.66	0.575	2.60	0.471	3.60	0.0955	5.90	0.295
0.76	0.353	1.68	0.5765	2.62	0.465	3.62	0.087	6.00	0.277
0.78	0.361	1.70	0.578	2.64	0.4595	3.64	0.079	6.10	0.256
0.80	0.369	1.72	0.579	2.66	0.454	3.66	0.070	6.20	0.233
0.82	0.3765	1.74	0.580	2.68	0.448	3.68	0.062	6.30	0.208
0.84	0.384	1.76	0.5805	2.70	0.442	3.70	0.054	6.40	0.182
0.86	0.3915	1.78	0.581	2.72	0.435	3.72	0.0455	6.60	0.125
0.88	0.399	1.80	0.5815	2.74	0.429	3.74	0.037	6.70	0.095
0.90	0.406	1.82	0.582	2.76	0.423	3.76	0.029	6.80	0.065
		1.84	0.582	2.78	0.416	3.78	0.021	6.90	0.035
				2.80	0.410	3.80	0.013	7.00	0.005
				2.82	0.403	3.82	0.005	7.01	0.000
				2.84	0.396	3.83	0.000		

β^2 as a Function
of Radius*

(r_c = radius of cathode; r = radius at any point P. β^2 applies to case where P is outside cathode, $r > r_c$. $(-\beta)^2$ applies to case where P is inside cathode, $r_c > r$.)

$\dfrac{r}{r_c}$ or $\dfrac{r_c}{r}$	β^2	$(-\beta)^2$	$\dfrac{r}{r_c}$ or $\dfrac{r_c}{r}$	β^2	$(-\beta)^2$
1.00	0.00000	0.00000	5.2	0.7825	10.733
1.01	0.00010	0.00010	5.4	0.7973	11.601
1.02	0.00039	0.00040	5.6	0.8111	12.493
1.04	0.00149	0.00159	5.8	0.8241	13.407
1.06	0.00324	0.00356	6.0	0.8362	14.343
1.08	0.00557	0.00630	6.5	0.8635	16.777
1.10	0.00842	0.00980	7.0	0.8870	19.337
1.15	0.01747	0.02186	7.5	0.9074	22.015
1.2	0.02875	0.03849	8.0	0.9253	24.805
1.3	0.05589	0.08504	8.5	0.9410	27.701
1.4	0.08672	0.14856	9.0	0.9548	30.698
1.5	0.11934	0.2282	9.5	0.9672	33.791
1.6	0.1525	0.3233	10.0	0.9782	36.976
1.7	0.1854	0.4332	12.0	1.0122	50.559
1.8	0.2177	0.5572	14.0	1.0352	65.352
1.9	0.2491	0.6947	16.0	1.0513	81.203
2.0	0.2793	0.8454	18.0	1.0630	97.997
2.1	0.3083	1.0086	20.0	1.0715	115.64

*Taken from I. Langmuir and K. B. Blodgett, Currents Limited by Space Charge Between Coaxial Cylinders, *Phys. Rev.*, 22:347–356 (1922).

$\dfrac{r}{r_c}$ or $\dfrac{r_c}{r}$	β^2	$(-\beta)^2$	$\dfrac{r}{r_c}$ or $\dfrac{r_c}{r}$	β^2	$(-\beta)^2$
2.2	0.3361	1.1840	30.0	1.0908	214.42
2.3	0.3626	1.3812	40.0	1.0946	327.01
2.4	0.3879	1.5697	50.0	1.0936	450.23
2.5	0.4121	1.7792	60.0	1.0910	582.14
2.6	0.4351	1.9995	70.0	1.0878	721.43
2.7	0.4571	2.2301	80.0	1.0845	867.11
2.8	0.4780	2.4708	90.0	1.0813	1018.5
2.9	0.4980	2.7214	100.0	1.0783	1174.9
3.0	0.5170	2.9814	120.0	1.0726	1501.4
3.2	0.5526	3.5293	140.0	1.0677	1843.5
3.4	0.5851	4.1126	160.0	1.0634	2199.4
3.6	0.6148	4.7298	180.0	1.0596	2567.3
3.8	0.6420	5.3795	200.0	1.0562	2946.1
4.0	0.6671	6.0601	250.0	1.0494	3934.4
4.2	0.6902	6.7705	300.0	1.0440	4973.0
4.4	0.7115	7.5096	350.0	1.0397	6054.1
4.6	0.7313	8.2763	400.0	1.0362	7172.1
4.8	0.7496	9.0696	500.0	1.0307	9502.2
5.0	0.7666	9.8887	∞	1.0000	∞

$(-\alpha)^2$ as Function

of Radius*

(r_c = radius of cathode; r = radius at any point P.)

$\dfrac{r_c}{r}$	$(-\alpha)^2$	$\dfrac{r_c}{r}$	$(-\alpha)^2$	$\dfrac{r_c}{r}$	$(-\alpha)^2$
1.0	0.0000	2.9	2.302	10.0	29.19
1.05	0.0024	3.0	2.512	12.0	39.98
1.1	0.0096	3.2	2.954	14.0	51.86
1.15	0.0213	3.4	3.421	16.0	64.74
1.2	0.0372	3.6	3.913	18.0	78.56
1.25	0.0571	3.8	4.429	20.0	93.24
1.3	0.0809	4.0	4.968	30.0	178.2
1.35	0.1084	4.2	5.528	40.0	279.6
1.4	0.1396	4.4	6.109	50.0	395.3
1.45	0.1740	4.6	6.712	60.0	523.6
1.5	0.2118	4.8	7.334	70.0	663.3
1.6	0.2968	5.0	7.976	80.0	813.7
1.7	0.394	5.2	8.636	90.0	974.1
1.8	0.502	5.4	9.315	100.0	1144
1.9	0.621	5.6	10.01	120.0	1509
2.0	0.750	5.8	10.73	140.0	1907
2.1	0.888	6.0	11.46	160.0	2333
2.2	1.036	6.5	13.35	180.0	2790
2.3	1.193	7.0	15.35	200.0	3270
2.4	1.358	7.5	17.44	250.0	4582
2.5	1.531	8.0	19.62	300.0	6031
2.6	1.712	8.5	21.89	350.0	7610
2.7	1.901	9.0	24.25	400.0	9303
2.8	2.098	9.5	26.68	500.0	13015

*Taken from I. Langmuir and K. B. Blodgett, Currents Limited by Space Charge Between Concentric Spheres, *Phys. Rev.*, 24:49–59 (1924).

APPENDIX E

Characteristics of

Standard Rectangular

Waveguides

Rectangular waveguides are commonly used for power transmission at microwave frequencies. Its physical dimensions are regulated by the frequency of the signal being transmitted. For example, at X-band frequencies from 8 to 12 GHz, the outside dimensions of a rectangular waveguide are 2.54 cm (1.0 in.) wide and 1.27 cm (0.5 in.) high designated as EIA WR(90) by the Electronic Industry Association, but its inside dimensions are 2.286 cm (0.90 in.) wide and 1.016 cm (0.40 in.) high. Table E-1 tabulates the characteristics of the standard rectangular waveguides.

TABLE E-1. CHARACTERISTICS OF STANDARD RECTANGULAR WAVEGUIDES

EIA Designation WR ( )	Physical dimensions		Cutoff frequency for air-filled waveguide in GHz	Recommended frequency range for TE_{10} mode in GHz
	Inside in cm (inches) Width a Height b	Outside in cm (inches) Width a Height b		
2300	58.420 29.210 (23.000) (11.500)	59.055 29.845 (23.250) (11.750)	0.257	0.32–0.49
2100	53.340 26.670 (21.000) (10.500)	53.973 27.305 (21.250) (10.750)	0.281	0.35–0.53
1800	45.720 22.860 (18.000) (9.000)	46.350 23.495 (18.250) (9.250)	0.328	0.41–0.62
1500	38.100 19.050 (15.000) (7.500)	38.735 19.685 (15.250) (7.750)	0.394	0.49–0.75
1150	29.210 14.605 (11.500) (5.750)	29.845 15.240 (11.750) (6.000)	0.514	0.64–0.98

TABLE E-1. CHARACTERISTICS OF STANDARD RECTANGULAR WAVEGUIDES (CONTINUED)

EIA Designation WR ( )	Physical dimensions				Cutoff frequency for air-filled waveguide in GHz	Recommended frequency range for TE$_{10}$ mode in GHz
	Inside in cm (inches)		Outside in cm (inches)			
	Width a	Height b	Width a	Height b		
975	24.765 (9.750)	12.383 (4.875)	25.400 (10.000)	13.018 (5.125)	0.606	0.76–1.15
770	19.550 (7.700)	9.779 (3.850)	20.244 (7.970)	10.414 (4.100)	0.767	0.96–1.46
650	16.510 (6.500)	8.255 (3.250)	16.916 (6.660)	8.661 (3.410)	0.909	1.14–1.73
510	12.954 (5.100)	6.477 (2.500)	13.360 (5.260)	6.883 (2.710)	1.158	1.45–2.20
430	10.922 (4.300)	5.461 (2.150)	11.328 (4.460)	5.867 (2.310)	1.373	1.72–2.61
340	8.636 (3.400)	4.318 (1.700)	9.042 (3.560)	4.724 (1.860)	1.737	2.17–3.30
284	7.214 (2.840)	3.404 (1.340)	7.620 (3.000)	3.810 (1.500)	2.079	2.60–3.95
229	5.817 (2.290)	2.908 (1.145)	6.142 (2.418)	3.233 (1.273)	2.579	3.22–4.90
187	4.755 (1.872)	2.215 (0.872)	5.080 (2.000)	2.540 (1.000)	3.155	3.94–5.99
159	4.039 (1.590)	2.019 (0.795)	4.364 (1.718)	2.344 (0.923)	3.714	4.64–7.05
137	3.485 (1.372)	1.580 (0.622)	3.810 (1.500)	1.905 (0.750)	4.304	5.38–8.17
112	2.850 (1.122)	1.262 (0.497)	3.175 (1.250)	1.588 (0.625)	5.263	6.57–9.99
90	2.286 (0.900)	1.016 (0.400)	2.540 (1.000)	1.270 (0.500)	6.562	8.20–12.50
75	1.905 (0.750)	0.953 (0.375)	2.159 (0.850)	1.207 (0.475)	7.874	9.84–15.00
62	1.580 (0.622)	0.790 (0.311)	1.783 (0.702)	0.993 (0.391)	9.494	11.90–18.00
51	1.295 (0.510)	0.648 (0.255)	1.499 (0.590)	0.851 (0.335)	11.583	14.50–22.00
42	−1.067 (0.420)	0.432 (0.170)	1.270 (0.500)	0.635 (0.250)	14.058	17.60–26.70
34	0.864 (0.340)	0.432 (0.170)	1.067 (0.420)	0.635 (0.250)	17.361	21.70–33.00
28	0.711 (0.280)	0.356 (0.140)	0.914 (0.360)	0.559 (0.220)	21.097	26.40–40.00
22	0.569 (0.224)	0.284 (0.112)	0.772 (0.304)	0.488 (0.192)	26.362	32.90–50.10
19	0.478 (0.188)	0.239 (0.094)	0.681 (0.268)	0.442 (0.174)	31.381	39.20–59.60
15	0.376 (0.148)	0.188 (0.074)	0.579 (0.228)	0.391 (0.154)	39.894	49.80–75.80

TABLE E-1. CHARACTERISTICS OF STANDARD RECTANGULAR WAVEGUIDES (CONTINUED)

EIA Designation WR ( )	Physical dimensions				Cutoff frequency for air-filled waveguide in GHz	Recommended frequency range for TE$_{10}$ mode in GHz
	Inside in cm (inches)		Outside in cm (inches)			
	Width a	Height b	Width a	Height b		
12	0.310 (0.122)	0.155 (0.061)	0.513 (0.202)	0.358 (0.141)	48.387	60.50–91.90
10	0.254 (0.100)	0.127 (0.050)	0.457 (0.180)	0.330 (0.130)	59.055	73.80–112.00
8	0.203 (0.080)	0.102 (0.040)	0.406 (0.160)	0.305 (0.120)	73.892	92.20–140.00
7	0.165 (0.065)	0.084 (0.033)	0.343 (0.135)	0.262 (0.103)	90.909	114.00–173.00
5	0.130 (0.051)	0.066 (0.026)	0.257 (0.101)	0.193 (0.076)	115.385	145.00–220.00
4	0.109 (0.043)	0.056 (0.022)	0.211 (0.083)	0.157 (0.062)	137.615	172.00–261.00
3	0.086 (0.034)	0.043 (0.017)	0.163 (0.064)	0.119 (0.047)	174.419	217.00–333.00

Notes: EIA stands for Electronic Industry Association; WR represents rectangular waveguide.

APPENDIX F

Characteristics of

Standard Circular

Waveguides

The inner diameter of a circular waveguide is regulated by the frequency of the signal being transmitted. For example, at X-band frequencies from 8 to 12 GHz, the inner diameter of a circular waveguide is 2.383 cm (0.938 in.) designated as EIA WC(94) by the Electronic Industry Association. Table F-1 tabulates the characteristics of the standard circular waveguides.

TABLE F-1. CHARACTERISTICS OF STANDARD CIRCULAR WAVEGUIDES

EIA Designation WC ( )	Inside diameter 2a in cm (inches)		Cutoff frequency for air-filled waveguide in GHz	Recommended frequency range for TE$_{11}$ mode in GHz
992	25.184	(9.915)	0.698	0.80–1.10
847	21.514	(8.470)	0.817	0.94–1.29
724	18.377	(7.235)	0.957	1.10–1.51
618	15.700	(6.181)	1.120	1.29–1.76
528	13.411	(5.280)	1.311	1.51–2.07
451	11.458	(4.511)	1.534	1.76–2.42
385	9.787	(3.853)	1.796	2.07–2.83
329	8.362	(3.292)	2.102	2.42–3.31
281	7.142	(2.812)	2.461	2.83–3.88
240	6.104	(2.403)	2.880	3.31–4.54
205	5.199	(2.047)	3.381	3.89–5.33
175	4.445	(1.750)	3.955	4.54–6.23
150	3.810	(1.500)	4.614	5.30–7.27
128	3.254	(1.281)	5.402	6.21–8.51
109	2.779	(1.094)	6.326	7.27–9.97
94	2.383	(0.938)	7.377	8.49–11.60
80	2.024	(0.797)	8.685	9.97–13.70
69	1.748	(0.688)	10.057	11.60–15.90
59	1.509	(0.594)	11.649	13.40–18.40
50	1.270	(0.500)	13.842	15.90–21.80
44	1.113	(0.438)	15.794	18.20–24.90
38	0.953	(0.375)	18.446	21.20–29.10
33	0.833	(0.328)	21.103	24.30–33.20
28	0.714	(0.281)	24.620	28.30–38.80
25	0.635	(0.250)	27.683	31.80–43.60
22	0.556	(0.219)	31.617	36.40–49.80
19	0.478	(0.188)	36.776	42.40–58.10
17	0.437	(0.172)	40.227	46.30–63.50
14	0.358	(0.141)	49.103	56.60–77.50
13	0.318	(0.125)	55.280	63.50–87.20
11	0.277	(0.109)	63.462	72.70–99.70
9	0.239	(0.094)	73.552	84.80–116.00

Notes: EIA stands for Electronic Industry Association; WC represents circular waveguide

APPENDIX G

Units of Microwave

Measurement

In field intensity measurements the units of measurement and the conversion of one unit to another are the essential parts of the process. A few widely used units are described below.

1. dB—The decibel (dB) is a dimensionless number that expresses the ratio of two power levels. It is defined as

$$dB \equiv 10 \log_{10} \left(\frac{P_2}{P_1} \right) \tag{G-1}$$

The two power levels are relative to each other. If power level P_2 is higher than power level P_1, dB is positive and vice versa. Since $P = V^2/R$, when their voltages are measured across the same or equal resistors, the number of dB is given by

$$dB \equiv 20 \log_{10} \left(\frac{V_2}{V_1} \right) \tag{G-2}$$

The voltage definition of dB has no meaning at all unless the two voltages under consideration appear across equal impedances. Above 10 GHz the impedance of waveguides varies with frequency, and the dB calibration is limited to power levels only. Table G-1 shows the conversion of voltage and power ratios to dB.

2. dBW—The decibel above 1 watt (dBW) is another useful measure for expressing power level P_2 with respect to a reference power level P_1 of 1 watt. Similarly, if the power level P_2 is lower than 1 watt, the dBW is negative.

3. dBm—The decibel above 1 milliwatt (dBm) is also a useful measure of expressing power level P_2 with respect to a reference power level P_1 of 1 milliwatt (mW). Since the power level in the microwave region is quite low, the dBm unit is very useful in that frequency range. It is customary to designate milli by a lowercase letter "m" and mega by an uppercase letter "M."

4. dBV—The decibel above 1 volt (dBV) is a dimensionless voltage ratio in dB referred to a reference voltage of 1 volt.

5. dBμV—The decibel above 1 microvolt (dBμV) is another dimensionless voltage ratio in dB referred to a reference voltage of 1 microvolt (μV). The field intensity meters used for the measurements in the microwave region often have a scale in dBμV, since the power levels to be measured are usually extremely low.

6. μV/m—Microvolts per meter (μV/m) are units of 10^{-6} volt per meter, expressing the electric field intensity.

7. dBμV/m—The decibel above 1 microvolt per meter (dBμV/m) is a dimensionless electric field intensity ratio in dB relative to 1 μV/m. This unit is also often used for field intensity measurements in the microwave region.

8. μV/m/MHz—The microvolts per meter per megahertz (μV/m/MHz) are units of 10^{-6} volt-second per broadband electric field intensity distribution. This is a two-dimensional distribution, in space and in frequency.

9. dBμV/m/MHz—The decibel above 1 microvolt per meter per megahertz (dBμV/m/MHz) is a dimensionless broadband electric field intensity distribution ratio with respect to 1 μV/m/MHz.

10. μV/MHz—Microvolts per megahertz per second of bandwidth (μV/MHz) are units of 10^{-6} volt-second of broadband voltage distribution in the frequency domain. The use of this unit is based on the assumption that the voltage is evenly distributed over the bandwidth of interest.

Table G-1 tabulates the conversion of voltage and power ratio to dB.

TABLE G-1.   CONVERSION OF VOLTAGE AND POWER RATIOS TO DECIBELS (dB)

Voltage ratio	Power ratio	− dB +	Voltage ratio	Power ratio	Voltage ratio	Power ratio	− dB +	Voltage ratio	Power ratio
1.000	1.000	0	1.000	1.000	0.596	0.355	4.5	1.679	2.818
0.989	0.977	0.1	1.012	1.023	0.589	0.347	4.6	1.698	2.884
0.977	0.955	0.2	1.023	1.047	0.582	0.339	4.7	1.718	2.951
0.966	0.933	0.3	1.035	1.072	0.575	0.331	4.8	1.738	3.020
0.955	0.912	0.4	1.047	1.096	0.569	0.324	4.9	1.758	3.090
0.944	0.891	0.5	1.059	1.122	0.562	0.316	5.0	1.778	3.162
0.933	0.871	0.6	1.072	1.148	0.556	0.309	5.1	1.799	3.236
0.923	0.851	0.7	1.084	1.175	0.550	0.302	5.2	1.820	3.311
0.912	0.832	0.8	1.095	1.202	0.543	0.295	5.3	1.841	3.388
0.902	0.813	0.9	1.109	1.230	0.537	0.288	5.4	1.862	3.467
0.891	0.794	1.0	1.122	1.259	0.531	0.282	5.5	1.884	3.548

TABLE G-1. (CONTINUED)

Voltage ratio	Power ratio	− dB +	Voltage ratio	Power ratio	Voltage ratio	Power ratio	− dB +	Voltage ratio	Power ratio
0.881	0.776	1.1	1.135	1.288	0.525	0.275	5.6	1.905	3.631
0.871	0.759	1.2	1.148	1.318	0.519	0.269	5.7	1.928	3.715
0.861	0.741	1.3	1.161	1.349	0.513	0.263	5.8	1.950	3.802
0.851	0.724	1.4	1.175	1.380	0.507	0.257	5.9	1.972	3.890
0.841	0.708	1.5	1.189	1.413	0.501	0.251	6.0	1.995	3.981
0.832	0.692	1.6	1.202	1.445	0.496	0.246	6.1	2.018	4.074
0.822	0.676	1.7	1.216	1.479	0.490	0.240	6.2	2.042	4.159
0.813	0.661	1.8	1.230	1.514	0.484	0.234	6.3	2.065	4.265
0.804	0.646	1.9	1.245	1.549	0.479	0.229	6.4	2.089	4.365
0.794	0.631	2.0	1.259	1.585	0.473	0.224	6.5	2.113	4.467
0.785	0.617	2.1	1.274	1.622	0.468	0.219	6.6	2.138	4.571
0.776	0.603	2.2	1.288	1.660	0.462	0.214	6.7	2.163	4.677
0.767	0.589	2.3	1.303	1.698	0.457	0.209	6.8	2.188	4.786
0.759	0.575	2.4	1.318	1.738	0.452	0.204	6.9	2.215	4.898
0.750	0.562	2.5	1.334	1.778	0.447	0.200	7.0	2.239	5.012
0.741	0.550	2.6	1.349	1.820	0.442	0.195	7.1	2.265	5.129
0.733	0.537	2.7	1.365	1.862	0.437	0.191	7.2	2.291	5.248
0.724	0.525	2.8	1.380	1.905	0.432	0.186	7.3	2.317	5.370
0.716	0.513	2.9	1.390	1.950	0.427	0.182	7.4	2.344	5.495
0.708	0.501	3.0	1.413	1.995	0.422	0.178	7.5	2.371	5.623
0.700	0.490	3.1	1.429	2.042	0.417	0.174	7.6	2.399	5.754
0.692	0.479	3.2	1.445	2.089	0.412	0.170	7.7	2.427	5.888
0.684	0.468	3.3	1.462	2.138	0.407	0.166	7.8	2.455	6.026
0.676	0.457	3.4	1.479	2.188	0.403	0.162	7.9	2.483	6.166
0.668	0.447	3.5	1.496	2.239	0.398	0.159	8.0	2.512	6.310
0.661	0.437	3.6	1.514	2.291	0.394	0.155	8.1	2.541	6.457
0.653	0.427	3.7	1.531	2.344	0.389	0.151	8.2	2.570	6.607
0.646	0.417	3.8	1.549	2.399	0.385	0.148	8.3	2.600	6.761
0.638	0.407	3.9	1.567	2.455	0.380	0.145	8.4	2.630	6.918
0.631	0.398	4.0	1.585	2.512	0.376	0.141	8.5	2.661	7.079
0.624	0.389	4.1	1.603	2.570	0.372	0.138	8.6	2.692	7.244
0.617	0.380	4.2	1.622	2.630	0.367	0.135	8.7	2.723	7.413
0.610	0.372	4.3	1.641	2.692	0.363	0.132	8.8	2.754	7.586
0.603	0.363	4.4	1.660	2.754	0.359	0.129	8.9	2.786	7.762
0.355	0.126	9.0	2.818	7.943	0.211	0.0447	13.5	4.732	22.39
0.351	0.123	9.1	2.851	8.128	0.209	0.0437	13.6	4.786	22.91
0.347	0.120	9.2	2.884	8.318	0.207	0.0427	13.7	4.842	23.44
0.343	0.118	9.3	2.917	8.511	0.204	0.0417	13.8	4.898	23.99
0.339	0.115	9.4	2.951	8.710	0.202	0.0407	13.9	4.955	24.55
0.335	0.112	9.5	2.985	8.913	0.200	0.0398	14.0	5.012	25.12
0.331	0.110	9.6	3.020	9.120	0.197	0.0389	14.1	5.070	25.70
0.327	0.107	9.7	3.055	9.333	0.195	0.0380	14.2	5.129	26.30
0.324	0.105	9.8	3.090	9.550	0.193	0.0372	14.3	5.188	26.92
0.320	0.102	9.9	3.126	9.772	0.191	0.0363	14.4	5.248	27.54
0.316	0.100	10.0	3.162	10.000	0.188	0.0355	14.5	5.309	28.18
0.313	0.0977	10.1	3.199	10.23	0.186	0.0347	14.6	5.370	28.84
0.309	0.0955	10.2	3.236	10.47	0.184	0.0339	14.7	5.433	29.51
0.306	0.0933	10.3	3.273	10.72	0.182	0.0331	14.8	5.495	30.20

TABLE G-1.   (CONTINUED)

Voltage ratio	Power ratio	− dB +	Voltage ratio	Power ratio	Voltage ratio	Power ratio	− dB +	Voltage ratio	Power ratio
0.302	0.0912	10.4	3.311	10.96	0.180	0.0324	14.9	5.559	30.90
0.299	0.0891	10.5	3.350	11.22	0.178	0.0316	15.0	5.623	31.62
0.295	0.0871	10.6	3.388	11.48	0.176	0.0309	15.1	5.689	32.36
0.292	0.0851	10.7	3.428	11.75	0.174	0.0302	15.2	5.754	33.11
0.288	0.0832	10.8	3.467	12.02	0.172	0.0295	15.3	5.821	33.88
0.283	0.0813	10.9	3.508	12.30	0.170	0.0288	15.4	5.888	34.67
0.282	0.0794	11.0	3.548	12.59	0.168	0.0282	15.5	5.957	35.48
0.279	0.0776	11.1	3.589	12.88	0.166	0.0275	15.6	6.026	36.31
0.275	0.0759	11.2	3.631	13.18	0.164	0.0269	15.7	6.095	37.15
0.272	0.0741	11.3	3.673	13.49	0.162	0.0263	15.8	6.166	38.02
0.269	0.0724	11.4	3.715	13.80	0.160	0.0257	15.9	6.237	38.90
0.266	0.0708	11.5	3.758	14.13	0.159	0.0251	16.0	6.310	39.81
0.263	0.0692	11.6	3.802	14.45	0.157	0.0246	16.1	6.383	40.74
0.260	0.0676	11.7	3.846	14.79	0.155	0.0240	16.2	6.457	41.69
0.257	0.0661	11.8	3.890	15.14	0.153	0.0234	16.3	6.531	42.66
0.254	0.0646	11.9	3.936	15.49	0.151	0.0229	16.4	6.607	43.65
0.251	0.0631	12.0	3.981	15.85	0.150	0.0224	16.5	6.683	44.67
0.248	0.0617	12.1	4.027	16.22	0.148	0.0219	16.6	6.761	45.71
0.246	0.0603	12.2	4.074	16.60	0.146	0.0214	16.7	6.839	46.77
0.243	0.0589	12.3	4.121	16.98	0.145	0.0209	16.8	6.918	47.86
0.240	0.0575	12.4	4.169	17.38	0.143	0.0204	16.9	6.998	48.98
0.237	0.0562	12.5	4.217	17.78	0.141	0.0200	17.0	7.079	50.12
0.234	0.0550	12.6	4.266	18.20	0.140	0.0195	17.1	7.161	51.29
0.232	0.0537	12.7	4.315	18.62	0.138	0.0191	17.2	7.244	52.48
0.229	0.0525	12.8	4.365	19.05	0.137	0.0186	17.3	7.328	53.70
0.227	0.0513	12.9	4.416	19.50	0.135	0.0182	17.4	7.413	54.95
0.224	0.0501	13.0	4.467	19.95	0.133	0.0178	17.5	7.499	56.23
0.221	0.0490	13.1	4.519	20.42	0.132	0.0174	17.6	7.586	57.54
0.219	0.0479	13.2	4.571	20.89	0.130	0.0170	17.7	7.674	58.88
0.216	0.0468	13.3	4.624	21.38	0.129	0.0166	17.8	7.762	60.26
0.214	0.0457	13.4	4.677	21.88	0.127	0.0162	17.9	7.852	61.66
0.126	0.0159	18.0	7.943	63.10	0.106	0.0112	19.5	9.441	89.13
0.125	0.0155	18.1	8.035	64.57	0.103	0.0110	19.6	9.550	91.20
0.123	0.0151	18.2	8.128	66.07	0.104	0.0107	19.7	9.661	93.33
0.122	0.0148	18.3	8.222	67.61	0.102	0.0105	19.8	9.772	95.50
0.120	0.0145	18.4	8.318	69.18	0.101	0.0102	19.9	9.886	97.72
0.119	0.0141	18.5	8.414	70.79	0.100	0.0100	20.0	10.000	100.00
0.118	0.0138	18.6	8.511	72.44		10^{-3}	30		10^{3}
0.116	0.0135	18.7	8.610	74.13	10^{-2}	10^{-4}	40	10^{2}	10^{4}
0.115	0.0132	18.8	8.710	75.86		10^{-5}	50		10^{5}
0.114	0.0129	18.9	8.811	77.62	10^{-3}	10^{-6}	60	10^{3}	10^{6}
0.112	0.0126	19.0	8.913	79.43		10^{-7}	70		10^{7}
0.111	0.0123	19.1	9.016	81.28	10^{-4}	10^{-8}	80	10^{4}	10^{8}
0.110	0.0120	19.2	9.120	83.18		10^{-9}	90		10^{9}
0.108	0.0118	19.3	9.226	85.11	10^{-5}	10^{-10}	100	10^{5}	10^{10}
0.107	0.0115	19.4	9.333	87.10		10^{-11}	110		10^{11}
					10^{-6}	10^{-12}	120	10^{6}	10^{12}

APPENDIX H

Microwave Coaxial

Connectors

For high-frequency operation the average circumference of a coaxial cable must be limited to about one wavelength in order to reduce multimodal propagation for elimination of erratic reflection coefficients, power losses, and signal distortion. The early standardization of coaxial connectors to interconnect electronic or electric system equipment during World War II was mandatory for microwave operation and low-reflection coefficient (or low-voltage standing-wave ratio—VSWR). Ever since, many modifications and new designs for microwave connectors have been proposed and developed. Figure H-1 shows six types of microwave coaxial connectors.

APC-3.5. The APC-3.5 (Amphenol Precision Connector-3.5 mm) was originally developed by Hewlett-Packard, but it is now manufactured by Amphenol. The connector provides repeatable connections and has a very-low-voltage standing-wave ratio (VSWR). Either male or female end of this 50-Ω connector can mate with the opposite type of an SMA connector. The APC-3.5 connector can work up to 34 GHz.

APC-7. The APC-7 (Amphenol Precision Connector-7) was also developed by Hewlett-Packard in the mid-1960s, but it was lately improved and is now manufactured by Amphenol. The connector provides coupling mechanism without male or female distinction and is the most repeatable connecting device. Its VSWR is extremely low in the range of 1.02 up to 18 GHz. Maury Microwave also has MPC series available.

BNC. The BNC (Bayonet Navy Connector) was originally designed for military system applications during World War II. The connector operates very well up to about 4 GHz; beyond that it tends to radiate electromagnetic energy. The BNC can accept flexible cables with diameter of up to 6.35 mm (0.25 in.) and characteristic impedance of 50 or 75 Ω. It is now the most commonly used connector under 1 GHz.

SMA.   The SMA (Sub-Miniature A) connector was originally designed by Bendix Scintilla Corporation, but it has been manufactured by Omni-Spectra, Inc., as the OSM connector and many other electronic companies. The main application of SMA connectors is on components for microwave systems. The connector is seldom used above 24 GHz because of higher-order modes.

SMC.   The SMC (Sub-Miniature C) is a 50-Ω connector and smaller than SMA. The connector is manufactured by Sealectro Corporation and can accept flexible cables with diameter of up to 3.17 mm (0.125 in.) for a frequency range of up to 7 GHz.

TNC.   The TNC (Threaded Navy Connector) is merely threaded BNC. The function of the thread is to stop radiation at higher frequencies so that the connector can work at frequencies up to 12 GHz.

Type N.   The Type N (Navy) was originally designed for military systems during World War II and is the most popular measurement connector for the frequency range of 1 to 18 GHz. It is a 50-Ω connector and its VSWR is extremely low, within 1.02.

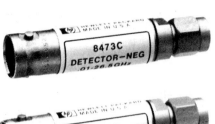

New coupling nut

APC-3.5 male

BNC female

Old coupling nut

(a) APC-3.5

(b) APC-7

BNC female

SMA female

BNC male

SMA male

(c) BNC

(d) SMA

APC-3.5
male

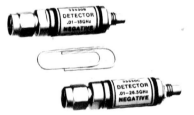

Type N female

SMC male
(plug)

Type N male

(e) SMC

(f) Type N

Figure H-1 Microwave coaxial connectors.

Index

Announcing. . . .

The Annual Prentice Hall Professional/Technical/Reference Catalog: Books For Computer Scientists, Computer/Electrical Engineers and Electronic Technicians

- Prentice Hall, the leading publisher of Professional/Technical/Reference books in the world, is pleased to make its vast selection of titles in computer science, computer/electrical engineering and electronic technology more accessible to all professionals in these fields through the publication of this new catalog!

- If your business or research depends on timely, state-of-the-art information, The Annual Prentice Hall Professional/Technical/Reference Catalog: Books For Computer Scientists, Computer/Electrical Engineers and Electronic Technicians was designed especially for you! Titles appearing in this catalog will be grouped according to interest areas. Each entry will include: title, author, author affiliations, title description, table of contents, title code, page count and copyright year.

- In addition, this catalog will also include advertisements of new products and services from other companies in key high tech areas.

SPECIAL OFFER!

- Order your copy of The Annual Prentice Hall Professional/Technical/Reference Catalog: Books For Computer Scientists, Computer/Electrical Engineers and Electronic Technicians for only $2.00 and receive $5.00 off the purchase of your first book from this catalog. In addition, this catalog entitles you to special discounts on Prentice Hall titles in computer science, computer/electrical engineering and electronic technology.

Please send me _____ copies of The Annual Prentice Hall Professional/Technical/Reference Catalog (title code: 62280–3)

SAVE!

If payment accompanies order, plus your state's sales tax where applicable, Prentice Hall pays postage and handling charges. Same return privilege refund guaranteed. Please do not mail cash.

- ☐ PAYMENT ENCLOSED—shipping and handling to be paid by publisher (please include your state's tax where applicable).
- ☐ BILL ME for The Annual Prentice Hall Professional/Technical/Reference Catalog (with small charge for shipping and handling).

Mail your order to: Prentice Hall, Book Distribution Center,
Route 59 at Brook Hill Drive,
West Nyack, N.Y. 10994

Name _____

Address _____

City _____ State _____ Zip _____

I prefer to charge my   ☐ Visa   ☐ MasterCard

Card Number _____ Expiration Date _____

Signature _____

Offer not valid outside the United States.

Dept. 1   D-PPTR-CS(9)